y

on

Zev Naveh
Arthur S. Lieberman

Landscape Ecology
Theory and Application

Second Edition

With 82 Figures

Springer-Verlag
New York Berlin Heidelberg London Paris
Tokyo Hong Kong Barcelona Budapest

Zev Naveh
The Lowdermilk Faculty
  of Agricultural Engineering
Technion-Israel Institute of Technology
Technion City, Haifa 32000
Israel

Arthur S. Lieberman
Cornell Abroad Program in Israel
University of Haifa
Mount Carmel, Haifa 31905
Israel

Front and back cover photographs courtesy of Almo Farina.

Library of Congress Cataloging-in-Publication Data
Naveh, Zev.
      Landscape ecology : theory and application / Zev Naveh, Arthur S.
  Lieberman. -- 2nd ed.
          p. cm.
      Includes bibliographical references and index.
      ISBN 0-387-94059-6 (New York : alk. paper). -- 3-540-94059-6
  (Berlin : alk. paper)
          1. Landscape ecology. 2. Landscape protection. I. Lieberman,
  Arthur S. II. Title.
  QH541.15.L35N38 1993
  333.7'ss--dc20   93-31792

Printed on acid-free paper.

Production managed by Karen Phillips, manufacturing supervised by Jacqui Ashri.
Typeset by MS Associates, Champaign, IL., and Asco Trade Typesetting Ltd., Hong Kong.
Printed and bound by Edwards Brothers, Inc., Ann Arbor, MI.
Printed in the United States of America.

9 8 7 6 5 4 3 2

ISBN 0-387-94059-6 Springer-Verlag New York Berlin Heidelberg
ISBN 3-540-94059-6 Springer-Verlag Berlin Heidelberg New York

This book is dedicated to the memory of
Dr. R. H. Whittaker—
The great ecologist and human being

Preface to the Second Edition

In the preface to the softcover edition of this book in 1989, we stated:

Since the publication of the first edition of this book, landscape ecology has made great strides. It has overcome its continental isolation and has also established itself in the English-speaking world. By attracting both problem inquiry and problem- solving-oriented scientists with different cultural, academic, and professional backgrounds from all over the world, it has broadened not only its geographical but also its conceptual and methodological scopes.

We are pleased to confirm in 1993 that the growth of landscape ecology continues, and to again express our gratification at the encouraging response to this first English-language monograph on the subject and its contribution to these developments. As before, we feel special satisfaction that it has reached not only the shelves of libraries and academic researchers, but that it has also appealed to professional practitioners, teachers, and their students from industrialized and developing countries, embracing the broad range of fields related to landscape ecology in the natural sciences as well as in the humanities.

Probably the best indication of this is the fact that each of the previous versions of this book exhausted their printing before we had time to prepare this updated edition. It is also apparent that the discipline of landscape ecology is perceived not merely as a spatial ramification of population, community, and ecosystem ecology for the study of the ecology of landscapes but also as an innovative, dynamic, and integrative field of environmental study and action in its own right, dealing with landscapes as tangible ordered wholes of natural and human systems at different scales and dimensions.

The above-mentioned trends in landscape ecology have continued with even greater intensity in recent years. Nevertheless, we believe that the basic conceptual and epistemological premises and their practical application presented in our first edition have retained their validity. However, we do realize the need to take recent developments into account, and we have updated each chapter with a summary of the advances we regard as most relevant for landscape ecology as a holistic and transdisciplinary science of landscape study, appraisal, history, planning and management, conservation, and restoration.

Since the publication of the earlier editions the rates of environmental degradation, pollution, and desertification; of depletion of natural resources; and of extinction of wild plant and animal species have further increased as the threats posed to life on earth by global changes have become even more alarming. At the same time, however, there is an increasing awareness of these dangers not only by ecologists and environmentalists but also by economists, politicians, and the public at large. One of the most outspoken and best informed of these politicians—Albert Gore—is the Vice President of the United States of America. These concerns found an overwhelming expression in the July 1992 Rio Earth Summit—alas, with results which did not come up to the great expectations for a new global partnership to resolve these crucial issues without further delay.

It has not been fully recognized that exponential growth of population and consumption are the major driving forces causing an overshoot of the capacity of our planet's biological, physical, and socioeconomic systems. As these forces are culturally deeply ingrained, this global environmental crisis cannot be resolved by technological, political, and economic means alone, but only by a far-reaching environmental and cultural revolution. As the third major wave of global change in human civilizations, after the agricultural and industrial revolutions, we believe this revolution should lead to a shift from consumption to conservation and from unrestrained quantitative growth to lasting qualitative improvement and development.

Recognition of the urgent need for this global environmental revolution and the feasibility of its realization, if humanity will accept this challenge in time, is no longer regarded as merely the utopian dream of "radical" environmentalists and "deep" ecologists. It is now shared by some of the most prominent economists, managers, and political decision-makers (Gore, 1992; King and Schneider, Club of Rome, 1991; Tolba, 1992). The team of systems analysts and modelers who wrote *Limits of Growth* (Meadows et al., 1972) concluded twenty years later in a new book (Meadows et al., 1992) that although these limits are real and close, the uncontrolled decline is not inevitable, if no further time is wasted. Their dynamic, cybernetic "World Model 3" indicates that "there is just exactly enough energy, enough material, enough money, enough environmental resilience, and enough human virtue to bring about revolution to a better world." This transition requires more than productivity: it also requires maturity, compassion, and wisdom.

Caring for the Earth, subtitled "A Strategy for Sustainable Living" has been published jointly by IUCN The World Conservation Union, United Nations Environment Programme (UNEP), and World-Wide Fund for

Nature, (WWF) (IUCN/UNEP/WWF, 1991). It provides a comprehensive and practical proposal for realizing this goal based on a strategy of mutually reinforcing actions at individual, local, national, and international levels.

In this book we are dealing with the environmental crisis and its resolution from a biocybernetic and hierarchical systems viewpoint within the context of landscape ecology. We emphasize that for this postindustrial environmental revolution a new symbiosis is required between modern man and nature at a higher organizational level—the "Total Human Ecosystem". Such symbiosis requires not only knowledge of the science of ecology but also ecological wisdom and ethics in order to recognize the place of humankind in nature and to behave in a manner which recognizes the limited resources on earth.

We hope that the conceptual and practical tools which we provide will help landscape ecologists play a useful role in this process. One of the first requirements for this purpose is to overcome the disastrous intellectual, professional, academic, and institutional fragmentation prevalent among those dealing with the current environmental crisis.

Within ecology, the means for such unifying tendencies have been presented in a very challenging and lucid way by Allen and Hoekstra (1992), who use a hierarchical systems approach as their basic premise. For the environmental movement at large, Norton (1991) has attempted to define a common interdisciplinary language and to propose an integrated theory of environmental management which centers around a better understanding of the human role in natural events. A similar tendency toward interdisciplinary convergence can be found also in the recently emerging branches of conservation biology (Brussard, 1991) and restoration ecology, which are closely related to each other and to landscape ecology (Allen and Naveh, 1993).

Of special significance in this respect is the emergence of the important synthetic field of ecology and economics. Its objective—as described in the announcement of the *Journal of the International Society of Ecological Economics*—is to transcend the normal conception of scientific disciplines seeking to integrate many different disciplinary perspectives in order to achieve ecologically and economically sustainable development.

Similarly, our concept of landscape ecology is one of a transdisciplinary, problem-solving, human ecosystem science. Landscape ecology can transcend the narrow, discipline-oriented paradigms of conventional ecological, geographical, and environmental management disciplines, thereby helping to bridge the gaps between them.

Applying the hierarchical view suggested by the late systems philosopher and educator Erich Jantsch (1970) to the levels of increasing coordination from multidisciplinarity to interdisciplinarity and transdisciplinarity, we regard the latter as the highest stage of multi-level and multi-goal coordination towards a common purpose. In ecological economics this purpose is the achievement of ecologically and economically sustainable development. In landscape ecology the purpose should be the highest attainable quality and health of our local, regional, and global landscapes, as an essential part of the quality of life on earth. The transdisciplinary purposes of both sciences are complementary: there can be no sustainable development without achieving full integration of "nature's household" (ecology) with "humanity's household" (economics); and healthy natural and cultural landscapes are among the major goals for sustainable development and the postindustrial environmental revolution."

If this updated version of *Landscape Ecology* contributes its small share towards the achievement of these goals, it will have fulfilled its purpose.

References

Allen, E., and Z. Naveh. 1993. Principles of Restoration Ecology: An Interdisciplinary Approach for Terrestrial Landscapes. Springer-Verlag, New York (in press).

Allen, T. H., and T. W. Hoekstra. 1992. Toward a Unified Ecology. Columbia University Press, New York.

Brussard, P. F. 1991. The role of ecology in biological conservation. Ecological Applications 1:6–12.

Gore, A. 1992. Earth in Balance—Forging a New Common Purpose. Earthwatch Publisher, London.

IUCN/UNEP/WWF. 1991. Caring for the Earth. A Strategy for Sustainable Living. Gland, Switzerland.

Jantsch, E. 1970. Inter- and transdisciplinary university: A systems approach to education and innovation. Policy Sciences 1:403–428.

King, A., and B. Schneider (Club of Rome). 1991. The First Global Revolution. Pantheon Books, New York.

Meadows, D. L., D. H. Meadows, and J. Randers. 1972. The Limits to Growth. Universe Books, New York.

Meadows, D. H., D. L. Meadows, and J. Randers. 1992. Beyond the Limits. Global Collapse or a Sustainable Future. Earthscan Publications Limited, London.

Norton, B. G. 1991. Toward Unity among Environmentalists. Oxford University Press, New York, Oxford.

Tolba, M. 1992. Saving Our Planet—Challenges and Hope. Chapman and Hall, London.

Preface to the First Edition

The notion of landscape ecology as an interdisciplinary science dealing with the interrelation between human society and its living space—its open and built-up landscapes—is a relatively recent one, having been conceived by geographers and ecologists in Central Europe following World War II. It is now being increasingly recognized as the scientific basis for land and landscape appraisal, planning, management, conservation, and reclamation. As such, landscape ecology is replacing many of the fields of applied ecology and geography to which the rather vague prefix "environmental" has been given.

These recent developments are being expressed by the growing interest in landscape ecology in the English-speaking world. While this book was being written, the first International Congress of Landscape Ecology was organized by The Netherlands Society of Landscape Ecology. This congress was attended by over 300 participants from 23 countries, representing a cross section of ecologists, geographers, conservationists, foresters, agronomists; regional, social, urban, and rural planners; and landscape architects from universities and governmental and private institutes of research, planning, and consulting. It led to the formation of the International Association of Landscape Ecology (IALE), which will have its inaugurating conference when this book goes to press.

During the International Congress of Landscape Ecology, various definitions were given to landscape ecology and different aspects were emphasized by the participants, according to their scientific and professional background. The broadest and most comprehensive definition was presented by the chairman of the organizing group, Professor I. S. Zonneveld (1982), regarding landscape ecology as both a formal biological, geological, and human science and as a holistic approach, attitude, and state of mind. In his opening lecture he stated:

Any geographer, geomorphologist, soil scientist, hydrologist, climatologist, sociologist, anthropologist, economist, landscape architect, agriculturist, regional planner, civil engineer—even general, cardinal, minister, or president, if you like, who has the "attitude" to approach our environment—including all biotic and abiotic values—as a coherent system, as a kind of whole that cannot be really understood from its separate components only, is a land(scape) ecologist.

In this book we have attempted to channel both these scientific and philosophical streams into a broad riverbed of transdisciplinary paradigms of landscape ecology. For this purpose we provide some conceptual and practical tools for synthesis and for systems thinking and acting, necessary to deal with the complex and closely interwoven physical, biological, and socio-cultural elements of landscapes, and to solve the severe problems arising from their present intensive human uses. But this book is neither a theoretical exercise nor a textbook or handbook of landscape ecology in the conventional sense. It is meant to institute a conceptual and epistomological basis for landscape ecology as a human ecosystem science and to show examples of its practical application in holistic land(scape) planning, management, and conservation.

The first section is devoted to the development of landscape ecology in Europe, and to its theoretical and conceptual foundation.

For this purpose, new frontiers are outlined in system theory and biocybernetics—the theory of self-organization and regulation in nature as applied to natural and human ecosystems—and human ecosystemology, as components of a new paradigm of a General Biosystems Theory. A central feature in this thesis is the recognition of the Total Human Ecosystem, integrating man and his environment. A redefinition is made of landscape as the concrete space/time-defined entities of this Total Human Ecosystem ecosphere. The role of landscape ecology is emphasized by providing scientific and educational information for the structural and functional integration of the biosphere, technosphere, and geosphere in the landscape. By furthering this new cybernetic symbiosis at the critical interface of resources and land use, landscape ecology can contribute to the urgently needed reconciliation between human society and nature.

In the second section, relevant examples are presented of practical scientific methods and tools and their application in holistic planning, management, and conservation, as guided by landscape ecological determinism. Methodologies from the areas of remote sensing, sensitivity analysis, and land-use capability analysis are used to focus on recent developments that can be employed in application of landscape ecological principles and knowledge in planning.

As an example of a holistic landscape approach to dynamic conservation planning, management, and education on a regional-ecological basis, the mediterranean climate sclerophyll forest zones have been chosen and practical solutions to these problems are offered.

By presenting these global, theoretical, and practical aspects of what is called, in general, "environmental management," as a unified task-solving and educational paradigm, we hope that this book will be of interest to a broad audience of professionals and decision-makers—scientists, teachers, and researchers and their students in both the basic and applied fields of general and human ecology, geography, and education, as well as in landscape architecture, regional planning, environmental engineering, agronomy, forestry, and landscape horticulture and in all related fields of conservation, wildlife, and resources management.

Reference

Zonneveld, I. S. 1982. Land(scape) ecology, a science or a state of mind. *In*: S. P. Tjallingii and A. A. de Veer (Eds.), Perspectives in Landscape Ecology. Contributions to research, planning and management of our environment. Proceedings of the International Congress of Landscape Ecology, Veldhoven, The Netherlands, April 6–11, 1981. Centre for Agricultural Publishing and Documentation, Wageningen, The Netherlands, pp. 9–15.

Acknowledgments for the Second Edition

Nine years have elapsed since the first edition of this book. Two individuals who showed enthusiasm and provided inspiration, Samuel D. Lieberman and Gustav Rosbasch, have since passed away. The absence of their counsel and support, ever available in preparing the 1984 edition, was keenly felt this time.

Ongoing appreciation and gratitude is conveyed to our wives, Ziona and Margot, as well as to Dora Lieberman and Gertrude Rosbasch.

In addition to the colleagues who were acknowledged in the first edition, we would like to express our thanks to those who assisted us by providing updated material from their own work and also made valuable suggestions for this second edition.

We sadly note the departure of two outstanding ecologists, Harold Biswell, a pioneer and leader in fire ecology in California, and Walter Westman, a disciple of the late Robert Whittaker, who achieved great professional distinction in carrying forth Whittaker's work. We were very much influenced by their approach, and they were among those who encouraged our proceeding with this second edition.

In the first edition it was deemed significant to include the visionary thinking of Arnold Schultz and Frank Egler concerning the evolving science of holistic Landscape Ecology. Their observations contributed greatly to the success of the first edition.

Acknowledgments for the First Edition

Both authors are deeply appreciative of the encouragement, patience, and cooperation of their wives, Ziona and Margot, and their children during the three-year period of writing this book. Special gratitude is also due to Samuel and Dora Lieberman, and to Gustav and Gertrude Rosbasch for their enthusiasm and interest throughout.

An expression of thanks is due to Dr. Duncan Poore, former Director of the International Union for the Conservation of Nature (IUCN) for allowing use, in Chapter 4, of the draft of a manuscript prepared by Zev Naveh for the IUCN World Conservation Strategies Source Book.

The authors gratefully note the diligent work of Vivian Rubinstein, who very ably typed the manuscript, and of Rosemarie Tucker.

We thank those individuals and publishers who willingly provided us with illustrations for the book, and especially to the MAB Secretariat of UNESCO, Paris.

We acknowledge planners, architects, landscape, and systems ecologists to whom we owe much for inspiration and encouragement and many of the ideas expressed in this book. Because of the great numbers of individuals involved, it is impossible to enumerate them, but we offer them a collective note of thanks for their assistance.

The Lady Davis Fellowship Trust of Jerusalem, Israel, enabled Arthur Lieberman to spend a sabbatical year at the Technion-Israel Institute of Technology in Haifa, by awarding him a fellowship as Visiting Professor at the Faculty of Agricultural Engineering. Both authors are indebted for that opportunity. Without this assistance and the chance to collaborate on location with his co-author, production of this book would have been much more difficult.

Contents

List of Figures and Tables

SECTION I

THE DEVELOPMENT
OF LANDSCAPE ECOLOGY
AND ITS
CONCEPTUAL FOUNDATIONS

1

The Evolution of Landscape Ecology

Landscape ecology is a young branch of modern ecology that deals with the interrelationship between man and his open and built-up landscapes. As will be shown in this chapter, landscape ecology evolved in central Europe as a result of the holistic approach adopted by geographers, ecologists, landscape planners, designers, and managers in their attempt to bridge the gap between natural, agricultural, human, and urban systems.

Definition of Landscape and Landscape Ecology

As the earliest reference to "landscape" in world literature, the Book of Psalms (48.2) can be cited. Here, "landscape" (נוֹף "noff" in Hebrew, probably etymologically related to יָפֶה "yafe," "beautiful") is the beautiful overall view of Jerusalem, with King Solomon's temple, castles, and palaces. This visual-aesthetic connotation of landscape is usually referred to in the English language as "scenery."

As Whyte (1976) has shown in his important book on land appraisal, the meaning of the term "landscape" has undergone great changes, but the original visual–perceptual and aesthetic connotation has been adopted in literature and art, and is still used by many persons involved in landscape planning and designing, and by gardeners. They are frequently more concerned with scenic-aesthetic landscape perceptions than with its ecological evaluation. This is well reflected in the great amount of English literature devoted to landscape assessment and reviewed by Arthur et al. (1977), Zube et al. (1975), and others. Unfortunately, the epistemological development of the landscape concept, leading to the new discipline of landscape ecology as described in this chapter, is almost unknown outside Europe. The term "landscape ecology" is virtually absent from North American literature. In a very comprehensive review of human ecology as an interdisciplinary concept (Young, 1974), a long discussion was devoted to related fields such as landscape architecture, land planning, nature conservation, and applied ecology in general, but landscape ecology was not mentioned, nor was a single non-English relevant literature source cited on this subject.

This does not mean that the scope of landscape ecology is not being addressed outside Europe. Nor does it mean that landscape planning and design professionals are not active in developing and implementing ecologically based analytical approaches and methodologies. Much initiation and improvement in land-use capability analysis and assessment approaches and methodologies has come from leading professional landscape architects (often connected with academic institutions) in the United States. When they deal with the landscape in a holistic sense, even though they may not formally employ the term "landscape ecology," they are in reality making use of the comprehensive or integrated approach it represents.

In the Germanic languages, landscape and its etymological equivalent, "Landschaft," also contain the geographic–spatial connotation of "land." Since the Renaissance, and especially in the 18th and 19th centuries, this spatial connotation has acquired a more comprehensive meaning in which the landscape is experienced as a spatial–visual whole reality of the total environment.

Landscape was introduced as a scientific–geographic term in the early 19th century by A. von Humboldt, the great pioneer of modern geobotanics and physical geography, who defined it as "Der Totalcharakter einer Erdgegend"– the total character of an Earth region. With the rise of classical Western geography, geology, and earth science, however, the term's meaning has been narrowed down to the characterization of the physiographic, geological, and geomorphological features of the earth's crust, as a synonym of "landform."

Russian geographers have again given it a much broader interpretation by including both inorganic and organic phenomena in the landscape concept, calling the study of its totality "landscape geography." These semantic and epistemological developments have been described in detail by Troll (1971), a leading German biogeographer, who defined landscape as "the total spatial and visual entity" of human living space, integrating the geosphere with the biosphere and its noospheric man-made artifacts. He regarded landscape as a fully integrated holistic entity, meaning a "whole" that is more than the sum of its parts and that should therefore be studied in its totality. As early as 1939, while studying problems of land use and development in East Africa, he coined the term "landscape ecology," realizing its great potential in the aerial photographic interpretation of landscapes. He hoped for a closer collaboration between geographers and ecologists, from which a unified earth and life research might develop–a new "ecoscience," as distinguished from "geoscience"–dealing only with the inanimate lithosphere and not the biosphere. In practice, landscape ecology combined the "horizontal" approach of the geographer in examining the spatial interplay of natural phenomena with the "vertical" approach of the ecologist in studying the functional interplay in a given site, or "ecotope."

The Division of Land Research (now named Land Use Research) of the Commonwealth Scientific and Industrial Research Organization (CSIRO) in Australia was one of the first important organizations to adopt this holistic approach in practical land surveying and evaluation for development. Under the leadership of C. S. Christian and A. G. Steward, the concepts of "land units" and "land systems" (Christian, 1958) were applied with great success in large-scale multidisciplinary integrated surveys (Christian and Stewart, 1968). Since then, with the new Division for Land Use Research, the scope of this methodology in producing land-use plans has been further broadened to include socioeconomic and ecological parameters.

Another important agency, working chiefly in developing subtropical and tropical countries, the International Institute for Aerial Survey and Earth Science (ITC) in The Netherlands, has further developed the holistic landscape concept. As described in detail by Zonneveld (1972) in his comprehensive textbook for photo interpretation, landscape ecology is the crucial subdivision of "land(scape) science," studying the landscape as a holistic entity made up of different elements, all influencing each other. These interpretations, as viewed by Zonneveld, are presented in Figure 1-1. The land as such, rather than living organisms, constitutes the central point of landscape ecology; it does not belong, in Zonneveld's opinion, to the biological sciences, like ecology, but is a branch of geography. He also states that any comprehensive physiographic or integrated approach to the survey of separate land elements is, in fact, making use of landscape ecology, even if the user has never heard of the actual term.

For the distribution of landscape units in space—their chorological relations— Zonneveld proposed the following hierarchical levels of increasing size:

1. The *ecotope* (or site) is the smallest holistic land unit, characterized by homogeneity of at least one land attribute of the geosphere—namely, atmosphere, vegetation, soil, rock, water, and so on—and with nonexcessive variations in other attributes.
2. The *land facet* (or microchore) is a combination of ecotopes, forming a pattern of spatial relationships and being strongly related to properties of at least one land attribute (mainly landform).
3. The *land system* (or mesochore) is a combination of land facets that form one convenient mapping unit on reconnaissance scale.
4. The *main landscape* (or macrochore) is a combination of land systems in one geographical region.

As in other scientific disciplines, there exist a great many different approaches to terminology and classification of landscape ecology. This is especially the case in theoretical geography, in which an almost overwhelming amount of semantic discussion can bury the major methodological and practical issues, as in the recent (German) book on landscape ecology by Leser (1976).

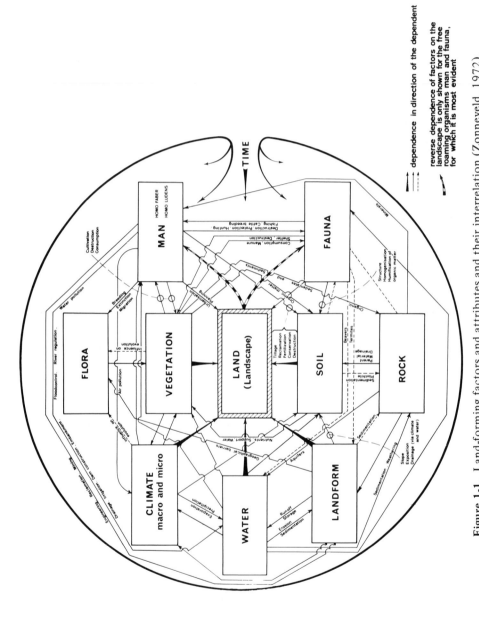

Figure 1-1. Land-forming factors and attributes and their interrelation (Zonneveld, 1972).

In West Germany, the Institute for Landscape Care and Nature Protection of the Technical University of Hannover was very instrumental in introducing landscape ecology as a scientific tool for landscape management and planning. Buchwald (1963), the head of this institute, defined the landscape as the total living space as a multilayered "Wirkungsgefüge"—interacting system—of both the geosphere and the biosphere, and he designated to landscape ecology the important task of helping to overcome the tensions between modern society and its landscapes resulting from the increasing demands of the industrial society and the natural land potentials. Langer (1970), from the same institute, has made an important first attempt at a system-theoretical interpretation of landscape ecology, which he defined as "a scientific discipline, dealing with internal functions, spatial organization and mutual relations of landscape-relevant systems." These regional ecological systems should be regarded as the highest level of ecological integration—above the aut-ecological and syn-ecological levels—with the ecotope as their smallest landscape element.

Langer (1973) continued to explore the differences between "Naturlandschaften" (natural landscape) and "Kulturlandschaften" (cultural landscape) that had first been noted by Schmithüsen (1963), another leading biogeographer and landscape ecologist; Langer stressed that the cultural landscapes are not related only to the natural sciences but also to the sociocultural sciences, as expressed by anthropogenic (human-caused) impacts through utilization. However, these anthropogenic elements in the cultural landscape do not just join the natural ones, but form units of a higher-level whole of "geosocial systems" which are relevant for the planning process.

These notions will be developed further in the next chapter. However, to define the highest level of integration of "man-plus-his-total-environment," we shall follow Egler (1964), who coined the phrase "total human ecosystem." Egler was one of the first North American plant ecologists to realize the holistic nature of vegetation and the active role of man (Egler, 1942) as an integral part of a higher ecosystem level, operating in the landscape. As early as 1942 Egler had criticized the conceptual and methodological failure of contemporary American plant ecology to take stock of this holistic nature of vegetation.

Another prominent North American ecologist with a then-pioneering holistic approach to ecology and the landscape was Dansereau (1957). In his groundbreaking *Biogeography: An Ecological Perspective,* he proposed the study of landscape as the highest integrative level of environmental processes and relations. He called this the "industrial level" and regarded human ecology, which is related not only to human geography, but also to anthropology, agriculture, forestry, sociology, and history, as the science studying the influence of man on the landscape. In this book he presented human landscape modification throughout history, through gathering, hunting, herding, agriculture, industry, and urbanization, as levels increasing intensity of human control. He claimed that through the present highest level of landscape modification by the elaboration of completely new ecosystems and deliberate molding of evolution, a new geological epoch in the exploitation of environmental resources—the "noospheric epoch" (from the Greek "noos," meaning "mind")—had been inaugurated. In a later critical essay on human ecology, Dansereau (1966) added a seventh level of landscape modification—atmospheric control and extraorbital travel. He claimed that we are now entering the *engineering phase*, in which virtually all landscapes in the world are under some kind of management. Apparently he was not aware of the rise of landscape ecology in Europe, but he mentioned the need for rational scientific ecological landscape management. In recent years Dansereau has further developed these ideas in two major directions: (1) He presented a comprehensive new system for ecological grading and classification of land occupation and land use mosaics, based on trophic levels of energy transfers in wild, rural, industrial, and urban ecosystems, using a three-dimensional model of these ecosystems which illustrates their relative trophic load in a very illuminating way (Dansereau, 1977). (2) He further elaborated on modern man's relationship with the landscape, defining landscape as the sum of its interacting ecosystems. In a penetrating analysis of the duality of human perception and human impact on the landscape, he regarded this relationship as a cyclic—or even cybernetic—process. In this "inscape/landscape" (Dansereau, 1975), there is a filtering inward from nature to man, upward from the subconcious to the conscious, and from perception to design and implementation. This process is occurring to the agriculturist, as well as to the forester, the engineer, and the town planner (and we could also add the landscape ecologist). In this way the inscape becomes a template for reshaping of the landscape. All of these original ideas and their lucid presentation have been summarized recently in an essay entitled "The Template and the Impact" (Dansereau, 1980).

Later we shall deal in more detail with the implications of this major "noospheric–cultural" evolutionary advance in which "homo faber," through the biological evolution of the brain cortex and consequent greater mental capabilities, has become a mighty geological agent, acting on the landscape in both a constructive and a destructive way. The geochemist Vernadsky (1945), who coined the term "noosphere," suggested that the noosphere, or "world dominated by the mind" of man, will gradually replace the biosphere, the natural evolving organic world. But as Odum (1971), in his now-classical textbook on ecosystem ecology, has pointed out, this is a dangerous philosophy. It is based on the assumption that man, through scientific and technological skill, can put himself above natural laws and is capable of living in a completely artificial world.

The term "noosphere" and man's role in its formation have been reevaluated by the great anthropologist and nature philosopher Teilhard de Chardin (1966). He believed in the active role of man in designing and furthering constructive evolution through self-reflection and human consciousness, and called this process "noogenesis." In this book we shall attempt to show how landscape ecology can contribute to this noogenesis through furthering those trends in ecological thinking and acting that may enable us to be at once both part of the biosphere and its modifier and caretaker.

A further step toward a system-theoretical definition of landscape and landscape ecology has been provided by Vink (1975), following the systems approach developed by Chorley and Kennedy (1971) in physical geography. They defined systems as "a structural set of objects or attributes, consisting of components or variables (i.e., phenomena that are free to assume variable magnitudes) that exhibit discernible relationships with one another and operate together as a complex whole, according to some observed pattern."

In discussing the role of landscape ecology in land use for advanced agriculture, Vink (1975, p. 135) emphasizes the above-discussed fact that landscapes—as carriers of ecosystems—are control systems in which the key components are controlled wholly or partly by "human intelligence" through land utilization and management. He therefore defined landscape ecology as "the study of the attributes of the land as objects and variables, including a special study of key variables to be controlled by human intelligence."

In this way landscape ecology can provide the functional ties of the objects and processes of the individual disciplines dealing with plants, animals, and man and their functional integration for present and future land uses.

Development of Landscape Ecology in Central Europe: Some Contributions

As was shown in the preceding section, landscape ecology was conceived in its formative stages chiefly as a biogeographic discipline bridging the gap between the spatial-chorological approach of the geographer and the functional and structural one of the ecologist. However, very soon the phytosociologists and geobotanists, because of their field orientation and their concern with the open landscape and its natural and man-modified vegetation, assumed a leading position in this endeavor. They were joined by applied ecologists—foresters, agronomists, and gardeners, as well as landscape architects and planners with ecological outlooks. The following is a short review of some of the major developments in this respect in central Europe, chiefly in West Germany and The Netherlands.

One of the central features in the theory of landscape ecology is the recognition of the dynamic role of man in the landscape and the quest for the systematic and unbiased study of its ecological implications. For this purpose it was essential to eliminate from any methodological and practical considerations preconceived Clementsian succession-to-climax dogmas. These not only dominated American ecology but were also accepted by most phytosociologists of the Braun-Blanquet school in central Europe. It was also important to broaden the scope of the phytosociological methods of this school and its sole reliance on floristic composition in the determination of plant communities by including causal ecological relations between vegetation and environment that were based on ecophysiological studies and insight. This has been achieved to a great extent through the contributions of some outstanding ecologists who educated a new generation of plant ecologists and others in related fields; members of this new generation now occupy leading positions in landscape ecology in West Germany.

A first important step in this direction was the replacement of the "climax" term by the more meaningful phrase, "potential natural vegetation." The latter was suggested by Tüxen (1956), who 20 years earlier had already expressed doubts about the validity of the monoclimax concept (Tüxen and Diemont, 1937). "Potential natural vegetation" was used also by the biogeographer Schmithüsen as "potentiale Naturlandschaft" (potential natural landscape) and by Kuchler (1967, 1975) in the United States. It means a conceptual abstraction and construction of the vegetation that would become established if man suddenly disappeared, and is based on current knowledge of actual existing vegetation potential, its developmental tendencies, and site relationships. At that time, Tüxen was the director of the Central Institute of Vegetation Mapping. Later he became the leader of a private Institute of Theoretical and Applied Phytosociology at Rintelen, where he organized challenging annual international symposia, including the first symposium on landscape ecology in 1968 (Tüxen, 1968).

Maps of potential natural vegetation have been completed in West Germany, as well as in other central European countries, and have yielded important information for landscape planning and management. In these maps, phytosociological methods of the Zurich–Montpellier school are used, but their interpretation has benefited much from the pioneering work carried out by Ellenberg (1950, 1974) on man-modified pastoral and weed associations and the introduction of "ecological species groups" as a criterion for plant community classifications. These could be used as indicators for climatic and edaphic conditions, thereby preparing the ground for fruitful synthesis with other classification and ordination methods that have become most important in integrated landscape-ecological surveys. In a recent attempt at an integrated synthesis of European and Anglo-American approaches to vegetation science (Ellenberg, 1956; Mueller-Dombois and Ellenberg, 1974), these methods have been made available to English readers. In a comprehensive monograph on the vegetation of central Europe (Ellenberg, 1978)[1] it was shown that in the old cultural landscapes of central Europe thousands of years of human impact have resulted in "not a single spot of the vegetation being retained in its original natural state" and have also caused far-reaching changes in ecological site conditions. Thus, in the so-called Urwald (pristine forest) traces of previous utilization can be found.

Ellenberg distinguished among *zonal* vegetation, which expresses the responses of the natural potential vegetation to climatic conditions; *extrazonal* vegetation, which responds to the local topoclimatic conditions of slope exposure; and *azonal* vegetation, which responds to the specific soil and moisture conditions occurring in different climatic regions, such as hydromorphic soils and sand dunes. Plant communities representing these three groups thus form complex mosaics in each landscape. An example of the far-reaching changes from the pristine Urlandschaft to the recent European Kulturlandschaft is presented in Figure 1-2.

---

[1]The English translation of this work is presently being prepared by Cambridge University Press.

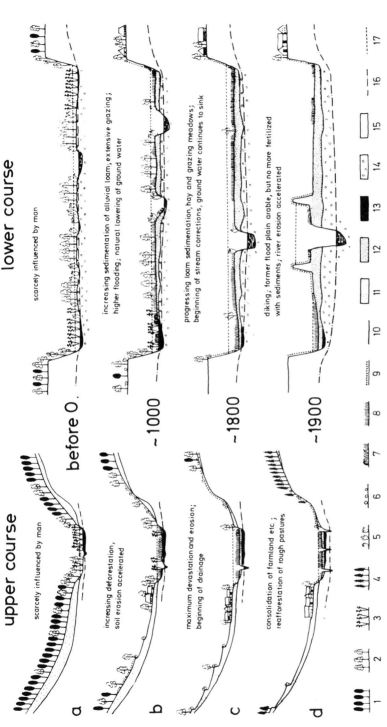

Figure 1-2. The conversion of a natural river-valley landscape in Central Europe into a cultural landscape in the course of 2000 years (Ellenberg, 1978). Major adverse landscape modifications were induced by deforestation, drainage, erosion, and siltation. (1) Beech forest, (2) oak a.o. mixed deciduous forests, (3) alder marsh, (4) conifer afforestation, (5) willow thicket, (6) other shrub thickets, (7) wet meadows, (8) fresh meadows, (9) dry meadows, (10) fields, (11) loess-loam, (12) meadow loam, (13) marshland, (14) gravel, (15) other soil types, (16) intermediate groundwater table, (17) intermediate floodwater table. (1)–(9) not in proper scale. (11)–(14) exaggerated in height.

In recent years, Ellenberg, as director of the Institute of Geobotany of the University of Göttingen, has further broadened the scope of ecology in central Europe by initiating integrated ecosystems studies, and especially the Solling project (Ellenberg, 1971), which became one of the most comprehensive and successful multidisciplinary forest and grassland ecosystem studies within the International Biological Program. He was also one of the founders of the German-Austrian Society for Ecology in 1971, an active multidisciplinary scientific organization in which landscape ecology occupies an important position. His perception of the ecosystem and man's dual part as both its internal and external component is presented in Figure 1-3. This approach and his functional ecosystem classification have served as starting points for our considerations, as described in the next chapter.

Another, even sharper, critic of the scholastic Clementsian and Braun-Blanquet dogmas was H. Walter, the former director of the Botanical Institute at the Agricultural University of Stuttgart-Hohenheim, who can be regarded as the father of modern ecology in Germany. He introduced ecophysiological methods into geobotanical studies, and in these he emphasized the need for a holistic treatment of ecological factors, including the anthropogenic one (Walter, 1960). In his monumental studies of the vegetation of the earth from such a holistic ecophysiological point of view (Walter, 1964, 1968)—summarized in English (Walter, 1973)—he provided a worldwide view of landscapes covered with intricate patterns of dynamic vegetation types, formations, and ecosystems. These are determined, not only in their floristic composition but also in structure, stability, diversity, and productivity, by regional climate, local site conditions, biotic interactions, and to a rapidly increasing extent by human modifications. In these studies, Walter (1964, p. 29) refuted the theories of climax and primary succession and stated that

> even the zonal vegetation, typical for distinct climatic zones, corresponds only in a very limited way to the climax concept. Furthermore, all attempts to save the climax concept by establishing a poly-climax term or by introducing climax groups or clusters—[as was proposed by Tüxen and Diemont (1937)]—are not satisfactory because they still include the concept of primary succession. A certain dynamic view of the vegetation is indeed justified. It should, however, not leave the ground of reality and lose itself in speculations.

For these reasons, he refrained from mentioning either climax or successional stages or hierarchical plant community classifications. He stressed, however, the importance of burning, grazing, and human interventions in shaping and maintaining vegetation types, and therefore in certain cases (such as that of *Calluna,* heather), the need for continued management as an ecological tool for their conservation.

Important cornerstones in the development of landscape ecology as an ecological science were the above-mentioned symposium on landscape ecology and phytosociology (Tüxen, 1968) and the publication of a comprehensive hand-

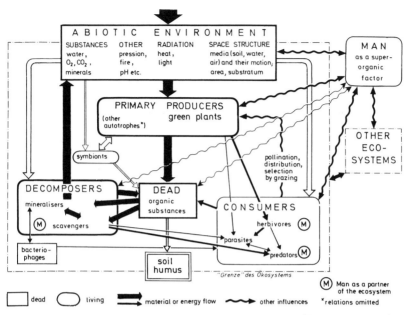

Figure 1-3. Simplified model of a complete ecosystem (Ellenberg, 1978).

book for landscape care and nature protection (Buchwald and Engelhart, 1968) that has been completely revised and is being republished. In these publications, landscape ecology was recognized as the scientific basis for both disciplines. Simultaneously, landscape ecology was also adopted by the landscape designing and planning professions. This acceptance is demonstrated in the publication of a condensed handbook on landscape ecology intended chiefly for architects (Woebse, 1975). Here, in contrast to the purely architectural–aesthetic designing approach, the ecological designing approach is emphasized. Thus, according to Woebse (1975), the criteria to be used for the evaluation of landscape quality should be oriented toward the maintenance of natural functions and structure and the landscape's ecological equilibrium, as conditioned by ecological site factors of climate and soil, by natural design material and not by artificial material, by the natural shapes of the land and not by geometric or other preconceived lines and shapes. They should take into consideration the dynamic changes occurring in the landscape, in contrast to the static end product of architectural designing. For these purposes, the designer should be well-versed in natural sciences and biology, so that he will be able to perceive the ecological principles, the natural potentials of the landscape, and the changes induced by economic utilization. At the same time, Woebse also stressed the importance of creating and maintaining the values of spiritual and emotional enlightenment that nature affords.

In recent years it has become more and more apparent that landscape care and planning—and thus also landscape ecology in the open landscape and the ecotones of rural and urban areas—are closely interwoven with all other interdisciplinary aspects of regional and urban planning. This realization leads to a further broadening of the basic landscape ecology concepts and the introduction of overall *ecological planning* as the central tool for landscape and regional planning and has replaced the distinction between care of the open landscape and the "greenplanning" of urban areas (Trent Forschunggruppe, 1973). Thus, as reported in the proceedings of the 1974 annual meeting of the German–Austrian Society for Ecology in Erlangen, many landscape–ecological studies are carried out as integral parts of landscape and regional planning projects, some on a rather large scale (e.g., not only in West Germany and East Germany, but also in The Netherlands—see following paragraph, Switzerland, Belgium, Sweden, and various Eastern European countries).

In West Germany, the Federal Institute for Nature Conservation and Landscape Ecology, headed by G. Olshowy at Bad Godesberg/Bonn, has played a leading role in the introduction of landscape ecology principles and methods and their practical application in regional planning and development (Olshowy, 1973). Some of these methods will be reviewed in more detail in Chapter 3. Olshowy (1975) has described the contribution of landscape ecology to planning by inventories (analyses) and evaluation (diagnoses) which are carried out both on larger-scale maps of 1:200,000 and in more detailed regional maps of 1:25,000, in the journal *Landscape Planning*. The latter now serves as an important international channel of communication in the field of landscape planning. Probably the most significant contribution of this institute is the now almost completed mapping of potential natural vegetation. These maps serve as valuable guides for landscape development, agriculture, reforestation and revegetation, the planning of highways, and mining operations, and they are used as indicators for the selection of trees and shrubs for these purposes (Trautmann, 1973).

A further important contribution to the development of landscape ecology as a scientific discipline was the establishment of special chairs for landscape ecology and related fields in some of the major universities of West Germany. Although in each university these chairs are attached to different departments and faculties and have different approaches and emphases in their teaching and research programs, they all deal with the complex interrelationships between man and his open, cultural, and industrial landscapes, and all aim at a compromise between the clashing natural, cultural, and socioeconomic demands and, at the same time, at the enrichment of the natural biotic environment. Thus, for example, at the Technical University of Aachen this chair, held by W. Pflug, is part of the Faculty of Architecture and its main achievements are in the field of landscape planning and design and in the education of ecologically trained and minded architects and planners (Pflug, 1973). At the Technical University of Hannover, landscape ecology is taught and practiced at the Institute for Landscape Care and Nature Protection, whose important contributions to the conceptual and methodological developments by Buchwald and Langer have already been mentioned. This work is oriented toward regional planning, as well as nature conservation and landscape reclamation. At the Technical University of Berlin there are two different groups. The Institute of Landscape Construction and Garden Art contributed much, under the leadership of H. Kiemstedt (1967, 1975), to the development and application of quantitative computerized methods for landscape ecology, which will be reviewed in Chapter 3. The other group, at the Department of Applied Botany, is carrying out important studies in urban ecology (Bornkamm and Grün, 1973; Sukopp, 1970, 1971), also dealing with the evaluation and management of nature reserves, the protection of endangered species, and the development of indicator species and plant communities for landscape and ecosystem classification applicable to landscape planning and management.

The tradition of the late G. Troll, who introduced landscape ecology as a geographic discipline, is continued by K. F. Schreiber, in the chair of landscape ecology of the Geographic Institute of the University of Münster. In an important lecture dedicated to the memory of Troll, Schreiber (1977a) outlined the conceptual and methodological development of landscape ecology and stated that, despite the present preoccupation with environmental conservation, the basic concepts of landscape ecology have remained the same. Among its most relevant contributions to planning and environmental protection, he mentioned the new concept of functionally oriented landscape classification and ordination. He emphasized the incorporation of experimental investigations in ecosystem research as an essential basis for landscape classification and application. He suggested simple and rapid methods of characterizing parameters relevant to the planning of the landscape-ecological structure of a region. Schreiber and his department have been instrumental in the development in West Germany of a dynamic approach to nature conservation through active intervention and in studying the controversial problems of "Sozial Brache"—the social fallow in derelict agricultural land in which agriculture is no longer profitable (Schreiber, 1977b). In order to decide whether to leave this land until it reaches a final "climax" stage or to preserve it by simulation of previous agricultural and pastoral practices, as done in The Netherlands and the United Kingdom in certain grassland types and in heather, or to transform it into a new recreational landscape, much more knowledge is required. For this purpose a series of long-term secondary succession studies have been established in West Germany (Schreiber, 1976) in which the effect of burning, grazing, herbicidal control, mulching, cutting, and the like, on vegetation and soil are carefully studied.

At the Faculty of Agriculture of the Technical University of München/ Weihenstephan, the chair of landscape ecology forms, together with the chairs for botany and landscape architecture, the Institute for Botany and Landscape Care, which trains more than 200 students each year, of whom about half specialize in landscape ecology. This chair, headed by W. Haber, has a very diverse research program. It deals with the interrelations of rural and urban ecosystems, land utilization systems and landscape structure, theory and practice of nature protection, biotope protection and development, the use of bioindicators for the evaluation of environmental burdens, landscape ecology in developing countries, and theoretical landscape ecology supported by mathematical and cybernetic models and methods. Some of the impressive work of this institute was presented at the eighth annual meeting of the German–Austrian Society for Ecology, which took place in September 1979 in Weihenstephan. Haber has also devoted much effort to the public education of decision makers in landscape ecology principles. This combination of theoretical planning-, conservation-, and management-oriented research and environmental education comes closest to the model of modern landscape ecology as presented in this book. In some recent lectures and articles, Haber (1979, 1980) has deepened the theoretical foundations of landscape ecology as part of a cybernetic and dynamic ecosystem theory and has further developed Odum's (1969) concepts of differential protection and production ecosystem patterns for the specific needs of regional planning, within the framework of a planning-oriented ecology as the major task of landscape care. These views have been incorporated in the following chapters of this book.

In The Netherlands, as a relatively small but very densely populated country with a severely impoverished flora and fauna, the appreciation of the importance of nature conservation on the one hand and the need for comprehensive landscape planning on the other is probably greater than in any other European country. This is reflected in actual land-use policies—in the application of integrated ecological approaches in large-scale physical planning and the dynamic scientific management of nature reserves, which also includes the creation of new habitats. Landscape ecology serves as the major scientific tool for these activities, and as reported by van der Maarel and Stumpel (1975), landscape ecological inventories of interdisciplinary teams are being carried out in 5 of 11 provinces. By 1974, 60 different projects had already been undertaken and an active working group for landscape ecology with about 100 members had been formed.

Landscape ecology is also firmly established in the extensive research in vegetation science, reviewed recently by Bakker (1979). As in Germany, it is based on phytosociological concepts of the Zurich–Montpellier school, but under the inspiring leadership of V. Westhoff and E. van der Maarel from the University of Nijmegen it has been developed into a dynamic and advanced quantitative methodology (Westhoff and van der Maarel, 1978). Landscape ecology has thus maintained close links with phytosociology and other scientific disciplines, such as geography, pedology, and hydrology, as well as with their application in agriculture, forestry, land reclamation, and drainage, especially in regional and town planning.

Because of their proficiency in major European languages—English, German, and French—Dutch ecologists have been able to bridge the language barrier and communication gap and have also incorporated into their own methodology other valuable concepts originating in Europe, the United States, and elsewhere. This was demonstrated by the initiative taken by landscape ecologists from The Netherlands Society of Landscape Ecology in organizing the First International Congress of Landscape Ecology at Vendhoven in April 1981.

An important contribution to the creation of a sound theoretical framework for landscape ecology and its practical application has been provided by the research on choice and management of nature reserves carried out in the special division of the State Forest Service for Nature Conservation and Landscape and by the State Institute for Nature Conservation Research (RIVON), which in 1960 merged with the Institute of Biological Research (ITBON) to form the present Research Institute for Nature Management (RIN), located at Leersom. In RIVON and RIN, C. G. van Leeuwen, together with V. Westhoff, developed the theoretical basis and practical guidelines for ecological management of nature reserves and landscapes, recently summarized by Bakker (1979). Its central feature is the recognition of the dynamic role of man in creating diversity and stability during many centuries of continuous land utilization with similar methods and gradual agricultural, pastoral, and other modifications in these seminatural landscapes. It resulted in the rejection of the nonintervention theory in nature reserves as being unrealistic, and even naive, in densely populated countries where human impacts throughout history have not necessarily been negative and impoverishing and have enlarged the variety in space (diversity) by creation of seminatural landscapes. Therefore, "nature techniques" must be applied, aiming at the development, construction, and improvement of the "natuurbouw" on one hand and the management and maintenance of the "natuurbeheer" on the other (van Leeuwen, 1973). External, or outward, management includes measures aimed at the prevention of excessive human influences that cause changes in environmental conditions (called "environmental dynamics"), especially urban and agroindustrial practices in the vicinity of natural reserves. Internal, or inward, management ensures the required level of human influences necessary to maintain or to create desirable situations in the nature reserve, such as excavations, treading, grazing, burning, mowing, and noninterference (Figure 1-4).

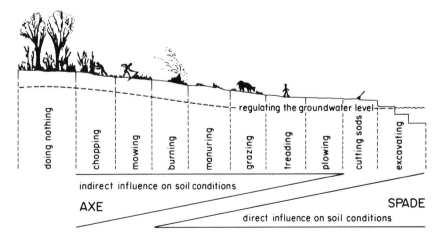

Figure 1-4. Types of inward nature techniques. Degrees of required minimal human influences (Bakker, 1979). Reproduced by permission of Dr. W. Junk BV, The Hague, The Netherlands.

The theoretical basis for these practices and for present landscape management in general in The Netherlands is the cybernetic relation-theoretical approach (van Leeuwen, 1966) to patterns (changes in space) and processes (changes in time) within ecosystems.

These principles and their application to conservation management in The Netherlands were first presented by Westhoff (1971) and van der Maarel (1971) in an important international symposium of the British Ecological Society in 1970 (Duffey and Watt, 1971). Westhoff (1971) classified landscapes according to their "naturalness," as judged by the composition of flora and fauna, the diagnostic value of pattern and process in vegetation and soil in relation to human influences, and distinguished four main types of open landscapes, presented in Table 1-1. Van der Maarel (1975) amplified this classification by also including Sukopp's (1972) distinctions of hemerobiotic degrees (from "hemeros," meaning "cultivated"), as measured by the proportion of neophytic species in the regional flora since 1500 (Table 1-2). This classification was applied in the mapping of ecotopes in Holland as related to the potential natural vegetation on a scale of 1:200,000 in a landscape-ecological study initiated by the Natural Regional Planning Agency of the Ministry of Housing and Physical Planning (Stumpel and Kalkhoven, 1978). In this study a general ecological model (GEM) was also prepared, with natural environment and human society as the main components of the ecosystem (van der Maarel, 1977). This GEM also follows van Leeuwen's cybernetic relation theory for the interpretation of the interrelations between the natural environment and society. Man's present interference is referred to as "anthropogenic environmental dynamics," leading to an increase in the dynamics (or total changeability of environmental factors) through time and to a decrease in the variety in space, notably the variety (or diversity) of biotic communities. This contradicts the natural development tendencies of ecosystems, but can be reduced to a conflict involving the fulfillment of production, carrier, and regulation functions of the natural environment.

This model is based on two theoretical concepts: (1) the ecosystem—the interacting system of biotic communities and their environment; and (2) the hierarchy of spheres of specific material energy influences and reactions, decreasing in dominance from the cosmosphere to the geosphere and the biosphere and noosphere of man and his influences. However, the same term, "ecosystem," is used for ecosystems as viewed by natural ecologists and for purposes of nature conservation, in which man is considered part of the (natural) environment of the biotic communities, and for ecosystems as usually viewed by ecological planners. For these, rightly, man is part of human society—as the main focus for planning—and the natural environment therefore includes both biotic communities and abiotic environmental factors.

As will be explained in the next chapter, these contradicting viewpoints result from the dichotomic position of man between noosphere and biosphere. This problem can be resolved by introducing the holistic landscape concept into this model and defining landscape as the "concrete space–time systems of the total human ecosystem," with the *ecosphere* as the largest global one and the *ecotope*

Table 1-1. Major Landscape Types Classified by Westhoff, According to the Degree of Naturalness

	Flora and fauna	Development of vegetation and soil	Examples
Natural landscapes	Spontaneous	Not influenced by man	Parts of the Wadden area (mud flats, coastal beaches, and salt marshes)
Subnatural landscapes	Completely or largely spontaneous	To some extent influenced by man	Parts of the dune landscape, most salt marshes, inland drift sands, deciduous woods with some cutting, final stages of succession in hydroseries in fens
Seminatural landscapes	Largely spontaneous	Drastically influenced by man (other formation than the potential natural vegetation)	Healthlands, oligotrophic grasslands, sedge swamps, reed swamps, inner dune grasslands, coppice, osier beds, many woods in which the tree stratum is arranged by man
Agricultural landscapes	Predominantly arranged by mar	Strongly influenced by man (soil often fertilized and drained; vegetation with ruderals, neophytes, and garden escapes)	Arable fields, sown grasslands, parks, conifer forests

Source: Bakker (1979). Reproduced with permission of Dr. W. Junk BV, The Hague, The Netherlands.

Table 1-2. Degrees of Naturalness in Ecosystems, Hemerobiotic State, and Some Characteristics of Vegetation and Soil

Naturalness	Hemerobiotic state[a]	Changes substrate	Changes vegetation structure	Changes floristic composition	Loss natives[b] (%)	Gain neophytes[b] (%)
Natural	A-hemerobiotic	No	No	No	0	0
Near natural	Oligo	Few	No	Most species spontaneous	<1%	5%
Semi-(agri-)natural	Meso-	Small, superficial	Other life form dominating	Most species spontaneous	1–5%	5–12%
Agricultural	Eu-	Moderate to drastic	Crops dominating	Few species spontaneous	6%	13–20%
Near-cultural	Poly-	Drastic artificial substrate	Open ephemeral	Few to no species	?	21–80%
Cultural	Meta-hemerobiotic	Drastic artificial substrate	–	–	–	–

[a]Hemerocultivated.
[b]Per 1000 km².
Source: van der Maarel (1975). Reproduced by permission of Dr. W. Junk BV, The Hague, The Netherlands.

as the smallest mappable landscape unit, serving as the basis for physical planning and management. Furthermore, there is need for a more concise epistemological definition of the term "ecosystem" and of the term "system" in general, distinguishing between concrete and abstracted systems and between natural–physical, or organismic, systems and cognitive mind–event systems, in order to comprehend the differing interrelations between the noosphere, the biosphere, and the geosphere. Finally, there is a need to introduce the new system–theoretical and human–ecological paradigms into our conceptual framework. This will be attempted in the following chapter.

Conclusion

We have shown that landscape ecology had its roots in Central and Eastern Europe, where biogeographers viewed the landscape not just as an aesthetic asset (as done by most landscape architects) or as part of the physical environment (as by most geographers), but as the total spatial and visual entity of human living space, integrating the geosphere with the biosphere and the noospheric man-made artifacts.

This holistic view of the landscape was first adopted by ecologists who were well versed in field-oriented vegetation science and/or trained originally as agronomists, foresters, gardeners, or planners. These abandoned their narrow professional outlook and dogmatic climax theories and modified the phytosociological methodology into dynamic and integrated field surveys and ecosystem studies.

Landscape ecology is presently viewed in Europe as the scientific basis for land and landscape planning, management, conservation, development, and reclamation. As such, it has overstepped the purely natural realm of classical bioecological sciences and has entered the realm of human-centered fields of knowledge—the sociopsychological, economic, geographic, and cultural sciences connected with modern land uses. This is reflected in the broad range of landscape-ecological planning-oriented and other studies on the complex interrelationships between modern man and his open, cultural, and built-up landscape, aiming at a compromise between the clashing natural, cultural, and socioeconomic demands and, at the same time, at an enrichment of man's biotic environment.

Several recent developments have inclined this trend even more toward a broader interdisciplinary human ecological basis for landscape ecology: The merging of landscape planning with ecological planning as integral parts of regional planning, emphasizing the quantitative assessment of the overall impact of man's land uses and his specific utilitarian requirements of the open landscape, especially in the field of recreation. This is making landscape ecology increasingly the major scientific basis for the creation of more balanced and far sighted policies and decision-making tools, such as those used up to now by economic planners, agronomists, foresters, engineers, and (above all) politicians.

Without doubt, many similar studies are also being carried out in other countries, and especially in the English-speaking world to which the rather vague terms "environmental conservation" or "environmental management" are sometimes given. However, in view of the great challenge of saving and improving man's total living space, there is urgent need for the convergence and consolidation of all these aims and methods toward a global human ecosystem science of landscape ecology. As a first step, this requires a central conceptual framework based on the unifying principles of ecological theory—or more specifically, *human ecosystemology*. This framework will be outlined in the following chapter.

References

Arthur, L. M., T. C. Daniel, and R. C. Boster. 1977. Scenic assessment—an overview. Landscape Planning 4:109–130.

Bakker, P. A. 1979. Vegetaton science and nature conservation. *In:* M. J. A. Werger (Ed.), The Study of Vegetation. Dr. W. Junk, The Hague, pp. 249–288.

Bornkamm, R., and W. Grün. 1973. Eine Versuchsanlage zur Frage der Flachdach Begrünung. Landschaft + Stadt 3:141–143.

Buchwald, K. 1963. Die Industriegesellschaft und die Landschaft. Beitr. z. Landespflege 1:23–41.

Buchwald, K., and Engelhart, W. (Eds.). 1968. Handbuch für Landschaftpflege und Naturschutz. Bd. 1. Grundlagen. BLV Verlagsgesellschaft, Munich, Bern, Wien.

Chorley, R. J., and B. A. Kennedy. 1971. Physical Geography, A Systems Approach. Prentice-Hall Intern., Inc., London.

Christian, C. S. 1958. The concept of land units and land systems. *In:* Proceedings of Ninth Pacific Science Congress 20:74–81.

Christian, C. S., and G. A. Stewart. 1968. Methodology of integrated surveys. *In:* Aerial Surveys and Integrated Studies. Proc. Toulouse Conf., UNESCO, Paris, pp. 233–280.

Dansereau, P. 1957. Biogeography: An Ecological Perspective. Ronald Press, New York.

Dansereau, P. 1966. Ecological impact and human ecology. *In:* F. Fraser Darling and J. P. Milton (Eds.), Future Environments of North America. Natural History Press, Garden City, New York, pp. 425–464.

Dansereau, P. 1975. Inscape and Landscape. The Human Perception of Environment. Columbia University Press, New York and London.

Dansereau, P. 1977. Ecological Grading and Classification of Land-Occupation and Land-Use Mosaics. Geographical Paper No. 58. Lands Directorate Fisheries & Environment, Ottawa, Canada.

Dansereau, P. 1980. The Template and the Impact. INTECOL Bull. 7/8:70–109.

Duffey, E., and A. S. Watt (Eds.). 1971. The Scientific Management of Animal and Plant Communities for Conservation (11th Symposium, Brit. Ecol. Soc., 1970). Blackwell Sci. Publ., Oxford, London, and Edinburgh.

Egler, F. E. 1942. Vegetation as an object of study. Philos. Sci. 9:245–260.

Egler, F. E. 1964. Pesticides in our ecosystem. Am. Sci. 52:110–136.

Ellenberg, H. 1950. Unkrautsgemeinschaften als Zeiger für Klima und Boden. Landwirtschaftliche Pflanzensoziologie 1. Ulmer, Stuttgart.

Ellenberg, H. 1956. Aufgaben und Methoden der Vegetationskunde. Ulmer, Stuttgart.

Ellenberg, H. (Ed.). 1971. Integrated Experimental Ecology. Methods and Results of Ecosystem Research in the German Solling Project. Springer-Verlag, Berlin, Heidelberg, New York.

Ellenberg, H. 1974. Zeigerwerte der Gefässpflanzen Mitteleuropas. Scripta Geobot., Göttingen.

Ellenberg, H. 1978. Vegetation Mitteleuropas mit den Alpen, 2nd edition. Ulmer, Stuttgart.

Haber, W. 1979. Theoretische Anmerkungen zur Ökologischen Planung. Verhandl. Ges. für Okologie 7, Weihenstephan.

Haber, W. 1980. Raumordungskonzepte aus der Sicht der Ökosystemforschung. Forschungs.-und Sitzungberichte der Akad. für Raumforschung und Landesplanung, Hanover.

Kiemstedt, H. 1967. Zur Bewertung der Landschaft für die Erhohlung. 1. Sonderheft d. Beiträge zur Landespflege, Stuttgart.

Kiemstedt, H. (Ed.). 1975. Landschaftsbewertung für die Erhohlung im Sauerland. Institut für Landes- und Stadtentwicklungs Forschung des Landes Nordhein-Westfalen. Band 1.008/1. Dortmund.

Kuchler, A. W. 1967. Vegetation Mapping. Ronald Press, New York.

Kuchler, A. W. 1975. Map of Potential Natural Vegetation of the Conterminous United States. 1:3,168,000. Special Publication No. 36, American Geographical Society, New York.

Langer, H. 1970. Die Ökologische Gliederung der Landschaft und ihre Bedeutung für die Fragestellung der Landschaftpflege. Landschaft + Stadt 3: 2–29.

Langer, H. 1973. Ökologie der geosozialen Umwelt. Landschaft + Stadt 5:133–140.

Leser, H. 1976. Landschaftsökologie. Ulmer (UTV Nr. 521), Stuttgart.

Mueller-Dombois, D., and H. Ellenberg. 1974. Aims and Methods of Vegetation Ecology. Wiley, New York.

Odum, E. O. 1969. The strategy of ecosystem development. Science 164:262–270.

Odum, E. O. 1971. Fundamentals of Ecology, 3rd edition. Saunders, Philadelphia, Pennsylvania.

Olshowy, G. 1973. Landscape planning in the Rhineland Brown Coal Area. *In:* D. Lovejoy (Ed.), Land Use and Landscape Planning. Aylesbury, England, pp. 249–251.

Olshowy, G. 1975. Ecological Landscape Inventories and Evaluation. Landscape Planning 2:37–44.

Pflug, W. 1973. Umweltschutz in Lehre und Forschung in der Technischen Hochschule Aachen. Landschaft + Stadt 5:98–115.

Schmithüsen, J. 1963. Der wissenschaftliche Landschaftsbegriff. Mitt. flor.-soziol. Arbeitgemeinschaft 10:9–19.

Schreiber, K.-F. 1976. Zur Sukession und Flächenfreihaltung auf Brachland in Baden-Würtenberg. Verhandl. Ges. für Ökologie 1976. Dr. W. Junk, The Hague, pp. 251–303.

Schreiber, K.-F. 1977a. Landscape planning and protection of the environment. The contribution of landscape ecology. Appl. Sciences and Development 9: 128–139.

Schreiber, K.-F. 1977b. Naturschutz und Flurbereinigung-einige Bemerkungen zu einem viel diskutierten Problem. Natursch. Landschaftpfl. 27:48–51.

Stumpel, A. H. P., and J. T. R. Kalkhoven. 1978. A vegetation map of The Netherlands based on the relationship between ecotopes and types of potential natural vegetation. Vegetatio 37:163–173.

Sukopp, H. 1970. Charakteristik und Bewertung der Naturschutzgebiete in Berlin (West). Natur und Landschaft 45:133–139.

Sukopp, H. 1971. Bewertung und Auswahl von Naturschutzgebieten. Schr. Reihe für Landschaftpflege und Naturschutz 6:183–194.

Sukopp, H. 1972. Wandel von Flora und Vegetation in Mitteleuropa unter dem Einfluss des Menschen. Ber. Landwirtschaft 50:112–139.

Teilhard de Chardin, P. 1966. Man's Place in Nature. Collins, London.

Trautmann, W. 1973. Vegetationskarte der Bundesrepublik Deutschland 1: 200,000 Potentielle natürliche Vegetation Blatt CC 5502 Köln. Schrift. für Vegetationskunde 6.

Trent Forschunggruppe. 1973. Typologische Untersuchungen zur rationellen Vorbereitung umfassender Landschaftsplanungen. Dortmund/Saarbruecken.

Troll, G. 1971. Landscape ecology (geo-ecology) and bio-ceonology—a terminology study. Geoforum 8:43–46.

Tüxen, R. 1956. Die heutige potentielle natürliche Vegetation als Gegenstand der Vegetationskartierung. Angew. Pflanzensoziologie, Stolzenau/Weser 13: 5–42.

Tüxen, R. (Ed.). 1968. Pflanzensoziologie und Landschaftsökologie. Intern. Symp. Intern. Ver. für Vegetationskunde Stolzenau und Rinteln.

Tüxen, R., and H. Diemont. 1937. Klimaxgruppe und Klimaxschwarm. Jber. Naturhist. Ges. Hannover 88/89:73–87.

van Leeuwen, C. G. 1966. A relation theoretical approach to pattern and process in vegetation. Wentia 15:25–46.

van Leeuwen, C. G. 1973. Oecologie en natuurtechniek. Natur en Landchap 27: 57–67.

van der Maarel, E. 1971. Plant species diversity in relation to management. In: E. Duffey and A. S. Watt (Eds.), The Scientific Management of Animal and Plant Communities for Conservation. Blackwell Sci. Publ., Oxford, London, Edinburgh, pp. 45–63.

van der Maarel, E. 1975. Man-made natural ecosystems in environmental management and planning. In: W. H. van Dobbin and R. H. Lowe-McConnel (Eds.), Unifying Concepts in Ecology. Dr. W. Junk, The Hague, pp. 263–274.

van der Maarel, E. 1977. Naar een globaal ecologisch model voor de ruimte ontwikkeling van Neerderland [Toward a global ecological model for physical planning in the Netherlands]. Ministry of Housing and Physical Planning, The Hague, The Netherlands.

van der Maarel, E. 1978. Ecological principles for physical planning. In: M. W. Holgate and M. J. Woodman (Eds.), The Breakdown and Restoration of Ecosystems, NATO Conf. Ser. 1 (Ecology), Vol. 3. Plenum Press, New York and London, pp. 413–450.

van der Maarel, E., and A. H. P. Stumpel. 1975. Landschaftsökologische Kartierung und Bewertung in den Niederlanden. Verhandl. Ges. für Ökologie 1974. Dr. W. Junk, The Hague, pp. 231–240.

Vernadsky, W. I. 1945. The biosphere and the noosphere. Am. Sci. 33:1–12.

Vink, A. P. A. 1975. Land Use in Advancing Agriculture. Springer-Verlag, Berlin, Heidelberg, New York.

Walter, H. 1960. Standortlehre (Analytische-ökologische Geobotanik). (Einführung in die Phytologie; Grundlagen der Pflanzenverbreitung.) Ulmer, Stuttgart.

Walter, H. 1964. Die Vegetation der Erde in öko-physiologischer Bertrachtung, Band 1: Die tropischen und subtropischer Zonen. E. B. Gustav Fischer Verlag, Jena.

Walter, H. 1968. Die Vegetation der Erde in öko-physiologischer Bertrachtung, Band 2: Die gemässigten und arktischen Zonen. Ulmer, Stuttgart.

Walter, H. 1973. Vegetation of the Earth in Relation to Climate and the Eco-Physiological Conditions. Springer-Verlag, New York, Heidelberg, Berlin.

Westhoff, V. 1971. The dynamic structure of plant communities in relation to the objectives of conservation. *In:* E. Duffey and A. S. Watt (Eds.), The Scientific Management of Animal and Plant Communities for Conservation. Blackwell Sci. Publ., Oxford, London, Edinburgh.

Westhoff, V., and E. van der Maarel. 1978. The Braun-Blanquet Approach, 2nd edition. *In:* R. H. Whittaker (Ed.), Classification of Plant Communities. Dr. W. Junk, The Hague, pp. 287–399.

Whyte, R. O. 1976. Land and Land Appraisal. Dr. W. Junk, The Hague.

Woebse, H. H. 1975. Landschaftsökologie und Landschaftsplanung. R. B. Verlag, Graz.

Young, G. L. 1974. Human ecology as an interdisciplinary concept: A critical inquiry. *In:* A. MacFayden (Ed.), Advances in Ecological Research. Academic Press, New York, pp. 1–105.

Zonneveld, I. S. 1972. Textbook of Photo-Interpretation, Vol. 7. (Chapter 7: Use of aerial photo interpretation in geography and geomorphology). ITC, Enschede.

Zube, E. H., R. O. Brush, and J. G. Fabos (Eds.). 1975. Landscape Assessment: Values, Perception and Resources. Dowden, Hutchinson, & Ross, Stroudsburg, Pennsylvania.

Recent Developments in Landscape Ecology

The Development of Landscape Ecology as a Global Science and the Contribution of the International Association of Landscape Ecology (IALE)

In recent years landscape ecology has undergone rapid development and has evolved from a strictly Central and Eastern European science into a global science. The International Association of Landscape Ecology (IALE) has served as an important catalyst for the consolidation of landscape ecology as an international and interdisciplinary science. The IALE, which has aproximately 1200 members in 42 countries from all continents, including several developing countries in Asia, Africa, and Latin America, is a major channel for communication through the bi-annual issue of the IALE news bulletin (published also in the **CSSR** *Ecologia* journal); the organization of international conferences, symposia, and seminars; and the establishment of international working groups.

A typical example for these important functions of IALE is the progress of landscape ecology in China, which was stimulated by earlier contacts of Chinese scientists with landscape ecologists in these meetings. This led to the establishment of an active branch of IALE (Chinese Association of Landscape Ecology—CALE), which is publishing a *CALE Bulletin* and is organizing meetings. A considerable amount of landscape ecological research is carried out already in different parts of China, sponsored chiefly by the Chinese National Foundation for Science. Two important centers for research and teaching are the Department of Landscape Ecology at the Institute for Applied Ecology, Academia Sinica, in Shenyang, under the leadership of Professor Xio Duning, and the Laboratory of Landscape Ecology at the Department of Geography of Peking University, under the leadership of Professor Chen Changdu.

An additional important achievement of IALE was the founding of the journal *Landscape Ecology* in 1987, thereby opening a special, permanent forum for current landscape-ecological studies. More and more articles devoted to landscape ecology are appearing in other scientific journals in a variety of languages, and there is a growing stream of books dealing wholly or partly with this subject.

Recent Important Publications on Landscape Ecology

It would be very difficult to cover in a comprehensive way the almost overwhelming diversity of topics which constitutes the present body of landscape ecology and to give full justice to the great strides made in recent years. We will attempt to provide a brief overview and for this purpose will rely chiefly on recently published papers and books and on the reports of meetings and other activities.

The journal *Landscape Ecology* is one important source. Wiens (1992) has recently attempted to answer the question, What is landscape ecology really? by screening the topics of the papers published in the first five volumes of this journal (in which contributions from North American authors still dominated). He found that about half of these dealt with landscape structure, 40% focused on land use, and 21% on plants; larger landscapes on the scale of hectares to kilometers were preferred by 73% and the medium scale by 20%. Issues of scaling occupied 24%; spatial pattern development, 21%; boundary flows, 9%; and disturbance, 8%. Although the papers were predominantly descriptive and conceptual, an increasing emphasis on quantification, computer modeling, and simulation was evident. Wiens concluded from his survey that landscape ecology is less concerned with theory and hypothesis-testing and more with problems addressing habitat fragmentation, reserve design, biological diversity, resource management, and sustainable development. For this purpose it should become more rigorous and quantitative and include investigations of a variety of organisms and systems over a range of spatial scales. Studies in landscape ecology should test formal propositions, especially through experimentation. These tendencies are already apparent in many of the European landscape-ecological studies, extensively reviewed by Leser (1991) (see below)—alas, in German.

In recent years contributions from European, Asian, and other countries in *Landscape Ecology* have increased considerably. Of relevance in this context is the special issue devoted to landscapes of France at multiple scales, edited by Decamps and Lefeuvre (1992). Another important issue of *Landscape Ecology*, dealing with coastal dune landscapes (van der Meulen et al., 1991) was based chiefly on presentations of European landscape ecologists at a conference on Landscape-Ecological Impact of Climatic Change (LICC).

The contents of the sequence of volumes devoted to landscape ecology and published by Springer-Verlag, for which the first edition of our book served as a pioneer, reflect many of the trends in concepts and methods which have emerged in recent years.

Landscape Heterogeneity and Disturbance, edited by Turner (1987), was the outcome of the first symposium of the US chapter of IALE. It represented approaches shared at that time by most American landscape ecologists. These approaches were based on the concepts put forth by Forman and Godron (1986) in their influential book on landscape ecology. They viewed landscape ecology as the study of spatially heterogeneous areas on scales of tens to hundreds of kilometers, composed of a cluster of systems containing patches of different shapes, numbers, kinds, configurations, and functions. This kind of "spatial landscape ecology" can be distinguished from "ecosystem ecology" chiefly by its focus on spatial heterogeneity, emphasizing ecological effects of the spatial patterning of ecosystems over large areas of landscape mosaics (Risser, 1987).

Further attempts to deal with spatial heterogeneity at different scales were presented in a special issue of *Landscape Ecology* (Dale et al. 1989), and also in the following Springer volumes.

Changing Landscapes: An Ecological Perspective edited by Zonneveld and Forman (1990), emerged from a symposium at the Fourth International Congress of Ecology at Syracuse, New York, in 1986. In contrast to *Landscape Heterogeneity, and Disturbance*, it contained an interesting mixture of different approaches. The "American" version of landscape ecology was presented in case studies from different, mostly coarse-grained landscapes. In general, the European landscape ecologists paid more attention to the smaller-scale dimensions of their fine-grained landscape units. These were treated not merely as "patches" but as "ecotopes" with distinct system properties, described in detail in Chapter 2 of this book. Zonneveld's chapter outlined the premises of holistic and transdisciplinary landscape ecology. He presented in a lucid way the closely interwoven three-dimensional scopes of topological (vertical) heterogeneity, the chorological (horizontal) heterogeneity, and the geospheric-global heterogeneity of landscapes. The close integration of landscape ecology with land-use planning and decision-making in Europe was shown in papers on the history of landscape ecology in Europe by Schreiber, on planning and management in Germany by Haber, and in Czechoslovakia by Ruzicka and Miklos. Some of these advances will be discussed further in the supplements of the relevant chapters.

Innovative approaches and methods to the study of agricultural landscapes were introduced by the chapters of Burel and Baudry in France and Merriam in Canada. Shifting the focus from homogeneous, individual fields and hedgerows to a view of landscapes as ecological systems, Burel and Baudry showed the important beneficial roles played by hedgerow network landscapes and their interacting and interconnecting biological and ecological functions. In spite of the great geographical distance separating them, both studies were closely related by the important concept of *connectivity*, applied by Merriam as a functional parameter of landscape connectedness by structural links and corridors such as woodland patches, hedgerows, fencerows, and roads. These links connect subpopulations of organisms into functional demographic units and into metapopulations; and subunits of nutrient pools are interconnected as fluxes into larger pools.

In an important international IALE symposium dealing with the theoretical and practical implications of landscape connectivity (Schreiber, 1988), Baudry and Merriam distinguished between connectedness as the quantitative measure of links between mappable elements, and connectivity as the measure of the process of animals moving among landscape elements.

Quantitative Methods in Landscape Ecology, edited by Turner and Gardner (1990), is a comprehensive presentation of new methods and approaches to quantitative analysis of landscape heterogeneity developed recently in North America. For this purpose efficient use has been made of recent advances in computer hardware, remote sensing, geographical information systems (GIS) information theory, hierarchy theory, percolation theory, fractal geometry, and model developing. For future directions, the editors regard as most important the testing of hypotheses in actual landscapes and the joint progress of theoretical and empirical work, ideally through an iterative sequence of model and field experiments. Most of these models are addressing basic questions about the influences of spatial patterning on ecological dynamics. Some of the methods and indices for quantifying spatial patterns described in this compilation are already used widely by researchers and land managers. This rich volume can be regarded as an important landmark in the development of what could be called "quantitative spatial landscape ecology." Those advances which are most relevant for further development of holistic, quantitative landscape ecology will be discussed briefly in Chapter 2.

The most recent Springer volume, *Landscape Boundaries*, edited by Hansen and di Castri (1992), presents a wealth of information on the influence of ecotones on biodiversity; on flows of energy, material, organisms, and water; and on innovative methods for their study. It is described by its editors as an attempt to integrate the concept of ecotone with patch dynamics theory and to enrich "landscape theory" (without clearly defining what is meant by this term), and to serve as a catalyst for the "next generation of landscape research and management" (Hansen et al., 1992). Probably the most important implication of this research is the new insight gained into ecotone dynamics in space and time (Delcourt and Delcourt, 1992), the use of innovative methods and models for the monitoring of ecotones to detect global change at different scales (Wein-stein, 1992), and the realization of the need for holistic landscape con-servation strategies to optimize biodiversity (Hansen et al., 1992).

Recent Developments in Europe

Recent conceptual and methodological developments of landscape ecol-ogy in Europe have been summarized by Leser (1991) in a very thorough manner in the third and completely revised edition of *Landschaftsoekolo-gie*. The introduction of the term "geo-ecology" as a synonym for *land-scape ecology* and the distinction between terms such as *biosystems* and *geosystems*, *biotopes*, and *geotopes* have created considerable confusion and controversy between European landscape ecologists and geog-raphers. Leser attempted to overcome these contradictions by coining the term *Landschaftoekosystem* ("landscape ecosystem") to signify a unifying concept treating landscapes as the total living space. According to Leser (1991, page 181), these concrete interaction systems of natural and anthropogenic components are almost identical with our definition of the Total Human Ecosystem landscapes described in Chapter 2. This is true also for his holistic system-theoretical interpretations of landscape ecolo-gy, which occupy a transdisciplinary position between the disciplines, with a problem-solving orientation in a very broad range of local, regional, and global environmental issues. The great advances achieved in recent years in landscape ecology in Central and Eastern Europe are very evi-dent in the second part of Leser's book, which deals with the methodolo-gy of landscape ecological analysis and synthesis and provides practical examples of their application in the field.

The teaching, research, and planning centers in Europe mentioned in Chapter 1 have continued and for the most part expanded their activities, and new centers have been established. One of the most encouraging results is the breeding of a new generation of landscape ecologists, well-trained and highly motivated, who will make their mark not only on Europe but also on the global scene.

This is very apparent from a recent special issue in honor of the 25th anniversary of the Chair in Landscape Ecology at the Technical Univer-sity of Munich in Weihenstephan under Professor W. Haber, published by his assistants and former students (Duhme et al., 1992). Its 25 con-tributions are an excellent demonstration of the broad spectrum and the high scientific standards reached in problem-solving-oriented landscape-ecolgical research and planning, inspired by the admirable leadership of Professor Haber.

In Great Britain a regional branch of IALE has been formed under the leadership of Dr. R. G. H. Bunce at the Merlewood Research Station. At Wye College of the University of London, a full postgraduate program in landscape ecology and management is offered by B. H. Green, the Sir Cyril Kleinwort Professor of Countryside Management, and his colleagues. This is the first of its kind in any English-speaking country, and we hope that this will soon be followed by the establishment of a global network in universities and professional training centers of full diploma and postgraduate curricula in landscape ecology and related fields.

In one of the younger and more dynamic centers for landscape ecology at the Institute for Geography, Socio-Economic Analysis, and Computer Science at the Roskilde University in Denmark, Dr. J. Brandt (the editor of the *IALE Bulletin*) and his colleagues have organized two successful international IALE seminars which became important further steps in the development of landscape ecology. The first seminar, held in 1984 (Brandt and Agger, 1984), was noteworthy for its attempt to provide broad coverage of large-scale holistic landscape-ecological orientations in research and planning. The second seminar (Brandt, 1991) concentrated on acute European problems. Its title, "Practical Landscape Ecology in Europe," indicated the strong tendency of European landscape ecologists to focus on the solution of pressing environmental problems, including rural land use, urban and suburban environments, landscape-ecological consequences of industrialization, environmental assessment, remote sensing applications, landscape restoration and forestry in less favorable areas, cultural aspects, among others. A special plenary session was devoted to the rapidly growing involvement of landscape ecologists in the research program of the European Community (EC). This has been recognized now by the ELEC, by including landscape ecology for the first time explicitly in its 1993 research program.

The involvement of European landscape ecologists in developing countries was demonstrated by a lecture on the application of remote sensing and GIS in Java, Sri Lanka, and Thailand by D. van der Zee (1991), who is continuing the important work of I. S. Zonneveld (see Chapter 1) at the International Institute for Aerospace Survey and Earth Science (ITC) in research and training landscape ecologists from developing countries. K.-F. Schreiber (1991), from the Institue of Geography at Muenster University, reported on his joint landscape-ecological studies with Israeli scientists on the ecotope structure of the Negev Highlands and their utilization. Another important study—one of many carried out under his guidance—has dealt with monitoring agrosystems at the African Sahel (Groten, 1991). As mentioned in Chapter 1, the chair for landscape ecology at Muenster University was established by the "father of landscape ecology," G. Troll, within the Institute of Geography. It has now been upgraded to an independent Institute of Landscape Ecology. The institute at Muenster has become one of the most active centers for landscape-ecology teaching and research and hosted the Second International Seminar of IALE on Connectivity (Schreiber, 1988).

In this symposium, many landscape ecologists from Eastern Europe reported on their intensive studies of the severe problems resulting from the land degradation, dereliction, and pollution that are the unfortunate "legacy" of the previous communist regimes.

The long tradition of holistic problem-solving and the active intervention of landscape ecologists in the planning and decision-making process in Czechoslovakia is evident in the reports of the international symposia which have been organized for more than twenty years by the Department of Landscape Ecology in the Institute of the Slovak Academy of Sciences. In a symposium dealing with spatial and functional relationships in landscape ecology (Ruzicka et al., 1988), Miklos et al. (1988) reported on the interdisciplinary ecological evaluation of the controversial Hrusov retentation reserve of the Danube River, carried out under the leadership of this institute. The study made use of the integrated method of landscape-ecological planning (LANDEP), which was conceived by the head of the institute, Dr. M. Ruzicka, jointly with Dr. L. Miklos and their colleagues (see also Ruzicka and Miklos, 1990). Dr. Miklos has been very active as Deputy Minister of Environment in Slovakia in the development of legislation for landscape-ecological planning in his country. He was instrumental in preparing and submitting a special statement of *IALE* at the Rio Earth Summit. In the symposium on spatial and functional relationships, several prominent landscape ecologists from North America presented their work and exchanged ideas with European participants. In a discipline which has had distinct differences in approaches on the two sides of the Atlantic, this was an important step in their gradual convergence into a unified, broad stream of theory and practice.

In recent years, landscape ecology has spread its wings in Europe further south to the Mediterranean Basin. In Italy, a dynamic IALE branch has been established by mostly young scientists in close cooperation with the Italian Ecological Society. These activities, led by Dr. A. Farina, the director of the Lunigiana Museum of Natural History, culminated in an international conference on "The Future of Mediterranean Landscapes," to which we will refer further in the supplement to Chapter 4.

There is a growing awareness of the importance of safeguarding not only rare and threatened plants and animals and unique ecosystems, but whole landscapes and their physical, ecological, and cultural assets. This challenge has been accepted by a special Working Group on Red Books for Threatened Landscapes (Naveh, 1992). This joint Working Group of the IUCN Commission on Environmental Strategy and Planning and IALE has broadened its scope and is called now Working Group on Landscape Ecology and Conservation and the Red Books will be called from now on Green Books for landscape conservation, to give them a more positive connotation. It consists of experienced landscape ecologists and prominent members of IUCN, who are submitting to IUCN a proposal for preparation of world-wide Red Lists of Endangered Landscapes, and initiating case studies of Green Books for Landscape Conservation.

The first case study has been carried out in western Crete (see also Chapter 4 supplement) and further examples are planned in several industrial and developing countries, in which a comprehensive methodology for "rescuing" endangered landscapes will be worked out. Based on integrated ecological and socioeconomic surveys and scenarios, these Green Books should serve as practical guidelines for political and professional decision-makers. They will also provide strategies for the preservation and restoration of "landscape ecodiversity": the natural and cultural landscape diversity, through multibeneficial, sustainable land-use practices (Naveh, 1993). Such Red Books were recommended for Mediterranean landscapes in the first edition of this book. They will now be put into practice on a global scale.

Finally, it is worthwhile to mention that European landscape ecologists are taking the lead in all active IALE Working Groups. These deal with landscape ecology of agrosystems and alluvial rivers, urban ecology, geographical information systems, cultural aspects of landscape ecology, ecological infrastructure, and landscape-ecological planning.

Landscape Ecology in Canada and the World Congress of Landscape Ecology in Ottawa—July 1991

In Canada, as in Europe, there is a long tradition of landscape-related work and a strong group of professional land planners and managers working closely with ecologists and geographers from universities. As a result, there is also a great demand for a discipline focusing on problem-solving at the scale of the landscape which requires knowledge of the complex relationship among socioeconomic, biological and physical components of our environment (Moss, 1988). This broader transdisciplinary view of landscape ecology was explicit in the central lectures by Rowe (1988) and by Rubec et al. (1988).

It was therefore not by chance that the IALE World Congress of 1991 became an important tool for promoting the convergence of these different trends.

The plenary lecture by Professor H. Decamps from the Centre des Ressources Renouvelables at Toulouse, France, emphasized the consequences of human-caused modifications on the land-water interface, and especially on the activity of river ecotones as buffers against nutrient fluxes and as corridors for species expansion along river networks.

A second plenary lecture was presented by Dr. P. Opdam of the Department of Landscape Ecology at the Research Institute for Nature Management in Leersum, The Netherlands, and chief editor of *Landschap*, the Dutch scientific journal of landscape ecology.

Opdam is leading an IALE Working Group on "Ecological Infrastructure" which is concerned with the spatial characteristics of landscape fragmentation, habitat patches, corridors, and barriers, and their effect on plant and animal populations as a basis for landscape conservation planning. In his plenary lecture he pointed out that landscape ecology must offer the basis for conservation policy by indicating problems in the early stages, predicting the outcome of measures to restore spatial relations, comparing spatial planning scenarios, and assessing impacts of further threats to landscape function. For this purpose it is essential to ensure close intergration of process-oriented and pattern-oriented investigations.

The themes in the IALE World Congress reflect the breadth of topics subsumed within the broad discipline of landscape ecology: the ecology of fluvial landscapes, river restoration, management of urban green space, suburban reforestation strategies, the use of satellites (SPOT) for prediction of barn owl breeding, the influence of boundary on habitat management by elk and mule deer, ecological regionalization in China, planning for nature restoration in The Netherlands, effects of land fragmentation in Australia, acid precipitation in Canada, and landscape-ecological effects of agricultural collectivization in Czechoslovakia. Some of these representations have been summarized in a lively way as "brief report for time-limited decision makers" (IALE, 1992). The strong trend towards practical, problem-solving-oriented landscape ecology was apparent and best expressed in the title of a lecture presented by three Australian wildlife ecologists (Hobbs et al., 1991): "Integrated Landscape Ecology— Doing Rather Than Describing."

The Eight Annual U.S.–I.A.L.E.–Symposium, Oakridge, March 1993

A similar trend was discernible in this meeting, sponsored by the Environmental Sciences Division, with the largest number of active landscape researchers in the USA. These had a very significant influence on the development of above-described quantitative spatial landscape ecology and also contributed much to the success of this meeting.

In addition to the almost overwhelming variety of innovative approaches and methods in spatial landscape ecology offered, more than half of the lectures were delivered by professional land managers, foresters, conservationists and restorationists who collaborated mostly with research scientists. Such close teamwork could serve as the most promising model for landscape-ecological research, distinguishing it from most other ecological disciplines, especially if these teams include specialists from the humanities and social sciences. This was indeed the case in some of the studies presented. There seems to be a growing realization now among American landscape ecologists that we are dealing mostly with semi-natural and cultural landscapes which have been shaped in the past and are driven presently by a closely interwoven network of natural and cultural forces. At least a third of the lectures were devoted to problems of human land uses and to agricultural and urban landscapes. In some of these, integrative ecological and socio-economic models were applied thereby fulfilling the plea expressed by Robert O'Neill, of Oakridge, in his opening plenary lecture "to incorporate more socio-economic theory." Several central speakers went even farther in their attempts to broaden the narrow bio-ecological conceptual and methodological framework which has characterized this American version of landscape ecology. This meeting has shown clearly that Amercian landscape ecologists have not only made further important advances in their quantitative spatial methods and their practical application, but they have also made an important step toward a broader transdisciplinary and holistic conception of landscape ecology. This reflects very well the emerging trend toward a gradual amalgamation of the different versions of landscape ecology into a broader and chiefly problem-solving oriented global science (which was noticable already at the Fourth World Congress in Ottawa, in 1992).

Cultural Aspects of Landscape Ecology

Cultural, social, aesthetic, and economic considerations occupied an important place in the Ottawa conference on landscape ecology. In a colloquium on the implications of social perceptions for landscape integrity, a strong case was made for the need to integrate the conservation of biodiversity with cultural and aesthetic values. The need to appreciate cultural perceptions of local people, especially farmers, and the need to gain their acceptance by communication and education for the implementation of landscape conservation and restoration plans was also discussed.

Interactions between culture and landscapes were also discussed in a well-attended workshop reflecting the growing interest of landscape ecologists in human ecological aspects, such as the role of intrinsic, traditional, and other "soft" landscape values. Is it possible to preserve these values together with landscape patterns and processes which have evolved in the past and are maintained by vanishing traditional land-use practices? These practices often clash with prevailing cultural perceptions of resourcism and utilitarian "economic rationality" in diverse ecological and cultural situations, such as the Mediterranean uplands, the Brazilian Amazon, and the Japanese coast.

Cultural aspects of landscape was the theme of 1989 conference organized in Barn, The Netherlands, by the IALE World Group on Culture and Landscape. The conference brought together a multidisciplinary group of landscape ecologists, planners and architects, ecologists, foresters and geographers, environmentalists, psychologists, anthropologists, artists, and museum curators (Svobodova, 1990). All were representatives of the "two cultures" which landscape ecology can bridge (Naveh, 1990). The organizer noted, "We need to link arms if we are to protect nature and landscape, it is both our moral duty and the duty of contemporary culture" (Svobodova, 1990).

The Place of Landscape Ecology in the Scientific Community

The recognition of landscape ecology as a distinct discipline by the scientific community, especially ecologists and geographers, has been slow. However, an important turning point in its international status occurred in 1986, when landscape ecology was chosen at the Fourth International Congress of Ecology (INTECOL) at Syracuse, New York, as one of the central themes of the plenary lectures (Naveh, 1986) and IALE was invited to organize a symposium. At the fifth INTECOL Congress in Yokohoma, Japan, 1990, it occupied an even more prominent place—four symposia were presented as well as a great number of lectures and posters.

Landscape ecology has found strong support in Japan from prominent ecologists, including Professor A. Miyawaki of Yokohama National University and Professor M. Numata, Director of the Natural History Museum and Institute at Chiba.

Internationally, there is a growing appreciation of the relevance of landscape ecology to the such pressing and complex ecological and sociological problems as land degradation and desertification, habitat fragmentation, and loss of biodiversity, attractiveness and recreation amenities and its importance to conservation and restoration ecology. Landscape ecologists are presenting their studies in growing numbers in other international meetings and are taking an active part in discussions, consultations, and policy-making concerning crucial environmental issues. For example, the 1989 European Conference "Landscape-Ecological Impact of Climatic Change" in Lunteren, The Netherlands (Boer and De Groot, 1990), resulted in an ongoing research program sponsored by the Dutch government and the European Community.

A statement by IALE to the Rio Earth Summit offered its capacities to the post-Rio need to develop projects aimed at achieving sustainability of regions and to educate researchers, planners, and managers.

In spite of these developments, a number of scientists claim that there is nothing original or unique about landscape ecology which would justify its recognition as a distinct scientific discipline. Among those criticizing landscape ecology and its holistic and transdisciplinary nature are those who Bohm and Peat (1987) have noted are focused only on "hard facts and logic." They have little interest in "soft" contents or intrinsic values and intangibles grounded in philosophy. Similarly, the "non-mathematical, and even non-scientific" humanities, such as arts and history, carry little professional value for them.

To such claims Bohm and Peat reply that "in order to sustain the creative activity of mind, and of the ongoing scientific development, it is necessary to remain sensitive to the ways in which similarities and differences are developing, and not to oversimplify and situation by ignoring or minimizing their potential importance."

They point out that this process is hampered by the tendency of scientists "to cling rigidly to familiar ideas in order to maintain a habitual sense of control and security and not to break the old patterns of thought." According to Kuhn (1970), these patterns have been established as the paradigm of "normal science": a way of working, thinking, communicating, and perceiving with the mind an unconscious or tacit form of consent. In biology and ecology this is especially true for those paradigms which are grounded in a narrow reductionist and positivist perception of science, ignoring the broader cultural contexts with which landscape ecology must deal.

Conclusions

This brief overview has summarized a number of indications that landscape ecology is coming of age. This can be explained by several mutually enforcing factors:

1. Landscapes can be recognized as tangible and heterogeneous but closely interwoven natural and cultural entities of our total living space. The health and integrity of landscapes are of vital importance for global survival.
2. As an approach to landscape study and management, the conventional, discipline-oriented, and mostly reductionist scientific paradigms can be replaced by more integrative, holistic, and transdisciplinary approaches and methods, based on a systems view.
3. There has been a dramatic rise in our ability to deal with landscapes holistically, thanks to advances in remote sensing and satellite images with finer and finer resolutions over larger and larger areas, combined with progress in processing larger masses of data in smaller and cheaper computers with more refined methods.

These achievements enable us to identify, analyze, synthesize, and (at least partly) quantify in more holistic ways the complex natural and cultural patterns and processes occurring on tangible stretches of land, ranging from a couple of square meters to regional and global scales. Landscape ecology has thereby acquired the capacity to grow from a descriptive science to a more rigorous prescriptive and anticipatory science.

This will enable it to act in a more efficient way as a goal-directed science for the study of the complexity of our Earth's landscapes, and for the safeguarding and improving of their intergrity, health, and natural and cultural diversity.

We believe that the three major challenges for further developments of landscape ecology, as presented in this volume, lie in the following mutually-reinforcing directions.

(1) We have to deepen our understanding, broaden our concepts, and improve our methods for the study, management and development of our local, regional, and global landscapes. For this purpose we have to capitalize on above-described developments for the application of integrative, quantitative methods and models which should enable us to cope in a more comprehensive way with the complex interactions between human and natural systems on landscape scales.

(2) In order to become such a transdisciplinary, problem-solving science, we have to try even harder and involve even more scientists and professionals with different backgrounds and outlooks in landscape study and management. The same time, we must avoid the dangerous split between theoreticians and practitioners, between so-called basic and applied researchers, and between purely biocentric and anthropocentric orientations. Unfortunately, this has not been achieved in any other branch of ecology, geography, or other environmental science. As will be shown in Chapter 2 and its supplement, landscape ecologists can accept this challenge by further developing sound transdisciplinary concepts and methods and by spending greater efforts in academic and professional education. Of greatest importance in this respect, would be the development of interdisciplinary teaching programs in which landscape ecology and management could become the bridging core course.

(3) As will be pointed out in the context of mediterranean landscapes (Chapter 4), we have to develop more efficient communication tools. This should enable us to transfer our scientific information into practical and usable information and help to persuade land manager, users, decision makers, and the public to choose the right options in order to achieve the final goal of environmental managment—namely a new symbiosis between man and nature in the post-industrial society.

(4) Because of the integrative and synthetic nature of landscape ecology, it occupies a unique position, and herein lies its greatest challenge. In our opinion, landscape ecology can accept this challenge by further developing sound transdisciplinary concepts and methods, in order to address its central theme: the study of the complexity of earth's Total Human Ecosystem landscapes and the safeguarding of their intergrity, health, and natural and cultural diversity.

References

Baudry, J., and Merriam, H. G. 1988. Connectivity in Landscape Ecology. Proc. 2nd Intern. Semin. of IALE, Muenster 1987, Muenstersche Geographische Arbeiten 29, pp. 23–28.

Boer, M. M, and R. S. De Groot. 1990. Landscape-Ecological Impact of Climate Change. IOS Press, Amsterdam.

Bohm, D., and F. Peat. 1987. Science, Order, and Creativity. A Dramatic New Look at the Roots of Science and Life. Bantam Books, New York.

Brandt, J. (Ed.). 1991. Practical Landscape Ecology. Proceedings of the European Seminar of the International Association for Landscape Ecology (IALE), Roskilde University Centre, Denmark, May 2–4, 1991. Roskilde Universitetsforlag GeoRue, Roskilde, Denmark.

Brandt, J., and P. Agger (Eds.). 1984. Methodology in Landscape Ecological Research and Planning. Proceedings of the First International Seminar of the International Association for Landscape Ecology (IALE), Roskilde University Centre, October 15–19, 1984. Roskilde Universitetsforlag, Roskilde, Denmark.

Dale, V. H., R. H. Gardner, and M. G. Turner (Eds.). 1989. Predicting across scales. Theory development and testing. Landscape Ecology 3. (Special Issue).

Decamps, H., and J. C. Lefeuvre (Eds.). 1992. Landscape of France at multiple scales. Landscape Ecology 5 (Special Issue 3).

Delcourt, P. A., and H. R. Delcourt. 1992. Ecotone dynamics in space and time. In: A. J. Hansen and F. di Castri (Eds.). Landscape Boundaries. Springer-Verlag, New York, pp. 19–54.

Duhme, F., R., Lenz, and L. Spandau (Eds.). 1992. 25 Jahre Lehrstuhl fuer Landschaftsoekologie in Weihenstephan mit Prof. Dr. h.c. W. Haber. Landschaftsoekologie Weihenstephan.

Forman, R. T. T., and M. Godron. 1986. Landscape Ecology. Wiley and Sons, New York.

Groten S. M. E., 1991. Satellitenmonitoring von Agrar-Oekosystemen in Sahel. Inst. Fuer Geographie der Westfaelischen Wilhelms-Universitaet Muenster.

Haber, W. 1990. Using landscape ecology in planning and management. In: I. S. Zonneveld and R. T. T. Forman (Eds.), Changing Landscapes: An Ecological Perspective. Springer-Verlag, New York, pp. 217–232.

Hansen, A. J., and F. di Castri (Eds.). 1992. Landscape Boundaries. Springer-Verlag, New York.

Hansen, A. J., P. G. Risser, and F. di Castri. 1992. Epilogue: Biodiversity and ecological flows across ecotones. In: A. J. Hansen and F. di Castri (Eds.), Landscape Boundaries. Springer-Verlag, New York, pp. 423–438.

Hobbs, R. J., D. A. Saunders, and G. W. Arnold. 1991. Integrated landscape ecology—doing rather than describing. IALE World Congress of Ecology 1991 Abstracts 56. Carleton University, Ottawa, Canada.

IALE. 1992. Scanning the Mosaic. Brief Reports from the IALE World Congress of Landscape Ecology 1991. Carleton University, Ottawa, Canada.

Kuhn, T. S. 1970. The Structure of Scientific Revolution. University of Chicago Press, Chicago.

Leser, H. 1991. Landschafts-oekologie. Eugen Ulmer GmbH & Co., Stuttgart. UTB, Uni-Taschenbuecher 521.3. Auflage.

Merriam, G. 1990. Ecological processes in the time and space farmland mosaics.

In: I. S. Zonneveld and R. T. T. Forman (Eds.), Changing Landscapes: An Ecological Perspective. Springer-Verlag, New York, pp. 121–126.

Miklos, L., M. J. Lisicky, and M. Kosova. 1988. Ecological evaluation of the special-interest area of the Hrusov retention reservoir. *In*: M. Ruzicka, T. Hrnciarova, and L. Miklos (Eds.), Structural and Functional Relationships in Landscape Ecology. Proceedings of the VIIIth International Symposium on Problems of Landscape Ecological Research, October 3–7, 1988, Zemplinaka Birava CSSR, pp. 9–22.

Moss, M. R. (Ed.). 1988. Landscape Ecology and Management. Proceedings of the First Symposium of the Canadian Society of Landscape Ecology and Management, University of Guelph, May 1987. Polyscience Publications Inc., Montreal, Canada.

Naveh, Z. 1986. Landscape ecology as a multidimensional science for global survival. Plenary Lecture, 4th International Congress of Ecology (INTECOL), Syracuse, USA. New York, August 1986 (manuscript).

Naveh, Z. 1990. Landscape ecology as a bridge between bio-ecology and human ecology. *In*: H. Svobodova (Ed.), Cultural Aspects of Landscape. Proceedings of the First International Conference of the Working Group "Culture and Landscape" of IALE, Castle Groeneveld, Barn, The Netherlands, June 28–30, 1989.

Naveh, Z. 1992. Workshop of the IALE-CESP-IUCN Working Group on Threatened Landscapes, Montecatini, May 1992. Working Paper, IUCN Commission on Environmental Strategy and Planning (CESP), Sacramento, California.

Naveh, Z. 1993. Biodiversity and landscape management. *In*: K. C. Kim and R. D. Weaver (Eds.), Biodiversity and Landscapes: A Paradox of Humanity. Cambridge University Press, New York (in press).

Opdam, P. 1991. The IALE- Working Group on ecological infrastructure—a report and summaries of current research programmes and projects. IALE Bulletin 9/1:5–18.

Risser, P. G. 1987. Landscape ecology: State of the art. *In*: M. G. Turner (Ed.), Landscape Heterogeneity and Disturbance. Springer-Verlag, New York, pp. 3–14.

Rowe, J. S. 1988. Landscape ecology: The ecology of terrain ecosystems. *In*: M. R. Moss (Ed.), Landscape Ecology and Management. Proceedings of the First Symposium of the Canadian Society of Landscape Ecology and Management, University of Guelph, May 1987. Polyscience Publications Inc., Montreal, Canada, pp. 35–42.

Rubec, C. D. A., E. B. Wiken, J. Thie, and G. R. Ironside. 1988. Ecological land classification and landscape ecology in Canada: The role of C.C.E.L.C. and the formation of the C.S.E.L.M. *In*: M. R. Moss (Ed.), Landscape Ecology and Management. Proceedings of the First Symposium of the Canadian Society of Landscape Ecology and Management, University of Guelph, May 1987. Polyscience Publications Inc., Montreal, Canada, pp. 51–56.

Ruzicka, M., T. Hrnciarova, and L. Miklos (Eds.). 1988. Structural and Functional Relationships in Landscape Ecology. Proceedings of the VIIIth International Symposium on Problems of Landscape Ecological Research, October 3–7, 1988, Zemplinaka Birava CSSR.

Ruzicka, M., and L. Miklos. 1990. Basic premises and methods in landscape ecological planning and optimization. *In*: I. S. Zonneveld and R. T. T. Forman (Eds.), Changing Landscapes: An Ecological Perspective. Springer-Verlag, New York, pp. 233–260.

Schreiber, K.-F. (Ed.). 1988. Connectivity in Landscape Ecology. Proc. 2nd Intern. Semin. of IALE, Muenstersche Geographische Arbeiten 29, Muenster, 1987.

Schreiber, K.-F. 1990. The history of landscape ecology in Europe. *In*: I. S. Zonneveld and R. T. T. Forman (Eds.), Changing Landscapes: An Ecological Perspective. Springer-Verlag, New York, pp. 21–34.

Schreiber, K.-F. 1991. Ecological gradients in the ecotope structure at the slopes of the Negev-Highlands—Their utilization to improve the primary production. *In*: J. Brandt. (Ed.), Practical Landscape Ecology. Proceedings of the European Seminar of The International Association for Landscape Ecology (IALE), Roskilde University Centre, May 2–4, 1991. Roskilde Universitetsforlag GeoRuc, Roskilde, Denmark, pp. 111–121.

Svobodova, H. (Ed.). 1990. Cultural Aspects of Landscape. Proceedings of the First International Conference of the Working Group "Culture and Landscape" of IALE, Castle Groeneveld, Barn, The Netherlands, 28–30 June 1989.

Turner, M. G. (Ed.). 1987. Landscape Heterogeneity and Disturbance. Springer-Verlag, New York.

Turner, M. G., and R. H. Gardner (Eds.). 1990. Quantitative Methods in Landscape Ecology. Springer-Verlag, New York.

Van der Meulen, F., J. V. Witter, and W. Ritchie (Eds.). 1991. Impact of Climate Change on Coastal Dune Landscapes of Europe. Landscape Ecology 6 (Special Issue 1/2).

Van der Zee, D. 1991. Remote sensing and Geographic Information Systems for land ecology studies in developing countries. *In*: J. Brandt. (Ed.), Practical Landscape Ecology. Proceedings of the European Seminar of The International Association for Landscape Ecology (IALE), Roskilde University Centre, May 2–4, 1991. Roskilde Universistetsforlag GeoRuc, Roskilde, Denmark, pp. 43–52.

Weinstein, D. A. 1992. Use of simulation models to evaluate the alteration of ecotones by global carbon dioxide increase. *In*: A. J. Hansen and F. di Castri (Eds.), Landscape Boundaries. Springer-Verlag, New York, pp. 379–393.

Wiens, J. A. 1992. What is landscape ecology, really? Editorial comment. Landscape Ecology 7:149–150.

Zonneveld, I. S. 1990. Scope and concepts of landscape ecology as an emerging science. *In*: I. S. Zonneveld and R. T. T. Forman (Eds.), Changing Landscapes: An Ecological Perspective. Springer-Verlag, New York, pp. 1–20.

Zonneveld, I. S., and R. T. T. Forman (Eds.). 1990. Changing Landscapes: An Ecological Perspective. Springer-Verlag, New York.

2

Conceptual and Theoretical Basis of Landscape Ecology as a Human Ecosystem Science

Some Basic Premises and Definitions

Having traced the development of landscape ecology as a scientific discipline in Central Europe, we shall now attempt to outline its conceptual and epistemological framework. In our view, this is derived from the following closely connected scientific theories.

1. *General systems theory* (GST): A holistic scientific theory and philosophy of the hierarchical order of nature as open systems with increasing complexity and organization and with living systems and ecological systems as their special biosystem subsets.
2. *Biocybernetics:* The theory of cybernetic regulation of these biosystems, enabling their self-stabilization and self-organization through deviation-counteracting (negative) and deviation-amplifying (positive) feedback couplings.
3. *Ecosystemology:* The theory of a transdisciplinary ecosystem concept with the Total Human Ecosystem (THE) as the highest level of ecological integration and with the ecosphere as its concrete space-time-defined global landscape entity.

New insights into the general systems theory, the holistic axiom, and the intermediate "holon" properties of biosystems at each integrative hierarchical level, recent findings on the nonequilibrium thermodynamic behavior of open systems and biosystems, and the formulation of basic biocybernetic rules have all greatly facilitated the amalgamation of the foregoing theories into a unified paradigm of human ecosystemology and its application in landscape ecology. As will be explained in this chapter, this paradigm culminates in the recognition of the Total Human Ecosystem as an open, nonequilibrium, self-realizing, and self-transcendent natural Gestalt system. This system cannot be grasped fully by formal language (expressed only in scientific terms and symbols), but only by another Gestalt system, namely, natural language.

The broad scope of this book as a synthesis of these basic premises, their recent ramifications, and their practical application allows only a condensed presentation of major principles and concepts with references to a small number of relevant original studies. However, in order to ensure their full comprehension even without further study, brief definitions of some of the most important terms and concepts, as related to general systems theory, biocybernetics, and ecosystemology, will be presented in the first part of this chapter. [For a more detailed outline of this theoretical background, see the relevant chapters in Bakuzis's (1974) lecture and research notes on the foundations of forest ecosystems. In these he provided a comprehensive compilation of citations from literature sources, on which we also relied in this chapter.]

D-1. According to Miller (1975), general systems theory (GST) is a set of related definitions, assumptions, and propositions that deal with reality as an integrated hierarchy of organization of matter and energy.

The terms "set," "class," and "system" are starting points for our further discussion, and they must first be defined more clearly:

D-2. A *set* is a collection of distinct objects or elements that, according to our view or knowledge, can be grouped together as a set (pencils, trees, numbers, etc.). In a linguistic way of speaking (the natural way of expression) General Systems Theory is a set of proper statements; in a general scientific way, general systems theory is an abstract analogy model of real systems; and in a set-theoretical way, general systems theory is a proper subset X_s (it has fewer members) of the set $X: X_s \subset X$.

D-3. A *class* is a set whose elements are distinguished by a common attribute. The choice of characteristic attributes and the classification of sets of objects according to these attributes are derived by a process of *abstraction:* A "horse" is a class of elements identified by distinct zoological attributes. In our ecosystem model (Figure 1-3), "producers, consumers," and the like have been grouped in distinct classes of elements—plants, animals—according to their *functional* attributes in energy/matter flow.

D-4. A *system* is a set of elements (or units) in a certain state, connected by relations that are closer than those with their environment. The set of relations among these elements and among their states constitutes the structure of the systems. Because of these relations, a system is always more than the sum of its elements; it is a *whole* (Sachsse, 1971). This concept of wholeness, having emergent qualities from the system's behavior of its elements as a whole, is the *basic holistic axiom.* Such "wholes" are, for instance, a melody or a poem, a water molecule, a planetary system, a rioting mass, or the system of rational numbers, irrespective of the logical aesthetic, physical, or psychological and other relations that constitute this system. However, we shall deal mainly with a special class of systems that is called in German "Wirkungsgefüge" and for which we propose the term "interaction systems."

D-5. An *interaction system* (in German, Wirkungsgefüge) is a special class of system whose elements are connected (or coupled) with each other by direct mutual influences. Therefore, if one element of such a system is affected, all others will be involved by these mutual influences, irrespective of the nature of the forces that effect the coupling of these elements (Sachsse, 1971). As mentioned in Chapter 1, an ecosystem is such an interaction system. The same is true for living systems such as bacteria, the human organism, a flock of geese, or a biotic community in a waterpool, as well as for a rural community, a military platoon, and the stock market, but it is also true for man-made servomechanisms, such as a watch or a refrigerator. Here, too, the interrelations between these elements and their states can be of a very different nature—they can be mechanical, natural, biochemical, biological, physical, social, cultural, and so on. In the English language, in general, no distinction is made between D-4 and D-5, and we shall deal mainly with the latter.

Cybernetics, in the broadest sense, can be called the science of interaction systems, and therefore it is closely connected with general systems theory. System concepts have many more ramifications, to which we shall refer later. First, however, there are certain basic physical terms of great significance that we shall attempt to define briefly.

D-6. Physical (geographic) space and abstracted space. *Physical space,* as defined in classical Euclidean three-dimensional geometry, is the extension surrounding a point. It is the only space in which concrete systems exist. Therefore, physical space is a *common space,* shared also by all observers, and all scientific data must be collected in it. However, scientific observers often view concrete systems, such as living systems (and ecosystems), as existing in space that they conceptualize or abstract from the phenomena with which they deal ("social space of pecking order," "niche space," etc.). Such *conceptual,* or *abstracted, spaces* may therefore be very useful if it is recognized that physical space is not the major determinant of certain processes in these systems, in spite of the necessary personal bias involved in observing or "measuring" them (Miller, 1975). In this book we are dealing chiefly with the physical–geographic space of ecosystems and landscapes. We should always be aware, however, that people live not only in physical space, but also in the noospheric-conceptual space of the total human ecosystem, of which, as will be outlined later, the landscape is a concrete entity. Jantsch (1975, p. 50) has provided a lucid description of this space: "When man emancipated from nature and embarked on his psychosocial revolution, he started to build for himself a conceptual space which is the realm of his mind as well as his feelings, of his imaginations and understanding, perception and conception. Man is the only animal on earth to have fully developed such a conceptual space." Jantsch stated that in addition to social and physical space, "there is also a *spiritual space,* which holds man's relations with the numinous; his quest for purpose, direction and meaning; his cultural inventions from values to religion, from the arts to philosophy and science." It is because of the impact of this spiritual space on our physical space in which we live—namely, our cultural landscapes—that these landscapes contain more than the measurable and quantifiable parameters in which space–time dimensions are expressed.

D-7. *Time* is the fundamental "fourth" dimension of the physical space–time continuum, since the particular instant at which a structure exists or a process occurs can be measured. A concrete system can move in any direction in the spatial dimensions, but only forward, never backward, in the temporal dimension (Miller, 1975). Because of the relatively short time in which ecological processes in landscapes have been studied systematically in the same location (in most cases for only a couple of years—the duration of a supported research project or a master's or doctoral thesis), there is a tendency to confuse these temporal and spatial dimensions. Therefore, two communities occurring side by side in one location are sometimes automatically regarded as successional stages.

D-8. *Energy and matter. Matter* is anything having mass and occupying physical space. *Energy* is the ability to do work. According to the first law of thermodynamics, energy can be neither created nor destroyed, but it can be converted from one form to another, including the energy equivalent of rest mass. Matter may have either *kinetic* energy (when it is moving and exerts a force on other matter), *potential* energy (because of its gravitational field), or *rest-mass* energy (the energy that would be released if the mass were converted into energy). Mass and energy are equivalent, and one can be converted into the other. Therefore, the joint term "matter/ energy" can be used in the general systems theory context (Miller, 1975). As we shall see later (D-11), in cybernetics the main emphasis has been shifted from the purely quantitative energetic and mass measurements to the qualitative aspects of the organization and steering of energy.

An important discussion on energy–matter relations in ecosystems took place at the first international congress of the International Association for Ecology (INTECOL) in The Hague (van Dobben and Lowe-McConnell, 1975).

One of the most important concepts, closely connected with energy but of a more abstract and complex nature, is *entropy.* We cannot provide a full formal interpretation here (one can be found in any modern textbook of thermodynamics), but we can stress some of its most relevant aspects from the systems point of view:

D-9. *Entropy* (from the Greek "entrepein," meaning "to turn into") is related to the second law of thermodynamics. This law states that all processes of spontaneous energy transformation in closed systems are moving into a more and more dispersed, and therefore more degraded and less available, form of energy, until a final state of complete homogeneity—a static equilibrium—has been reached in which all differences in energy gradients have been equalized. Since no energy potentials are left, the system becomes "lazy" and no further changes will occur without additional input of potential energy. Entropy is the measure of the state of the system, its *homogeneity* and its *irreversibility,* because this process goes in only one direction (i.e., coffee and milk mix but never separate themselves again).

In statistical mechanics, the meaning of the second law has been broadened, and entropy has become an important concept as the measure of *structural disorder;* its opposite, the decrease of entropy—negative entropy or *negentropy*— has become the measure of *order.* This has been achieved mainly by Boltzmann, who expressed the phase-space distribution of gas molecules in physical systems at different times as their *thermodynamic probability* (*D*) by determining the ratio between their macroscopic state (the distribution of *all* molecules or elements in different classes) and their microscopic configuration (the distribution of *different* molecules or elements within each class).

This probability is calculated by permutations:

$$\frac{N!}{(p_1 N)! \cdot (p_2 N)! \cdot \ldots \cdot (p_k N)!}$$

where N is the total number of elements—the observed gas molecules—and p_1, $\ldots$, p_k their percentage of distribution in cells $1, \ldots, k$ of the phase space. Entropy (*S*) can now be calculated as being proportional to the natural logarithm ln (log to the base *e*) of *D*:

$$S = k^* \ln D$$

where k^* is Boltzmann's constant (1.38×10^{-16} ergs/°C). This "entropy function" means the tendency to move toward the most probable state in which all possibilities have been exhausted by the realization of *one* macroscopic state of the molecules—namely, that of equal distribution or complete randomness.

If *D* is a measure of atomic disorder, then $1/D$ is a measure of order. Therefore

$$S = k \ln D \quad \text{or} \quad \text{negentropy} = k \ln 1/D$$

In more general terms, entropy can be defined as the measure of the degree to which the components of an aggregate are mixed up in an unpredictable (or random) way, and its opposite, negentropy, as the degree to which these components are ordered in a predictable (nonrandom) way.

Open systems (explained in D-17) differ from closed systems (D-18) in regard to entropy by their capacity to move toward a *dynamic equilibrium,* or steady state. In this, the state parameters are kept constant in spite of matter/energy exchange with the environment, and the conservation of structure is enabled by minimum entropy production and by maximum order or negentropy. This purposeful behavior of living systems toward an equal final steady state that is conditioned not by the environment but by their structure has been defined by Bertalanffy as the "principle of equifinality in organic systems."

In man-devised systems that are regulated by automatic feedback control (also called servomechanisms), such as a refrigerator, the constant state of low entropy—the (low) undertemperature—can be ensured only by input of energy from a larger system, namely, the electric power station. This stage of low entropy and high negentropy finds its expression in natural ecosystems by the high organizational and structural order and diversity of food chains and food webs of well-developed biotic communities. In these, the high-quality potential and chemical energy, and therefore low-entropy-producing energy, derived from solar radiation and photosynthesis is dissipated into high-entropy (low-quality) metabolic heat and respiration. According to Odum (1971a), natural "mature" ecosystems, by pumping out disorder, are gradually reaching such a steady-state equilibrium, or *climax,* in which energy efficiency [or in Margalef's (1968) terms, "information" (D-10)] is maximized. Later, we shall discuss the fallacy of this energetic version of Clementsian climax theories, based on the axiomatic assumption that nature must strive always toward a steady-state equilibrium.

The following definitions are some of the major concepts of information theory and cybernetics, presented as far as possible without much formal mathematical treatment.

D-10. *Information,* as defined in information theory by Shannon and Weaver (1949), is the measure of the uncertainty of an event and its removal by a message. Such a *message,* in its broadest sense, is any effect, regardless of its kind and physical nature; as such it is independent of the mechanism of its transmission. The amount of information contained in such a message is mathematically expressed by the number of choices between two equally probable possibilities, or *binary decisions* ("bits"). Thus, for example, if we have to choose a specific card from a set of 16, the probability of choosing the right one before receiving any information is only 1:16 ($P_0 = 1/16$). But if someone tells us the right card to choose, we can make the proper choice at once, and then $P = 1.0$, and the ratio between the probabilities, P/P_0 becomes 16. As shown in Figure 2-1, two binary choices would have been necessary for the correct choice of one card out of four, and in our case, according to the information (I) formula:

$$I = \log_2 P/P_0 \quad \text{therefore} \quad I = \log_2 16 = 4 \text{ bits}$$

Had the choice been made from 32 cards, then since $\log_2$ of 32 is 5, $2 \times 2 \times 2 \times 2 \times 2$ bits of information would have been needed.

The information contained in one volume of the *Encyclopedia Britannica,* with 1000 pages, each with 10^6 bits, is 10^9 bits; and that of a library with

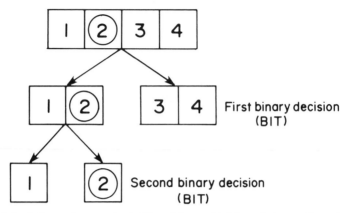

Figure 2-1. Information content (I) as binary decisions between four signs (N): $I = \log_2 N - 4 \log_2 = 2$ bits (from Klaus, 1969).

1,000,000 books would be 4×10^{12} bits, whereas the information storage capacity of a magnetic tape of 800 m with 5550 bits/cm would amount to 4×10^8. In a very useful book on biological energy, Lehninger (1965) calculated the approximate information content of a protein molecule with a molecular weight of 115,000, containing 1000 amino acid residues of 16 different kinds, on the basis of their specific sequence of order, as 4×10^3 bits, and that of a bacterial cell as 10^{12} bits. According to Sachsse (1971), the information storage capacity in the DNA molecules of human chromosomes is about 10^6 and that of the human brain 10^{12}. However, we cannot infer the functional significance of bits from their number. This should be kept in mind in our attempt to estimate the "valued" information of living systems and ecosystems resulting from natural evolution.

As a measure of the removal of uncertainty and randomness and increase in order and organization, information has been considered an equivalent of neg-entropy (D-9), and Shannon, the founder of information theory, has used the Boltzmann entropy function as a measure of information. However, this is merely a formal identity, and as mentioned earlier, information is more than a physical dimension expressed by this formula. The information content of an ecosystem or a plant community can be expressed by diversity measurement, such as the "Shannon–Weaver average information index" of evenness and homogeneity in distribution of plant and animal species, namely:

$$H' = - \sum_{i=1}^{N} \log_2 P_i \text{ bit}$$

However, this diversity index should not be equated with entropy measure as a thermodynamic parameter, but merely as a biological parameter of "alpha diversity" (Whittaker, 1975). For a more detailed discussion of these concepts in ecology, see Margalef (1968).

Schaefer (1972) has used a simple cybernetic model to show the epistemological relations between information and other system concepts. Two concepts that are positively related (+) increase each other (the more of A, the more of B) and two that are negatively related (−) decrease each other (the more of A, the less of B). Thus, two negative relations cancel each other (minus × minus = plus) and any logical sequence can be detected by the unequal number of relations (see also D-15). This is illustrated in Figure 2-2.

Information transmission has been described by Shannon and Weaver (1963) as the selection of a desired message out of a set of possible messages by an information source. The physical carrier of this message is a *signal* (a word, the color or smell of markers, electrical impulses, etc.) that is conveyed by a transmitter as a combination of signs (letters, mathematical symbols, electrical oscillations, etc.) to a receiver, which changes the signal back to a message or decodes it. Any unwanted additions, by disturbance from the environment, to these signals are called *noise*. The selection of useful information and the elimination of unwanted noise are most important adaptive and regulative functions in living systems and ecosystems and part of their natural evolution. The signs of their biophysical information are the results of physical laws and/or the result of biological processes. In cultural information these signs may have evolved naturally, like our day-to-day natural language; or they may have been created artifically by conventional symbols as formal languages, such as mathematics, traffic signs, or the Morse code. Weizsäcker (1974) has stressed the differences between *syntactic information* (the objective structures of arrangement of signs), *semantic information* (attaching a meaning to this syntactic information, as interpreted by its receiver), and *pragmatic information* (which becomes meaningful by its effect on the receiver and is expressed in his action, which may in turn affect the sender of the information). Of this pragmatic information he said that it "is what generates information potential," and this is essentially the biological and

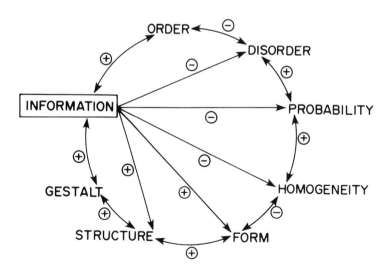

Figure 2-2. Epistemological relations between information and other system concepts (from Schaefer, 1972).

cultural information with which we are dealing here, in contrast to the mechanistic, mathematically derived Shannon–Weaver information of communication science, which deals primarily with syntactic information. According to von Weizsäcker, pragmatic information is composed of two complementary aspects, namely, *novelty* and *confirmation.* However, pure novelty does not contain any information and pure confirmation does not bring anything new. Therefore, somewhere between both we must realize an optimization of balanced pragmatic information.

The Shannon–Weaver information is characterized by a high degree of confirmation and very little novelty, because information is considered mainly to reconfirm and strengthen already existing information structures, geared (like the thermodynamic Boltzmann ordering principles from which it has been derived) toward equilibrium.

There is therefore a fundamental difference between both information concepts, and as we shall see later, for our purpose, the nonequilibrium, novelty-seeking, pragmatic ecological information is most relevant.

D-11. *Cybernetics* (from the Greek "cybernetes," meaning "steermanship") was originally defined by Wiener as "the science of control and communication in the animal and the machine" and by Ashby as "the art of steermanship." In GST, however, it has received a broader definition as the *science of behavior of interacting systems.* According to Ashby (1964), cybernetics does not ask "What is this thing?" but "What does it do?" Cybernetics is not interested in the qualitative and quantitative effects of relations between systems and their elements, but in their structural relations. In this way a very high level of abstraction can be reached, enabling one to explore and describe, in a similar way, the structural features of such different realms as technical organizations, machines and servomechanisms, biotic communities, ecosystems, and human relations. However, abstraction of interaction systems does not mean their reduction to mechanical systems. An organism or an ecosystem has features analogous to those of a machine, but organisms or ecosystems are not identical with machines and can be described, as was done by Ashby, as "adaptive cybernetic machines."

Only a few brief definitions of some cybernetic concepts and relations most relevant to our discussion can be presented here. For further study, see Ashby's (1964) excellent introduction from a biological point of view. A good, simple introduction to the technological aspects has been provided by Porter (1969), and a broader, interdisciplinary GST approach has been presented by Milsum (1968). Very helpful are a (German) comprehensive dictionary of cybernetics by Klaus (1969) and the books by Schaefer (1972) and Sachsse (1971), emphasizing biological, didactic, and interdisciplinary scientific and philosophical aspects of cybernetics. Examples of the application of cybernetics as a formal mathematical method in system ecology has been provided by Patten (1972); but for an informal, lucid presentation of the cybernetic relations in the tundra ecosystem, see Schultz (1969). Of greatest relevance to our discussion are the books by Vester dealing with biocybernetics, to which we shall refer in more detail later. Unfortunately, his most challenging (newly edited) book on our present "cybernetic revolution" (Vester, 1980) is not yet available to the English-speaking reader.

D-12. *Structure and coupling of black boxes of elements of systems.* According to Klaus (1969), in the cybernetic sense the structure (or organization) of a system is the set of relations that combines its elements with all their isomorphic relations (D-13). Cybernetics deals with dynamic systems in which energy, material, and information relations between elements (or systems) are described as *couplings.* Those elements are perceived as *black boxes* from which only their input and output relations are considered. They are coupled if a certain output of one element (or system) is at the same time the input of another, as shown in Figure 2-3. In general, all inputs of elements are considered as components of an input vector $\mathbf{I}$: $\mathbf{I} = (x_1, x_2, \ldots, x_n)$; and all outputs of these elements as components of an output vector $\mathbf{O}$: $\mathbf{O} = (y_1, y_2, \ldots, y_m)$. Thus, in Figure 2-3, three components of $\mathbf{O}E_i$ are also components of $\mathbf{I}E_k$ coupling conditions. These coupling conditions between two elements can be described mathemathically by the relations: $\mathbf{I}^{(k)} = K_{ik}\mathbf{O}^{(i)}$, in which the matrix K_{ik} shows which components of the output vector of one element are coupled with those of the input vector of the other. Couplings are noted by the number 1 and their absence by 0. Such a matrix can have, for example, the following form:

$$K_{12} = \begin{pmatrix} 0 & 0 & \ldots & 1 \\ 0 & 1 & \ldots & 0 \\ & & \ldots & \\ 1 & 0 & \ldots & 0 \\ 1 & 0 & \ldots & 0 \end{pmatrix}$$

If we assume, in addition, that no element can be coupled with itself ($K_{11} = 0$, $K_{22} = 0$, etc.), then the totality of all couplings (i.e., their structure) can be described by a *structure matrix:*

$$S = \begin{pmatrix} 0 & K_{12} & \ldots & K_{1N} \\ K_{21} & 0 & \ldots & K_{2N} \\ & & \ldots & \\ K_{N1} & K_{N2} & \ldots & 0 \end{pmatrix}$$

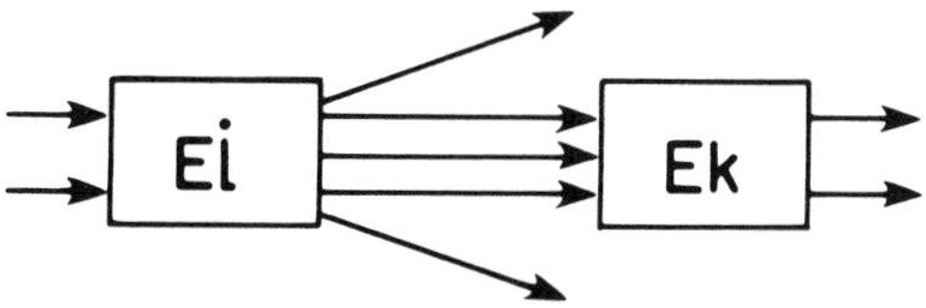

Figure 2-3. Coupling of two elements (from Klaus, 1969).

Such a structure matrix is not the formal description of a distinct concrete dynamic system, but in each case it is the class of all those systems with similar coupling matrices K_{ik} between their elements. Therefore, cybernetic structure models of all systems of this class can be constructed. These can be described in a less formal way as a graph or, in the case of isomorphic models (D-13), as a flow chart, shown in Figure 2-4.

D-13. *Homomorphic and isomorphic equivalence relations.* Equivalence relationships are analogies in the structure of a whole class of systems whose elements are coupled by similar coupling matrics (D-12), irrespective of different energetic and material realizations of these couplings in each matrix:

If we arrange two systems A and B with sets of elements $A = b_1, b_2, b_3$ in such a way that one is the image of the other, then *homomorphic* equivalence relations (or *homomorphy*) exist if the image of each coupling product of A is reflected in a similar coupling product of B. But if this arrangement is possible in both directions, and thereby these equivalence relations of the structural properties of both systems are even closer and more perfect, we call these *isomorphic* relations, or *isomorphy*.

Thus, for instance, the map of a certain landscape is isomorphic to the actual spatial relations between the different elements of this landscape. But the trophic relations in food webs between predators and their prey are homomorphic, and only mutual cannibalism could be described as isomorphic.

In mathematical set theory, homomorphic relations are reflexive: $(A1 \equiv A2)$ and transitive $(A1 \equiv A2)(A2 \equiv A3)$, but isomorphic relations are also symmetric $(A1 \equiv A2)$ and $(A2 \equiv A1)$. For a detailed description of isomorphic cybernetic relations, see the excellent introduction to cybernetics by Ashby (1964).

These equivalence relationships can be used for structural ecosystem modeling. In a homomorphic model of an ecosystem, its elements can be presented at the desired level of discrimination as classes of compartments and subsystems that are presented as symbols, letters, and the like; but in isomorphic models their dynamic relations are mapped in flowcharts. The ecosystem model in Figure 1-3 is a combination of both. Schultz (1969), in his illuminating paper on

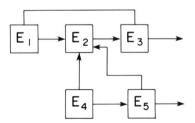

Figure 2-4. Isomorphic model of structure.

Figure 2-5. Diagram of information coupling (from Klaus, 1969).

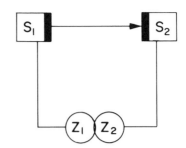

the arctic tundra ecosystem (one of the first biocybernetic ecosystem studies ever published), has presented such models with more detailed explanations. The structural properties of community webs as isomorphic models have been described in detail by Gallopin (1972).

D-14. *Information coupling* takes place in cybernetically regulated systems that differ from physical systems by their capacity for information processing and exchange, in addition to energy/matter conversion and exchange. It should be stressed that these informational processes, although bound to energy/matter transfer, have different qualities and are therefore not reducible to the latter. They also include the conversion of physical stimuli into perception, pattern recognition, formation of concepts, deductive and inductive processes, learning, and theory making as part of cultural information processed in human–inventive systems, to which will be referred below.

As can be seen in Figure 2-5, the effective connection between two systems, or elements, S_1 and S_2 is realized biochemically, physically, or otherwise, by signals that cause a change in system behavior. But this change is conditioned by an additional bridge between S_1 and S_2, created by a common pool of signs Z_1 and Z_2, which can be coded and decoded again. For the steering and regulation of biological, technological, and other processes, this information must be realized in an energetic and/or material way.

This means, in other words, the structuring of energy with the help of information (D-10). In this process, however, small inputs of steering energy may lead to manifold outputs of steered energy. Thus, for instance, in the steering of a car, a ratio of $1:75^2$ is reached, and in that of biological processes, ratios of even $1:10^{12}$ may be achieved (Schaefer, 1972).

One important means for this cybernetic regulation is by information feedback couplings, which we now discuss in more detail.

D-15. *Feedback coupling.* In contrast to the causal linear relations in which one part affects the other but not vice versa, as expressed in physical cause $\rightarrow$ effect or psychological stimulus $\rightarrow$ response, cybernetics is chiefly interested in *mutual causal effects*, $\leftrightarrows$, where each part affects the other. Thus in cybernetically regulated systems the output values again affect (or are fed back into) the input values of the system, thereby creating feedback loops. We can distinguish between positive and negative feedback couplings (or loops):

In positive feedback couplings the effect and its countereffect act in the same direction and amplify each other, either by increasing (symbolized by a +) or decreasing (symbolized by a −) each other. The more people of fertility age in the population (*A*), the more babies (*B*) can be conceived; and if more children are being born, the faster the population will grow again, because more people will again reach fertility level.

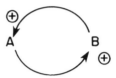

Such population growth curves are typical nonlinear *exponential growth curves*. The same is true for interest-bound capital investments, physiochemical chain reactions (such as atomic explosions), armament races, inflation rates, bank crashes, and the like. But in the latter we deal with *exponential extinction* rates, and in this case the lower the security against bank crashes (*A*), the less money its customers will deposit (*B*), but the lower this capital input of such an endangered bank, the lower its security level will drop, until the final collapse is reached.

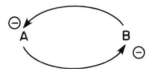

In negative feedback couplings, on the other hand, the input effect and the output effect are opposed to each other and thus control and cancel each other out. A simple example is the dynamic equilibrium in the population dynamics of foxes and hares. The more hares (*A*), the more foxes (*B*) will be able to prey on them; but if there are more foxes, then more hares are hunted, their number will decrease, and consequently that of the foxes will decrease, again enabling the reestablishment of a greater hare population.

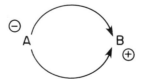

A more complex but very accurate example of negative feedback loops is the present tendency to lower population growth rates because of density stresses operating in a negative feedback. Increase in fertility and life expectancy (*A*), coupled with depopulation of rural areas, has led to greater population densities, especially in the larger cities (*B*); but these, in return, increase density stresses (*C*), which adversely affects fertility and health, thereby slowing down the rates of population increase and densities.

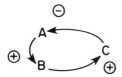

In developing countries, the extreme stresses of malnutrition and starvation in the poorest countries and the "cultural" negative feedbacks of birth control in some of the highly urbanized and more advanced developing countries are having similar effects. In general, a feedback loop with an even number of negative influences (−) is deviation amplifying and a loop with an odd number is deviation counteracting and stabilizing. The terms "positive" and "negative" therefore have mathematical connotations and should not be confused with the desirability or undesirability of the final results of such couplings. Many times these are undesirable in positive feedback couplings but desirable in negative ones. This is especially the case when such negative feedback loops ensure the maintenance of an optimum state, or "set point," in critical system variables of man-made cybernetic systems or servomechanisms, such as the set point of the desirable temperature in a heat thermostat system or the water level in hydrological control systems or the gasoline level in a carburetor. In living systems, too, the maintenance of optimum values, such as body temperature and CO_2 or hormone levels in the blood, or the optimum population density in natural ecosystems, are ensured by negative feedback couplings. These counteract any deviation or "error" induced by disturbances from the internal or external environment by again changing the input values, either increasing or decreasing them. We can therefore also call negative feedbacks deviation-counteracting or -attenuating mutual causal loops, or compensating or equilibrium feedback couplings, as opposed to positive feedbacks, which are deviation-amplifying mutual causal loops and sometimes also disequilibrating runaway or cumulative feedback couplings.

In living systems the capacity to maintain a dynamic equilibrium, or *homeostasis,* as a whole is ensured by a great number of closely interrelated cybernetic neural, endocrine (hormonal), and neuroendocrine feedback regulation mechanisms. These are hierarchically ordered, and chemical or electrical energy (or both) is used as feedback information signals. In defining the laws of cybernetics, Wiener, who worked in close collaboration with the physiologist Rosenfeld (a student of Cannon, the "father" of homeostasis), was very influenced by these biological processes of autoregulation. Figure 2-6 is a simple structural model of the principal mechanisms of such cybernetic autoregulation by a closed information system of negative feedback loops. The mathematical description of this feedback process is based on the assumption that it proceeds without delay. This is not generally the case in biological and ecological processes, where these control mechanisms are much more complex and closely interlinked, or "nested," in a hierarchical order. Thereby, their mathematical, and even graphical, description can become only a very imperfect model of reality. [For detailed formal descriptions, see Milsum (1968) and Bennet and Chorley (1978).]

As shown in Figure 2-6, each physical, chemical, demographic, or other variable has a desired reference input value, or set point, provided by a higher hierarchical control system as a guiding value (in the case of servomechanisms, such as a thermostat, this is human regulated). This desired value (t) can be held constant by a *sensor* (called in living systems a *receptor*) responding to any deviation or error introduced from the environment by disturbing factors and activating a regulation center, the *controller* (K) by an *error signal* as the difference between the actual and desired value $X(t) - by(t)$ and transduced into a *control signal.* This is received by an *effector,* or feedback transducer, sending its output back to the sensor via feedback elements, and thus by successive approximation correcting the error. This is recorded as the summation, or as the difference $(-b)$ between the two signals, so that the final output $y(t)$, instead of being $K \times (t)$, has "gained" from the feedback coupling.

The steering of a ship in a desired direction, in spite of environmental disturbances (or "errors") introduced by wind and currents, has been presented by Bennet and Chorley (1978) as a man–machine hierarchical (or nested) negative feedback control system. As mentioned in D-11, the term "cybernetics" was coined to fit this closed-loop feedback mechanism, *steermanship.*

A different way of presenting cybernetic feedback regulations of populations as "demostats" is shown in Figure 2-7. Here, the dynamic flow equilibrium in an annual mediterranean pasture (or the grass understorey of a mediterranean woodland) and its disruption by either increasing or decreasing grazing pressure is demonstrated as an adaptive mixed control system. This is maintained both by natural feedback loops of climatically induced seasonal and annual fluctuations in pasture productivity and composition and by cultural negative feedback loops of rotational pasture management. We shall refer to this example again in our discussion of nonequilibrium systems and in Chapter 4, which is devoted to the management of mediterranean ecosystems.

Whereas in the earlier stages of cybernetics, emphasis was chiefly on negative feedback couplings as deviation-counteracting, equilibrating, and

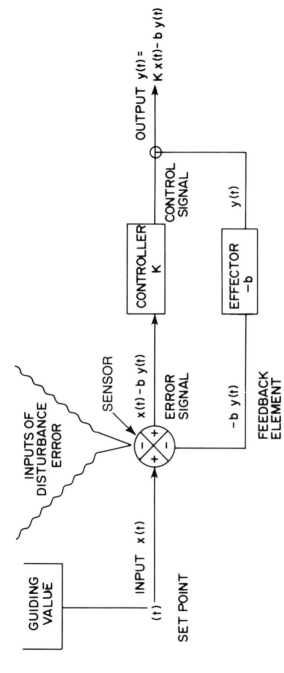

Figure 2-6. Closed negative feedback control system. Without feedback coupling the output $y(t)$ from input $x(t)$ would have been $Kx(t)$, but through the feedback gain by the effector, the output $(t) = Kx(t) - b(t)$.

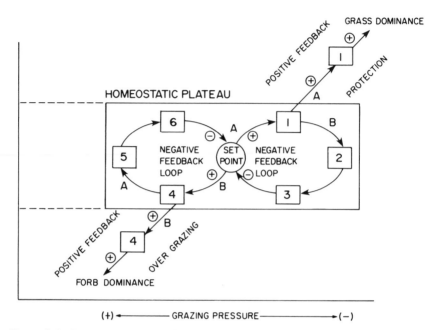

Figure 2-7. Demostat model of mediterranean grassland, maintained by cybernetic feedback control of grazing pressure. A set point of high spatial diversity and high long-term, global stability induced by climatic and rotational fluctuations is maintained by a nomeorhetic flow equilibrium, ensuring "order through fluctuation." (+A) Lower grazing pressure: (1) increased grass density, reduced forb and legume density. (−B) Higher grazing pressure: (2) increased defoliation stress on grass, increased competition by legumes & forbs; (3) control of vigor of grasses, mixed grass–legume–forb population. (+B) Higher grazing pressure: (4) reduced grass density, increased forb and legume density. (−A) Lower grazing pressure: (5) decreased defoliation stress on grass, decreased competition by grass; (6) control of vigor of legumes and forbs.

therefore stabilizing, mutual causal processes, attention has shifted more recently to the importance of deviation-amplifying mutual causal processes of positive feedback loops. Maruyama (1963), in an important article, has called these "second cybernation" and has shown their importance in the developmental and growth processes of organisms, as well as of social and ecological systems for increasing differentiation and complexity in structure. In this way, *self-organization* is achieved and can be maintained by deviation-counteracting processes of *self-stabilization*. This basic principle of self-organization in the evolution of biosystems can therefore be explained by positive feedback loops which increase nonhomogeneity, and thereby also negentropy and information (D-10), in contradiction to the postulates of the second law of thermodynamics. More recently Maruyama (1976) has also shown their great importance in cultural evolution and in processes of symbiosis–cooperation and integration as a basic evolutionary mechanism in the organic world and in human systems. He concluded that "the principle of the biological and social universe is increase of diversification, heterogeneity, and symbiotization. What survives is not the strongest but the most symbiotic." Later in this chapter we shall discuss these important principles of self-organization and stabilization in their relation to the epistemology of the total human ecosystem as the highest level of ecological integration and symbiosis.

General Systems Theory—Promises and Dangers

GST was conceived by Bertalanffy in the late 1930s when he realized that the concepts of closed systems isolated from their environment, as used in classical physics, are misleading when applied to open biological systems. He presented a lucid overview of his theories at an important symposium, "Beyond Reductionism" (Bertalanffy, 1969); the following has been abbreviated from that presentation. [Most of his important essays have been published in a special volume (Bertalanffy, 1968).]

The discontent with mechanistic and reductionistic tendencies and the need for new paradigms and a reorientation in scientific thinking in order to deal with the complexities of whole systems, organizations, and the like, is shared today by almost all branches of the natural and human sciences. Even physics, which inspired these mechanistic and reductionistic approaches, is no longer "mechanistic," since matter has been dematerialized in modern theory. Determinism has reached its limit in quantum theory, and ultimate particles are defined only by highly abstract mathematical expressions. Not too long ago, classical physics was concerned solely with *linear causality* in terms of cause and effect, stimulus-response, and the like. Then physics confronted the problem of *unorganized complexity,* which is essentially answered by statistical laws, its paradigm being the second law of thermodynamics (see D-9). Now, however, we are confronted with problems of *organized complexities* at all levels of the hierarchical structure of the universe. For these, new paradigms of systems and organization are essential within a conceptual framework which transcends that of traditional science.

Thus, in contrast to conventional disciplines, which tend to isolate and reduce the elements of their study into chemical compounds, enzymes, or cells, or into elementary sensations, stimulations, energetic modules, and others, and expect that putting them together again, conceptually or experimentally, will cause the whole to reemerge, the systems approach explores many systems of the universe as wholes. In the words of Bertalanffy (1969, pp. 60-61):

> The models and principles thus developed are intended to be interdisciplinary, that is, applicable to systems of various kinds and encountered in different scientific disciplines. To give a simple well-known example: the feedback model originally came from technology (exemplified by the thermostat, feedback control in radio receivers, and automation in general) but the same feedback scheme can be applied to many physiological regulations [see D-15] to contain psychological and

social phenomena, etc. Such models are therefore *isomorphic* [D-13] in different branches of knowledge and can often be transferred from one realm to another . . . . There appear to be general system laws applying to systems of various kinds, irrespective of what their nature, components and interactions may be.

GST therefore cuts across narrowly defined borders kept separate in traditional scientific disciplines, and as a *transdisciplinary metatheory*, or a "conceptual over-arching global theory" (Grinker, 1976), it links and integrates cultural and ideological barriers, quantitative and normative approaches, and qualitative and descriptive approaches. In the third L. von Bertalanffy memorial lecture of the Society for General Systems Research, Rapoport (1976) called GST a "bridge between two cultures." He pointed out its significant contribution to a revitalization of philosophy in a "world intellectually dominated by science." In his opinion, it has revived the role of speculative analogy and organismic thinking as a complement to analytical thinking, and has suggested concepts appropriate to a holistic approach to both a conception of reality and a theory of cognition. In addition, by emphasizing the interdependence of this planet as an interaction system, this systems view—as promoted by GST—also has most profound ethical implications.

Laszlo (1972), in his important work *Introduction to Systems Philosophy*, has attempted such an expansion into systems philosophy as a holistic philosophy of nature that also encompasses traditional philosophical fields. He applied GST as a "general system language" or metalanguage to the reassessment of traditional philosophical concepts in reference to these newly emerging system concepts, as an attempt at a *"new transdisciplinary paradigm of contemporary thought."*

As pointed out by Kuhn (1970) in his brilliant analysis of the structure of the scientific revolution, the appearance of such new paradigms of conceptual schemes, which are not perceived, or are sometimes even suppressed, by "normal science," is marking the beginning of the scientific revolution. This is presently the case with what we can call the "antireductionistic" scientific revolution.

At the same time, however, GST has also inspired the development of rigorous quantitative approaches and studies. It now embraces a spectrum of pure and applied systems sciences, from mathematical and applied systems theory to technology and systems engineering, management, and operation research. In these, systems concepts are translated into operational terms by "software" (operation models) and "hardware" (control technology for the solution of practical problems), including environmental problems and resource utilization from small-scale (local and regional) to large-scale (global) dimensions of world models.

Closely related to GST is a whole family of mathematical theory that made possible the formalization and quantification of the systems approach. These include set theory, the theory of automata, game and queuing theory, and most recently, catastrophe theory—but especially cybernetics and information theory, the sciences of control and communication.

Of great importance in this respect are the "state-determined cybernetic machine" and the formal theory of homeostasis, as developed by Ashby, and their application to biological systems. Another, most important, formal mathematical systems approach is the state–space approach developed by Zadeh as a generalization of electric circuit theory. It has been extended into "fuzzy systems," whose quantities may acquire values from fuzzy sets (i.e., sets that are not precisely defined and can be handled by special grammar language) instead of formal numbers and algebraic symbols (Zadeh, 1965).

These theories and their successful application, together with further advances in GST elaborated on later, have made the holistic viewpoint more rigorous and explicit and have opened the way for a new scientific paradigm of studying and comprehending wholes as organized complexities, and our world as the most highly organized and complex one.

The great biologist and systems philosopher C. H. Waddington (1977) left, after his death in September 1975, the manuscript of a popular guide to all these new tools of thought about complex systems. This is a very enjoyable and highly recommendable introduction for those who have neither sufficient mathematical background nor the time to study in depth GST, operations research, cybernetics, and the other fields mentioned here briefly. It also contains a full list of suggested reading.

In contemporary ecology these developments are chiefly reflected in the rapidly expanding field of ecosystem ecology, which has benefited much from the advances in system computerizing, modeling, and simulation. However, greatest advances have been achieved in system analysis and not in system synthesis. Linear and deterministic energy/material flow models, tractable by differential equations, can describe only some functional quantitative aspects; but for a truly holistic comprehension of the system as a whole, the structural dynamic, qualitative, and regulative aspects are also necessary. Some of the new approaches to these problems will be outlined further. The danger, in our opinion, lies chiefly in the present trend to produce quantitative "ecosystem models" as a fashionable and prestigious novelty, without realizing the great discrepancies between these models and ecological reality. No model can be better than the data and the conceptual and methodological sophistication used in its construction. Therefore, if this trend continues unhampered, it may lead to a situation in which, instead of one good modeler–mathematician handling the data of ten or more field ecologists, physiologists, and the like, ten mediocre modelers will work aimlessly with the poor data supplied by one lonely and frustrated field ecologist.

Thus, for instance, the development of very ambitious models of the American biome studies within the International Biological Program (IBP) was hampered by inadequate field data and the lack of coherent integration and overall synthesis. For these reasons, the objectives of these large-scale modeling projects (to give direction to research, advance understanding of ecosystems, expand the data base of whole systems, etc.) were hardly achieved. According to Mitchell et al. (1976), who evaluated three of these studies, large grants for large models did not necessarily accelerate progress in ecosystem research. As stated by Egler

(1970, p. 82), system analysis is merely a new mathematical and mechanical technology added to some basic concepts: "There is the danger of the abstract concepts withering, and only the dry technology being left. The advent of the computer has often encouraged the trivialization of scholarship and the belief that the things that count are those that can be counted."

A most striking recent example of such a "system technology" is the book on environmental systems by Bennet and Chorley (1978). As a comprehensive exposition of these technologies it is highly valuable. However, their technological multidisciplinary approach, culminating in "environmental modeling," without the holistic, scientific, conceptual, and epistemological dimension of GST and its recent ramifications, fails to provide a sufficiently broad base for "interfacing man with nature," which is claimed by the authors to be their book's chief object.

In his above-mentioned lecture, Bertalanffy (1969, p. 5) also warned that a purely technological and mechanistic systems approach may have further, even more menacing, consequences: "Systems can become a Janus face of possibly both beneficial and menacing consequences, leading to further dehumanization and making human beings even more into replaceable units and, what Mumford so pungently called, the 'megamachine of society'."

Young (1974), in discussing the relationship between GST and human ecology, further warned that "a general system approach could return us to a discarded mechanistic focus, ignoring the reality of man." He also doubted whether methods of definition and measurement utilized in biological ecology were acceptable to human ecology and believed that—if these methods were mathematical system analyses, as in natural ecosystems—they might not be directly applicable to human systems.

These dangers seem to be very real, and one can already recognize such a reductionistic tendency in ecology in the depiction of energy exchange as the only basic process, not only in natural ecosystems, but also in human systems, as when Odum (1971b) represents all phenomena, including human culture and religion, by energy flow diagrams and models. As Ellenberg (1973) rightly remarked, this approach would be harmful to both ethics and ecosystem research.

However, these warnings and criticisms should not lead us into a wholesale rejection of GST, but should guide us in avoiding the pitfalls of system approaches: too broad generalizations on the one hand and too narrow reductionistic and mechanistic interpretations on the other. Thus, GST has opened new vistas for contemporary ecology, which Halfon (1978) calls "theoretical systems ecology"—new ways of looking at ecological systems and of building bridges between field ecologists, systems ecologists, and systems scientists in general. Some of these contributions are contained in the volumes on system analyses and simulation in ecology (Patten, 1972).

As will be shown in this chapter, for landscape ecology as a human ecosystem science, GST has even more far-reaching implications. Its greatest merit lies in providing the conceptual framework to bridge the gap not only between the two cultures of science and the humanities, but also between these and the technoeconomic and political "culture" in which decision making on actual land uses are carried out. Recent developments in GST have also provided some powerful intellectual and practical tools for a better comprehension of the complex inter-relationships between man and his total environment. Some of the most relevant of these developments will be outlined in the following section.

General Living Systems Theory and Some Further Definitions

One of the most significant attempts at the development and expansion of GST was made by J. G. Miller in his monumental book *Living Systems* (Miller, 1978). He had presented his ideas earlier, in a much more condensed form, in the journal *Behavioral Science,* of which he is the chief editor and from which we have cited some of his definitions (Miller, 1975). He summarized his theory at the second Bertalanffy memorial lecture during the annual meeting of the Society for General Systems Research. (This international organization has as its aim the further development of theoretical systems and especially the applica-tion of GST to various organizational and societal problems.) Miller distin-guished among conceptual, concrete, and abstracted systems, and within each of these, certain subsets of systems:

D-16. *Conceptual systems* were defined by Miller (1975) as units of terms, such as words, numbers, or other symbols, like mathematical systems or computer simulation programs. Their relationships are sets of pairs of units, each pair being ordered in a similar way. In linguistic systems they are expressed by words, in conceptual models by symbols and figures, and in mathematical systems by logical or mathematical symbols. All are concepts that always exist in one or more concrete systems, living or nonliving, such as a matrix, a textbook, or a computer.

In mathematical systems, for instance, the variables are the elements and their relations are functions that constrain the possibilities of change or their degree of freedom. A system *state* is a set of variables with given values and its change is a change in the value variables. A *mathematical structure* is the description of these dynamics—the *sequence of states* (Pankow, 1976).

If the functions permit only one direction of change, the system has only one degree of freedom and then we are dealing with a *deterministic,* or *reversible,* system. But if it is not subject to functional constraints of the elements and we have, with a large number of elements, an indefinite number of degrees of freedom, then we speak of a *stochastic,* or *irrever-sible,* system. These have many more possibilities than they are capable of realizing and their system states are then no longer defined in a micro-scopic way for each variable and value, but in a macroscopic way by classes, grouping microscopic states. The values of their variables are, then, replaced by probabilistic values of classes, according to statistical calculations of these probabilities of states, as in the determination of entropy (D-9). The introduction of feedback relations (D-15) between elements in cybernetic systems has made possible entirely new methods for the structural description of such functions.

There is, however, a great difference if these systems are expressed in everyday, or natural, language, as opposed to the logical, graphical, or mathematical symbols of formal languages. According to Pankow (1976), natural languages are defined by their capability of representing not only objects but also themselves, they are self-transcendent and any self-transcendent system is a Gestalt system. This is a further, and most important, system concept not used by Miller.

D-17. *Gestalt* is the German word for form, figure, configuration, or pattern, but also the quality that forms, figures, and patterns have. Thus, Gestalt is both form and form-ness, pattern and pattern-ness. The Gestalt term was first adopted by a special school of psychology founded by Koehler in 1912, and more recently by the sociologist Lewin in the idea of Gestalt structures of social relations. It has now become an important notion for the wholeness of systems. There cannot be two absolutely identical Gestalt systems, but they may be adequately represented by Gestalt languages as the organs of consciousness, and they are related through homology (D-13). Formal languages, on the other hand, cannot represent themselves, but only other objects. They are non-self-transcendent and are related to each other through analogy. These important concepts will be used later on in our distinctions of two basic classes of open systems.

D-18. According to Miller (1975), a *concrete system* is a nonrandom accumulation of matter/energy in a region in physical space–time that is organized into interacting interrelated subsystems or components. The units of these systems and their relationships, including spatial, temporal, spatiotemporal, and causal ones, are empirically determinable by an observer.

Most concrete systems are *open systems,* whose boundaries are at least partially permeable, permitting energy/matter and information transmission to cross them. Such inputs can repair system components that break down, replace energy that is used up, regulate the functions of the system, and maintain its organizational structure.

Closed systems, on the other hand, have impermeable boundaries, and only their internal elements are coupled. No interaction or transmission from and into the environment occurs. In reality, a concrete system is never completely closed; it is either relatively open or relatively closed. But in the latter systems we can ignore their interaction with their surroundings. Their energy is gradually used up and the matter gradually becomes disorganized. Such closed systems are much simpler to handle, and they have played an important role in classical physics, as well as in defining the thermodynamic laws of entropy (D-9).

D-19. *Abstracted systems,* according to Miller, are sets of relationships abstracted or related according to the interest, viewpoint, or philosophical view of the observer. In contrast to concrete systems, the boundaries of abstracted systems may be conceptually established at regions that cut through the units and relationships in physical space occupied by concrete systems. Thus, the Sea of Galilee is a concrete (limnological) ecosystem, but "limnological ecosystems" are abstracted ecosystems.

D-20. *Living systems,* according to Miller (1975), are a special subset of all possible concrete systems that are composed of plants and animals, using inputs of foods or fuel to restore their own energy and repair breakdown in their original structure. They contain genetic material composed of deoxyribonucleic acid (DNA) and they are largely composed of protoplasm. They can exist only in a certain environment in a relatively narrow range on the surface of the earth, according to their tolerance amplitudes.

In his efforts to extend GST by a general theory of living systems as concrete, energy/matter information processing systems, Miller identified the evolutionary development of levels of increasingly complex living systems, namely, cells, organisms, groups, organizations, societies, and the supranational systems. At each of these levels, living systems have as components systems of their level below and are components of systems at the level above. Miller (1975) described 19 subsystems processes which he considered as critical for all living systems. The reproducer system processes both matter/energy and information and all others process either matter/energy or information. Each of the critical subsets keeps a set of variables in the steady state by negative feedback loops, and the system as a whole maintains the steady state of other variables by adjusting relationships among subsystems. He regarded the biosphere as the "suprasystem of all the human systems" and as a global national system. In this chapter we shall attempt a more rigorous and epistemologically sounder definition of the biosphere within a general ecosystem classification.

Miller's impressive and formidable attempt to treat living systems at a new level of generalization (but also with careful, detailed documentation of these across-species and across-level phenomena and isomorphies in a 7 X 19 matrix) leads to a *taxonomy of physiological systems* or, in the words of Boulding (1980), a "universal physiology." His system is derived essentially from the description of organisms, especially of the human body, and this is where it works best. It seems a little too complex as we move toward the cell or the organ, and not complex enough as we move upward to the group, the organization, the society (by which essentially he means the national state), and the supranational system, which never quite gets to the world as a total system.

In considering the relevance of General Living Systems Theory for landscape ecology, one should realize that all the above-described relationships are directed inward, to internal physiological processes and relations. This is in contrast to the ecological approach, in which the major focus is on the relations of living systems outward, with their environment and their mutual influences. In fact, all definitions of ecosystems emphasize their nature as interaction systems between living organisms and their nonliving environment. Tansley (1935), in introducing this term, stated that ecosystems comprise "not only the organismic complex but also the whole complex of physical factors forming what we call the environment," while Lindeman (1942), in his trophic–dynamic approach, regarded the ecosystem as composed of physical–chemical–biological processes acting within space–time units of any magnitude. Therefore ecosystems, integrating these living systems and their environment, should be considered as a higher level of integration, above the organismic living systems level. As we shall see in our further discussions, this functional–structural integration of living systems and their physical environment results in the emergence of new system qualities, yielding new isomorphies. These have not been included in Miller's scheme because he treated his higher levels of "living systems" not as concrete ecological entities but as concrete biological, sociopolitical, geographic, and economic ones. At the same time, however, much could be gained by including his findings and the revealing isomorphies of "critical subsystems," together with other new concepts, in an emerging general theory of biosystems.

The Hierarchical Organization of Nature and the Holon Concept

The holistic axiom that "the whole is more than the sum of its parts," first stated by Smuts (1926, 1971) and introduced to ecology by Egler (1942) as the concept of the hierarchical organization of nature, has become a basic philosophical presumption of GST. According to this concept, the universe is regarded as an organization—an ordered whole of a hierarchy of multilevel stratified systems, each higher level being composed of lower levels of systems with additional emerging qualities. This rule of hierarchical organization is displayed by all complex structures and processes of a relatively stable character, from the subatomic and atomic physical and chemical levels, to the suborganismic and organismic biological levels, and to the superorganismic ecological and social levels of integration, up to the world and galactic systems. For further elaboration of these points, see Laszlo (1972) and an excellent essay on the architecture of complexity by Simon (1962). More recently, Bakuzis (1974), in the above-mentioned *Foundations of Forest Ecosystems,* has provided a very comprehensive summary of recent advances in the epistemology of these concepts from the biological and ecological point of view. Of great importance in this respect were two symposia, both conducted in 1969: The Alpach symposium, "Beyond Reductionism" (Koestler and Smithies, 1969), dealt chiefly with the biological aspects, whereas in the other one (Whyte et al., 1969) hierarchical structures were treated in philosophically and formally broader ways. In the latter, Bunge (1969) suggested some useful working definitions. His basic definition is that of a "level structure," which is taken as a family of sets, having a relation between the sets that represents emergence or novelty-generating processes. The emergence relation that holds *between* the sets does not hold *within* the sets, whose elements are taken to be belonging to a level structure. A hierarchical structure, or hierarchy, is a set equipped with a relation of domination or its converse, subordination. Arthur Koestler, who was a famous novelist as well as a scientific writer and philosopher [he summarized his ideas in a very readable scientific popular work (Koestler, 1978)],[1] reassessed these hierarchical concepts in a challenging lecture, "Beyond Atomism and Holism," at the above-mentioned symposium (Koestler, 1969). He rightly remarked that the term "hierarchy" is often wrongly used to refer simply to order of rank on a linear scale or ladder. Its correct symbol is a living tree: a multileveled, stratified, outbranching pattern of an organizational system, branching into subsystems that branch into subsystems of lower order, and so on; a structure encapsulating substructures, and so on; a process activating subprocesses, and so on. (See the hierarchical tree diagram of ecological levels in Figure 2-11, which was designed after this model.) Koestler (1969) went even further and claimed that, contrary to deeply ingrained habits of thought, neither "parts" nor "wholes" in this absolute sense exist anywhere, whether in the domain of living organisms or of social organizations. What we find, instead, are intermediary structures on a series of levels in ascending order of complexity, each of which has two faces looking in opposite directions. The face turned toward the lower levels is that of an autonomous whole, the one turned upward is that of a dependent part. He coined the word "holon" for these Janus-faced subassemblies (from the Greek "holos," meaning "whole," with the suffix "-on"; cf. "neutron" or "proton"), suggesting a particle or part. Using the human organism as an example, Koestler showed the great epistemological value of this holon concept for bridging the missing link between atomism and holism and supplanting the dualistic way of thinking in terms of "parts" versus "wholes."

The structure and behavior of an organism, as well as any other hierarchically ordered whole, cannot be explained or "reduced to" its elementary parts. But it can be "dissected" into its constituent branches of holons. Each member of this holon hierarchy—or, for short, *holarchy*—on whatever level, is a subwhole, or holon, in its own right—a stable, integrated structure equipped with self-regulatory devices and enjoying a considerable degree of autonomy or self-government. Thus, in the human organism, cells and organs are subordinated as parts to the higher centers in the hierarchy—the circulatory system, the nervous system, and so on—but at the same time they function as quasi-autonomous wholes.

This term may be applied not only to biological but also to ecological, sociological, and cognitive holarchies which display rule-governed behavior and/or structural Gestalt constancy. According to Koestler (1978), this dichotomy between an integrative, or self-transcendent, tendency, resulting from the function as part of the larger whole, and the self-assertive tendency to preserve

---

[1] After suffering from severe, incurable illness, he committed suicide in March 1983. His last book (Koestler, 1982) was another—even more comprehensive—selection of his writings with comments.

individual autonomy in the holarchy, has far-reaching implications for the physical and biological, as well as human, realms. We shall use this holon concept in all our further discussions, especially in our attempt to conceptualize the unique place of man in nature—as a holon of the biosphere—and the important role of landscape ecology in the attempt to reconcile this dichotomy, at least in the field of land use.

In mathematical terms of set theory, this holarchic relativity can be expressed by stating that the holon a is included ($\subset$) in its superholon but at the same time includes its subholon, c: $(c \subset a) \subset b$, and in this case c is a subsubholon relative to b. In a more general way it can be stated that if a given holon a has component subholons $c_1, c_2, c_3, \ldots, c_n$ linked by their sum R, then a is part of a holarchy when $[a = (c_1, c_2, c_3, \ldots, c_n)R] \subset b$ in which a is a superholon (see also Pattee, 1973).

A further, most relevant, contribution to a more rigorous reformulation of these holistic paradigms was the leading lecture in the Alpach symposium by the eminent biologist Weiss (1969). Many years earlier, in 1925, he revealed his farsightedness and "pioneering spirit"; he was not only the first to replace the mechanistic cause-reflex principles by a "general system theory of animal behavior"; he was also the first to coin the term "molecular biology," in 1952 in his proposals for restructuring the biological sciences. To the recent reductionistic trends in this science he remarked (Weiss, 1969, pp. 10-11), "There is no phenomenon in living systems that is not molecular, but there is none that is *only* molecular either. It is one thing not to see the forest for the trees, but then to go on to deny the reality of the forest is a more serious matter; for it is not just a case of myopia but one of self-inflicted blindness." In his lucid overview of the hierarchical organization of living systems, he provided convincing proof that this is a real phenomenon "presented to us by the biological object and not the fictions of a speculative mind." He argued that biology has made spectacular advances by adopting the principles and methods of the inorganismic sciences and mathematics, but has not widened its conceptual framework in equal measure. For this purpose it relays the courage of its own insight into living nature, recognizing that organisms are not just heaps of molecules but dynamic systems with stratified determinism. The basic flaw, in his opinion, lies in equating science with the doctrine of microprecise causality, or microdeterminism, derived by a hypothetical downward extension–atomization of empirical macrodeterminacy. Since this is not explainable in terms of aggregates of microdeterminate events for those phenomena in living systems which defy the description purely in terms of micromechanical cause-effect chain reactions, the systems concept and the principles of systemic organization are applicable.

> The fact that the top level operations of the organism thus are neither structurally nor functionally referable to direct liaison with the processes on the molecular level in a steady continuous gradation, but are relayed step-wise from higher levels of determinacy (or "certainty of determinability") through intermediate layers of greater freedom or

variance (or "uncertainty of determinability") to the next lower levels of again more rigorously ascertainable determinacy, constitutes the principle of *hierarchic organization* (Weiss, 1969, p. 33).

He further stressed the dynamic features of systems as expressed in the mutability of systemic patterns in evolution, ontogeny, maturation, learning, and the like, as well as in the capacity of recombination into what then appear as supersystems with emerging properties of novelty and creativity. These features are of greatest importance in our discussion and will be developed further later on.

Weiss demonstrated the essential character of the invariance of a system as a whole, in comparison with the more variant fluctuability of its constituents, by a simple mathematical formula. This shows that in a system the sum of deviations in measurable physical, chemical, or other parameters, expressed as the variance V of its elements $a, b, c, \ldots, n$, are greater than the variance of the total complex S:

$$V_s \ll (V_a + V_b + V_c + \ldots + V_n).$$

This is a result of the systems behavior under the internal constraints in the degrees of freedom of its components by coordination and control. It is realized by the capacity of cybernetic self-regulation through negative feedback loops, known in biological systems as homeostatic control (as discussed in D-15). These feedback couplings enable the system to compensate for changing conditions in the environment in the system's internal variables as "adaptive self-stabilization." Through these self-regulative properties the system becomes more than its parts, not in a quantitative-summative way but in a qualitative-structural way. Weiss redefined this holistic principle of emergent qualities as follows: "The information about the whole, about the collective, is larger than the sum of information about its parts."

We conclude by stating that "ordered wholeness" implies *nonsummative* system properties emerging from the relation between the parts. These can be detected only by considering the whole, and they become meaningless in reference to these isolated parts. If these were summative properties, the complex, even if sharing a joint space–time region, would have only a cumulative character—like a heap of sand or any aggregate lacking interaction between the parts. It would then be sufficient to sum up their properties in order to obtain those of the whole, and we could remove or add sand grains without any change in their structure and organization; the sand heap would remain the same sand heap, although its total volume might have changed.

In this way the almost mystical interpretation of "whole more than the sum of its parts" has been restated in a more rigorous way which lends itself to scientific, empirical speculation and experimentation. As will be shown later in regard to ecosystems and landscapes, at least part of these nonsummative relations can be expressed as functions of state factors of dependent and independent variables.

Holograms and System Perception

One of the major claims of the holistic systems approach, namely, the importance not of the details of the elements but of their interrelations for the perception of the whole, can be demonstrated by the "hologram" block portrait of President Lincoln as a computer-generated image, prepared by Harmon (1973) and presented by Vester in his illuminating exhibition, "The World as an Interlaced System" (Figure 2-8). For the recognition of this face, even the most painstaking study of each individual square—its degrees of greyness and dimensions and the preparation of an exact table of the distribution percentages of the differently shaded squares—is not sufficient. On the other hand, the face can be made out by the process of "pattern recognition," even from a relatively small number of squares, as soon as we view it from a distance or blur its outlines. In this way the squares fall into the archetypal pattern of the face stored in our memory as that of Lincoln and thereby we compensate the missing parts through perception of the whole from the fuzzy structural image of its reality.

This pattern of recognition is based on the "hologram" storage and retrieval of information in our brain. Unlike the localized information storage of a library, a photo album, or computer, there are no separate compartments for each "piece" of memory stored in the brain. Therefore, even if our memory is weakened or part of the brain is damaged, this memory will still be projected as a whole—as in a laser-beam-produced three-dimensional holograph—if only in a more and more blurred and less accurate way.

According to Vester (1980), this example can teach us a major lesson in systems approach: For the recognition, assessment, and management of complex systems as wholes, not the detailed study of individual data—or in the case of a landscape, the professional and scientific disciplines and departments dealing with piecemeal information (biology, meteorology, hydrology, construction, engineering, sociology, economy, etc.)—but the identification of their cross-linkage and interfacing in reality, their structural relationships, are imperative for its comprehension. Therefore, in this process of system-thinking we must sacrifice the accuracy of these details and satisfy ourselves with the "fuzzy set" of the whole, learning about the most crucial couplings and their functional interrelations.

The major challenge lies in the recognition of these "hologram" structural patterns in a world of increasingly complex and increasingly interwoven systems. For this reason a revolution in scientific thinking is urgently needed—a scientific revolution in the terms of Kuhn (1970), in which the paradigms of discipline-oriented "normal" science and their solely analytical, reductionistic, and linear approaches to reality have to be replaced by systems-oriented, synthetic, integrative, and cybernetic approaches. But this is a matter not only of theoretical scientific interest but of great practical importance for the solution of the pressing problems in our ecological relation with the biosphere and the global natural resources and for all those human-ecological, psychosociological, cul-

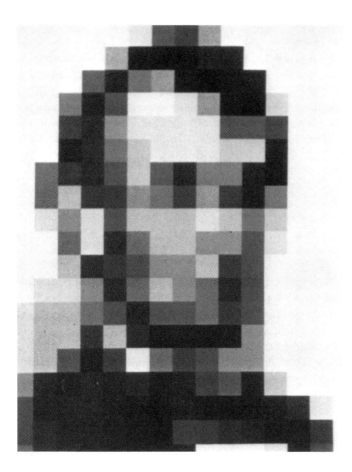

Figure 2-8. Block portrait of President Lincoln. Even from this fuzzy picture, composed only of a few squares, we can recognize the full features of his face if we view it from a distance with half-closed eyes. Courtesy of L. D. Harmon and Bell Laboratories.

tural, medical, economic, and political problems for which the holistic axiom has to be turned into a *holistic imperative.*

The difficulties arising, for those trained in discipline orientation, in coping with complex systems have been discussed in a penetrating study by the cognitive psychologist Dörner (1976). Following Piaget's structuralistic approach (discussed later) to the human cognitive apparatus based on assimilation and accommodation processes, Dörner distinguished between *epistemic* and *heuristic* structures. The epistemic structure (from the Greek "episteme" meaning "knowledge") is the specific knowledge about reality necessary to solve the problem—the information data base in our mind. Thus, for instance, having seen a picture or the statue of Lincoln before enables us to recognize his image from the squares, even in a blurred way. This thinking process can be guided by subconscious motivations (like the id guidance in Freud's primary processes). The heuristic structures (from the Greek "heurism" meaning "searching method"), however, to which most of Dörner's book is devoted constitute the construction method necessary for the solution of the problem. This cannot be retrieved from memory but can be treated as information processing events. In these we can state not only what the processes are accomplishing, but also how they produce the product. Some of these heuristic principles, dealing with the solution of specific problems in specific reality regions (like those dealing with complex systems), can be acquired through training and "rethinking."

In a computerized simulation game, Dörner asked 12 professionals from different relevant disciplines to propose an integrated development plan for the overall improvement of an imaginary African country, Tana. The results achieved were very disappointing; if these proposals were carried out they would worsen the lot of the people, destroy the agricultural–economic base, and create new, even more severe, problems. In the discussion of the outcome of this trial, Dörner stressed the preconceived epistemic structuring of the situation by the specialists and their faulty heuristic strategies—their thinking in affectation chains and not in the required affectation networks and couplings.

Self-Transcendent and Formal Openness

Much of the discussion at the Alpach symposium (Koestler and Smithies, 1969) was devoted to the criticism of reductionism, which the eminent neurologist and psychotherapist Frankl (1969) regarded as a mask for nihilism, camouflaged as "nothing-but ness," and as a "kind of projectionism. It projects the human phenomena into a lower dimension" and it can be counteracted by what Frankl calls "dimensional anthropology, to preserve the one-ness and humanness of man in the face of the pluralism of the sciences. Pluralism, after all, is the nourishing soil on which reductionism flourishes." As a major argument against the reductionistic approach, Frankl (1969) referred to the intrinsic openness of human existence, using the term "*self-transcendent* quality." It is pointing beyond itself rather than being a closed system of physiological reflexes or psychological responses to various stimuli. This openness of human existence, once we have projected man out of his own dimension into dimensions lower than his own, necessarily disappears.

In describing the differences between formal and natural languages, as defined by Pankow (1976; see D-16), we have mentioned the self-transcendent qualities of natural Gestalt systems. Pankow (1976), in the important book *Evolution and Consciousness* (Jantsch and Waddington, 1976), has further developed this concept of self-transcendence by distinguishing between formal openness to the flow of energy/matter and information (described by analogy with the help of formal languages) and the self-transcendent openness of living systems and other natural Gestalt systems, which have the capability to represent themselves or can be described by homology with the help of other Gestalt systems, such as natural language. Pankow (1976) illustrated this by the following parable:

> Somebody draws a landscape with the pencil. In this way a picture takes form consisting of lines, points, and gaps. By exploiting further the potential of various nuances of grey, the likeness of the image can be greatly enhanced. However, the interplay of different colors cannot be improved any further, but has to be questioned itself.

This means that the representation (model, image) of ecosystems or landscapes depends not only on words, but also on the *means* of representation—on language as the "organ" of consciousness. Only if I experience self-transcendence in my own thinking and speaking can I also recognize self-transcendence in the being and consciousness of other living systems. Only if my means of representation are themselves self-transcendent can I represent other self-transcendent objects, and my ultimate, most direct, means of representation, which is also accessible to them, is my use of language. By means of sensitive feeling (i.e., through the feeling of entering and recognizing the outer world), a homologous representation of this self-transcendence is possible.

Jantsch (1976, p. 9), in this same book, carried this thought further, regarding self-transcendence as "a new light on the evolutionary paradigm." He argued that evolution, or order of process, is not only a paradigm for the biological domain, but also a holistic view of how totality "that hangs together in all of its interactive processes moves, as an interacting, dynamic process, spanning a vast spectrum from subatomic process to biological, social, mental, and psychic processes to 'noogenesis'—the evolution of man's cultural/mental world, which created the noosphere."

In an evolutionary view, process and structure become complementary aspects of the same evolving totality. But it is not sufficient to characterize these systems simply as open, adaptive, nonequilibrium, or learning systems. They are all that and more. They are self-transcendent—capable not only of representing and realizing themselves, but also of transforming themselves. They are "evolution's vehicle for qualitative change and thus ensure continuity. For them, Being falls together with Becoming." In this respect, life becomes a much broader concept than just survival, adaptation, and homeostasis. By self-realization through self-transcendence, life constitutes, in Jantsch's words, "the creative joy of reaching out, of risking and winning, of differentiating and forming new relations at many levels, of recognizing and expressing wholeness in every living system. Creativity becomes self-realization in a systemic context. In the evolutionary stream, we all carry and are carried at the same time," (Jantsch, 1976, p. 10).

As the key to a better understanding of man and his societal systems, as well as of nonhuman living systems and ecosystems, self-transcendence becomes, in Pankow's (1976, p. 17) words "the common interdisciplinary beginning and end for the humanistic, as well as the natural, sciences. Interdisciplinarity through self-transcendence does not require the formalization of disciplines, but unifies the disciplines while preserving the variety of their ways for thinking and speaking (points or angles of view)." This unification of disciplines, leading toward transdisciplinarity as the final goal, will be discussed in more detail later. But it is important to emphasize, with the help of the self-transcendent concept, the realization that living systems and ecosystems and the total human ecosystem, as represented by its concrete landscapes or ecotopes (see p. 82) cannot be fully grasped by their formal openness as black box inputs–outputs, even if the most sophisticated cybernetic methods are used for this representation of their analogies and isomorphies. In the realm of environmental education, in the public, academic, and professional schools, this has far-reaching implications. It puts additional weight on the need for broadening our educational basis from the purely cognitive to the affective realm—from perception and intellectual comprehension to the perception of perception, namely, consciousness, and from knowing and understanding to loving and caring for.

In their above-mentioned book, Bennet and Chorley (1978) in their attempt at a "multidisciplinary synthesis of socioeconomic and physico-ecological systems" distinguished between "hard" systems, susceptible to rigorous specifications, qualification, and mathematical prediction in their responses (i.e., control systems and space–time systems), and "soft" systems, not tractable by such methods (i.e., cognitive systems and decision-making systems). These might be useful systems–technological distinctions, but they are not sufficient for "interfacing man with nature," which in our opinion should be attempted with the help of new, transdisciplinary paradigms of the total human ecosystem. Thus, the distinction between formal openness and self-transcendence is a much more fundamental one that cuts across all up-to-now-mentioned systems concepts and opens new vistas for such transdisciplinary systems paradigms in the realm of landscape ecology, as well as human ecosystem sciences. In the following section we shall provide some examples which show that such natural transcendent Gestalt systems can be treated in more formal ways as open systems far from equilibrium.

New Concepts of Stability:
Catastrophe Theory and Homeorhesis

In the conventional use of stability by mathematicians and engineers, it is recognized as denoting conditions very near to equilibrium points. In recent years, however, the concept of stability has been broadened by ecologists. Thus, Holling (1973, 1976) has derived the concept of nonequilibrium ecosystems empirically and by demonstration with topological models. He defined the notion of *resilience* as a system's capacity to absorb changes of state variables and parameters and to persist in a globally stable dynamic regime, far from equilibrium. Stability, on the other hand, is a system's ability to return to an equilibrium state after temporary perturbations. The more rapid the return and the fewer the fluctuations, the more "stable" the system is. The balance between resilience and stability is, according to Holling, the product of the system's evolutionary history and the fluctuations it has experienced. This is true, according to our findings (Naveh, 1982; Naveh and Whittaker, 1979; Naveh and Kinski, 1975), for mediterranean woodlands and grasslands, which are very resilient and still fluctuate greatly and therefore have low "stability" as per Holling. He pointed out, rightly, that this evolutionary interplay between resilience and stability might resolve the conflicting views of the roles of diversity and stability, if the latter is not meant as an equilibrium-centered mathematical term. May (1973) applied such an equilibrium-centered mathematical approach in a much-cited analysis of stability and diversity, and showed that complex systems may fluctuate more than less complex ones. But if there is more than one domain of attraction (the region within which stability occurs in topological models), then the increased variety can simply move the system from one domain to another (as will be shown in the next section, which deals with dissipative structures).

There is also more and more evidence from recent ecological studies, including our studies in the Mediterranean region, that, as reasoned by Holling, instabilities in numbers can result in more diversity of species and in spatial patchiness, and hence in increased resilience. Westmann (1978), in an important article, has further discussed resilience and its measurement and suggested slightly different definitions.

Another very promising formal approach to nonequilibrium systems is catastrophe theory, a new field in mathematical topology that makes possible the formulation of comprehensive qualitative models of systems with the help of three-dimensional graphs dealing with discontinuous and divergent phenomena and sudden changes that have previously eluded rigorous mathematical formulations. It was originated by Thom (1970), who also employed it in topological models of biology. In an endeavor to understand how language is created, Zeeman applied it to sociological and psychological problems and published a most illustrative article on catastrophe theory in *Scientific American*. According to Zeeman (1976), this method has the potential for describing the evolution of forms in all aspects of nature, and hence it embodies a theory of great generality. It can be applied with particular effectiveness to situations where gradually changing forces or motivations lead to abrupt changes in behavior (which is why the method is called "catastrophe theory"):

> Many events in physics can now be recognized as examples of mathematical catastrophes. Ultimately, however, the most important applications of the theory may be in biology and the social sciences, where discontinuous and divergent phenomena are ubiquitous and where other mathematical techniques have so far proved ineffective. Catastrophe theory could thus provide a mathematical language for the hitherto 'inexact' sciences. (Zeeman, 1976)

This theory has been recently explained very illustratively, by Woodcock and Davis (1978), using nontechnical language, as "a revolutionary new way of understanding how things change." They provided many examples of its application in physics, chemistry, biology, sociology, psychology, economics, politics, and public opinion. These included three-dimensional models of "cups catastrophe" of the conflict between nuclear power lobby pressure and the "ecology" lobby, and the emergence of compromise by a more comprehensive appraisal of national energy needs, on the one hand, and ecologically sound policies through an "internal lobby" of state agencies, and emphasis on alternative energy sources, on the other hand.

As shown by Jones (1975), catastrophe theory may also provide models for dynamic ecological processes such as pest outbreaks. He claimed that "it satisfies the call for an orientation that is qualitiative, structural and global and it provides a starting point for generating hypotheses when answers are needed and information is scarce."

In a similar vein, Waddington (1975), the late eminent geneticist and biologist–philosopher, urged the turning away from linear and deterministic mathematical, biological, and ecological models, which cannot do justice to adaptive evolutionary dynamics of bio- and ecosystems. He applied catastrophe theory to multifactoral development in morphogenesis and embryology. In this, he used topological models of attractor surfaces with three dimensions, presenting the multidimensional space occupied by the innumerable variables simultaneously controlling these dynamic processes. These surfaces have the shape of valleys with a stabilized stream as a canalized pathway of change, or "chreod" (from the Greek for "necessary path"), and the diversification processes through time are simulated by the branching of the single first valley into several smaller streams, thereby creating the *"epigenetic landscape"* shown in Figure 2-9. The points of measurement of the components at any given time are symbolized by little balls rolling on the valley bottom. As explained in more detail by Waddington (1977), the changes going on through time in the system controlling the development can be represented by bending the attractor surfaces. Such a branch point becomes a "catastrophe" when it leads to the creation of a fold in the attractor

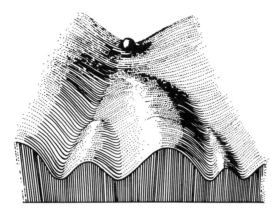

Figure 2-9. The epigenetic landscape. Biosystems move through time in multi-dimensional space along "chreods" of the river flow plain as long as the homeorhetic flow equilibrium of stable pathways of change is maintained and the systems continue to alter in the same way that they have altered in the past (from Waddington, 1975).

surface, thereby displacing the ball and letting it fall either into a lower chreod or out of the chreod. Thus, bio- and ecosystems move through time in multi-dimensional space as locally unstable but globally stable systems. But great environmental and/or human-induced perturbations can push these systems out of the chreods over the "watershed."

In illustrating feedback couplings (D-13), an example of such locally unstable, fluctuating, but globally stable and therefore resilient, *metastable* ecosystems was provided in Figure 2-7, using the annual mediterranean grassland-woodland. These ecosystems meandered in a broad river floodplain of natural (climatic) and cultural (management) fluctuations (in addition to the great spatial variability) throughout many centuries of traditional agropastoral management and their negative feedback loops. However, they are being presently pushed out and over their watersheds by either overgrazing or complete protection, which causes positive runaway feedbacks.

Waddington coined the important term "homeorhesis" (from the Greek meaning "preserving the flow") to denote evolutionary stability (as opposed to the stationary stability of homeostasis), or the preservation of a system's flow process as a pathway of change through time. As a goal, this concept means to keep systems altering in the same way as they have altered in the past.

As will be shown in more detail in Chapter 4, the maintenance of such a homeorhetic flow equilibrium should be the major objective of dynamic conservation management. But as described later, in a much broader sense such a dynamic homeorhesis, and not a stationary homeostasis, should also be the goal of the new symbiosis between man and nature in his postindustrial landscapes.

Self-Organization and Symbiosis in Biosystems and Human Systems

Order Through Fluctuation: The Theory of Dissipative Structures

Of greatest importance for this objective of dynamic homeorhesis is the recent breakthrough in the field of nonequilibrium thermodynamics and its applications to ecological and human systems. Until recently, thermodynamic nonequilibrium was treated as a temporary disturbance of equilibrium, but Prigogine and his co-workers from the "Brussells group" showed that nonequilibrium may be also a source of order and of organization. They founded a nonlinear thermodynamics of irreversible processes, enabling the description of the spontaneous formation of structures in open systems that exchange energy and matter with their environment and lead to the evolution of new, dynamic, globally stable systems. This new ordering principle of self-organization, creating "order through fluctuation" is of fundamental importance for our conceptual framework; we will present it here shortly, without its formal thermodynamic mathematical derivations. [For these, see the recent comprehensive book by Nicholis and Prigogine (1977) on self-organization in nonequilibrium systems; it includes a special chapter on ecosystems. A condensed version has been published by Prigogine (1976), in the above-mentioned book, *Evolution and Consciousness* (Jantsch and Waddington, 1976), which presents a sweeping view of these concepts as the basic nonequilibrium ordering principles governing the dynamic aspects of evolving coherent systems, from the physical–chemical "prebiotic" level of trimolecular models and polymerization to the biological level of aggregation processes of ameba and social insects, to the human-systems level of social systems.]

As Prigogine (1973) pointed out in another important paper on biological order, summarized in more detail by Bakuzis (1974), biosystems operate outside the realm of the equilibrium of classical thermodynamics, and therefore we have to broaden their framework to include nonequilibrium situations. Such a state of nonequilibrium is brought about by the effects of environmental constraints on the system. When these constraints are weak, the state of nonequilibrium attained by the system is simply an extrapolation of its state of equilibrium and it thus exhibits the same qualitative properties. But when increasingly powerful constraints remove the system further from its equilibrium, a new phenomenon can appear—that of *ordered structures,* which, as opposed to those in equilibrium, are maintained only by permanent energy exchange with the environment. They are called *dissipative structures* because they maintain continuous entropy production and dissipate accruing entropy. This entropy, however, does not accumulate in the system, but is part of the continuous energy exchange with the environment. With the help of this energy and matter exchange, the system maintains its inner nonequilibrium and this, in turn, maintains the exchange processes. With this "coherent" system behavior, order is maintained beyond instability thresholds. But if the fluctuations from outside or—as positive "evolutionary" feedbacks—from inside the system exceed a critical size and no longer can be absorbed by energy exchange, these structures are driven beyond a threshold to a new regime, and thus a *qualitative* change in the system's dynamics is introduced. Thus, their tendency is to move through a sequence of mutating transitions to new regimes which, in each case, generate the conditions of renewal of higher entropy production within a new and higher regime of organization and order. They thus create "order through fluctuation."

There is much similarity between the topological approach of Waddington to the development of evolving systems along chreods of epigenetic landscapes and the replacement of homeostasis by homeorhesis for the maintenance of this dynamic process. At present, however, the mathematical approach of catastrophe theory can represent, by the changes in the attractor surfaces, only the switches from one equilibrium stage to another as a new dynamic regime, and not the coherent behavior of dissipative structures and their self-organization by these fluctuations.

Prigogine and his collaborators showed how such dissipative structures can lead to a whole spectrum of characteristic dimensions linked to chemical reactions and transport phenomena. They pointed to examples of how the formation of these structures can be accompanied by symmetry breaking and the appearance of new forms and shapes. Dissipative structures can be considered as giant fluctuations, but their coherent character links their dynamic mechanisms to their spatiotemporal organization, since both have their origin in these fluctuations. Thus, the three levels of description can be represented by the following scheme:

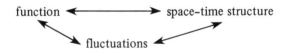

Prigogine described the deterministic character of such systems by kinetic equations, and at the same time, represented the systems' stochastic character by nonlinear stochastic equations (D-16). Therefore, contrary to the views of contemporary biologists like Monod (1970), expressed in his often-cited book, *L'Hasard et la Nécessité* (Chance and Necessity), in the process of evolutionary self-organization, biological systems are no longer in conflict but combine and complement each other. Living systems are thus not "strange products of chance" nor "the improbable winner in a huge lottery"; life does not appear to correspond to an ongoing struggle by an army of Maxwellian demons (who prevent the random mixture of "hot" and "cold" molecules between two chambers, and thereby flout the second law of thermodynamics of increasing entropy and randomness) to maintain the highly improbable condition of low entropy and high negentropy, order and information. In the words of Nicolis and Prigogine (1977, p. 14): "Far from being outside of nature, biological processes follow from the law of physics, appropriate to specific nonlinear interactions and to conditions far from equilibrium. Thanks to these conditions the flow of energy and matter may be used to build and maintain functional and structural order."

As stated rightly by Jantsch (1976), these principles form the core of the most comprehensive approach to evolutionary systems so far developed. "Order through fluctuation" can be regarded as a new description of self-organization of matter in conditions far from equilibrium, and this has opened the way for realizing that evolution toward increasing complexity and organization is the result of structural fluctuations and innovations that can appear suddenly in a previously stable system and drive it, subsequently, to a new regime. These fluctuations are not represented by the older cybernetic principle of "ultrastability," based on the stepwise adaptation of systems to their environment by deviation-reducing mutual causal processes of negative feedback loops, until the final equilibrium is reached. They require a new concept of metastability of non-equilibrium systems, achieved through self-reinforcement and self-organization of fluctuations through internal, deviation-amplifying mutual causal processes of positive feedback loops, mentioned also as the "second cybernation" by Maru-yama (1963) in our discussion of feedback couplings (D-15). Many ecosystems exposed to strong environmental fluctuations show such characteristics of meta-stable, self-organizing, and self-renewing dissipative structures, open to the exchange of solar and heat energy (and therefore also to entropy) with their environment. They therefore defy the notion that the sole ordering principle of natural ecosystems is the steady-state equilibrium, reached in mature "climax" ecosystems through maximizing of energy and information efficiency.

These assumptions are supported by recent findings in studies on the dynamics of marine and terrestrial ecosystems (Beddington et al., 1976; Bernstein and Jung, 1979), which show that such unstable, constantly perturbed and stressed systems can be quite persistent. This is also the case in mediterranean shrub-, wood-, and grasslands, which (as mentioned earlier) are in a metastable stage, with their threshold values depending on a complexity of natural and cultural parameters and climatic fluctuations. Apparently, these provide the means for energy exchange with the environment and return the system's capacity for renewal of entropy production through rest periods, such as grazing, cutting, and burning rotations, until a new peak of entropy is reached in each dynamic regime.

Self-Organization and Planning in Human Systems

These thermodynamic principles of nonequilibrium systems have even more far-reaching implications for the evolution of human systems and for our paradigm of the emerging total human ecosystem.

Jantsch (1975), in his important book on self-organization and planning in the life of human systems, pointed out that "order through fluctuation" seems to be a basic mechanism permeating all hierarchical levels of human systems, organizations, institutions, and cultures, as well as the overall dynamic regimes of mankind at large and their evolution from hunting and gathering to primitive agriculture, to our present global neotechnological society. He maintained that social sciences, following biology, have recognized only external Darwinian factors, and he tried to shed some light on internal factors in the evolution of human systems. In this evolutionary process, human systems are perceived as those systems in which, and through which, human life evolves and organizes itself. For this process of self-organization the theory of dissipative structures provides a theoretical basis, not only for social and cultural organization, but also for self-organization toward a higher state of system organization. Such a theory could also deal with the effect of introducing or adding instability to human systems. It may become capable of explaining cultural mutations that until now have seemed to defy any explanations in terms of energy flow (or the exertion of power). These "autocatalytic" processes or positive feedback loops bear on economic, as well as physical, ecological, social, and cultural, aspects.

In his recent book *The Self-Organizing Universe* (Jantsch, 1980), which became his last because of his unfortunate premature death in 1981, Jantsch's paradigm of self-organization in nature was carried even further. He viewed the cosmic, biological, and sociocultural evolution and their scientific and humanistic implications as a synthesis of Prigogine's concepts with other, new concepts of dynamic micro- and macroevolution, especially "autopoiesis" as the capacity for continuous renewal of biosystems (Maturana and Varela, 1975) and of self-reproducing hypercycles (Eigen and Schuster, 1979). He thereby created a comprehensive framework that reaches far beyond the presently fashionable post-Darwinian and sociobiological interpretations of evolution. Because they are outside the scope of this book, these ideas will not be discussed here further, but we shall refer to his earlier descriptions of self-organizing systems.

Jantsch (1975) distinguished among the following three basic types of internal self-organizing behavior:

Mechanistic systems, which do not change their internal organization;

Adaptive (or organismic) systems, which adapt to changes in the environment through changes in their internal structure in accordance with preprogrammed information (e.g., engineered or genetic templates);

Inventive (or human action) systems, which change their structure through internal generation of information (invention) in accordance with their intentions to change the environment. Such information is generated within the system in feedback interactions with the environment.

Both adaptive and inventive systems play their role in the unfolding of evolution in their own way and in the scale of their proper time. The evolutionary time scale for adaptive systems is the biological domain and corresponds to what Teilhard de Chardin called "the unfolding of *biogenesis,*" whereas for inventive systems in the human domain it would correspond to noogenesis.

Jantsch (1975) realized that wherever social systems aim at changing their own internal organization, the political processes for planning and enacting such changes become an integral aspect of their regulation. The *external* behavior of the system—its ecological feedback relationships with the environment and their chiefly deleterious effects—ought to become part of the political structure. But

unfortunately, these evolutionary and anthropological aspects of feedback interactions belong, at least partly, to cultural change, which is still beyond the grasp
of the conventional political process. Here again, we might add, the role of transdisciplinary environmental education of decision makers and politicians is
crucial. In their way of interacting with and acting upon the environment, their
"external self-organization," Jantsch distinguished two major types:

1. *Finalistic* (or teleological) *systems,* which are rigidly controlled, deterministic, purposive, or heuristic systems and are pursuing aims at different levels that are inherent in their present dynamic structure.
2. *Purposeful systems,* changing their behavior in cybernetic interaction with a change in the environment. To these belong "education for self-renewal" eco-systemic views of world dynamics and policies geared to stability rather than growth. Such systems are, therefore, normative in their planning of the future. Such tasks of operational normative planning in the realm of purposeful systems, namely, the invention of new operational targets, can be fulfilled only by human inventiveness.

We shall elaborate on these new operational targets in the realm of land use
and resources planning and management and the scientific and educational feedback loops that landscape ecology can provide for their achievement. But without doubt, these can be realized only within the context of such "purposeful
systems."

Bennet and Chorley (1978), in their above-mentioned book, have rightly
realized that for what they called "interfacing of physico-ecological and
socio-economical systems" any technical development in modeling ability and
information bases is overshadowed by this all-embracing need to develop our
understanding of the psychological and sociological organization of society and
its epistemological bases, especially in terms of the meaning of nature to man,
in terms of better-understood human goals, and indeed of a better understanding
of the meaning of humanity itself. Not providing such an epistemological base
and putting all their faith in the systems technology of modeling, they concluded that "the real world situations which environmental systems attempt to
model are complex" and that there is "no easy answer for this ultimate control
problem except to exhort an increasing realism through system modelling."

We hope that by providing broader systems concepts and human ecosystemology paradigms, and by not putting our faith in finalistic systems, we shall be
able to provide some new operational targets for purposeful, normative system
planning and management.

Guiding the Train of History

Bennet and Chorley (1978, p. 549) claimed that "most ecological writings ignore
the essential dilemmas of how transition to the Utopian society can be reached
or the lag of time which is required for painful changes in attitudes and the

much needed re-education of planners and government." Among others, they blamed Schumacher (1971) and his intermediate technology as being "divorced from reality."

The well-known biophysicist and futurologist, J. R. Platt, expressed very different views on this subject of "guiding the major train for history" (Pratt, 1974). [His very significant books and essays, as well as most other contemporary systems concepts, approaches, and studies mentioned here, have not been cited by Bennet and Chorley (1978).] Platt relies on very similar basic premises of what he calls "new biological principles" (namely, GST, biocybernetics, and ecology for social planning), giving us a new view of the self-conscious design of social change. He distinguished among three periods of the future: the *inertia* period, for which social science is deterministic and fairly accurate predictions may be provided if we understand the laws of continuity in society; the *choice* period, in which happenings are governed by cybernetic rules and our predictions become "if–then" prophecies, advice, or warnings, so that the alert society can avoid foreseen changes or choose desirable directions that have just become visible. For this period the problem is not (as has been attempted in some of the deterministic world models) to predict but to change the future. The third period is the *uncertainty* period, for which it is fruitless to plan, except in a very general way guided by the same moral rules that usually worked in the past. These three periods are not sharply separated and their lengths are different for different problems. The length of the inertia period is very often overestimated, and as Pratt shows by several examples, the time required for social changes is shorter than is generally realized. Higher education, fast communication, and television make our societies far more plastic and responsive to social changes than we believed was possible in the past. According to Pratt, the dynamics of society and social change in the choice period is rooted not in physical science but in the biological sciences, which are characterized by cybernetic interactions between every organism and its environment.

Although the momentum of world population, with all its pressing problems, can hardly be leveled off in less than 10–25 years, Pratt believes that the period within which our present choices can create fairly predictable changes might be 10–20 years. Within a choice period of 20 years or so, we may be able to reach some of the goals of well-being that we now deem necessary for our whole world population. But for this purpose there is urgent need for an improved social design for the future of mankind if we are not to be destroyed by the crises of the next few years. Platt (1974, p. 300) states:

> Today the human race is passing through a far greater change than the Industrial Revolution or the Protestant Reformation. In the last 30 years, our great jumps in energy, weapons, speeds of communication, and in such biological areas as contraception and agriculture, are forcing mankind to new and more intense levels of interaction, on a shrinking planet. Our population pressure and our pollution, our impact on plants and animals, are felt around the world. If we are to survive these enormous problems we have created, we can only survive by moving to new

and more stable forms of global organization and responsibility. Such a change for the whole human race will be a new jump in the evolution of mankind—like the appearance of eyes or speech, or like the metamorphosis of a caterpillar into a butterfly.

Platt sees the role of technical elites in "map-making" for the future, for "lookouts" to see what possibilities lie ahead and how we can reach them through social inventions, but in the end it must remain for all the people, the participants in our planetary wagon train, to decide whether they want to go in one direction or another. For this group decision it is essential that all social decisions be participatory ones. It must also be recognized that we are in what is called a "non-zero-sum game" in which one man or group does not simply win what another one loses, but in which we all win together or lose together, and that we understand the historical and evolutionary processes of change, which cannot be instantaneous and are rarely smooth.

This "wagon-train model" of history gives us a more coherent picture of the collective process of social design and change as one of shaping history in an ongoing cybernetic process that is powered by practical Utopian plans. The economic and political power necessary to achieve change comes from the fact that a Utopia can be profitable for everybody, in economic as well as human terms. "What we mean by a future Utopia of future improvement in society is a better-operating system, a new kind of vision and hope that is tangible enough to be worth working for and worth investing our effort or money. The problem is only to carry forward this process for a better world society even more rapidly" (Platt, 1974, p. 30).

The Biocybernetic Symbiosis between Biosystems and Human Systems

Among those few systems scientists who are deeply involved in preparing such practical guidelines based on the above-mentioned "new biological principles" for the "wagon train of history" are F. Vester and his interdisciplinary study group. His guidelines are not based on preconceived ideological or theoretical notions derived from isolated disciplines and they do not regard this regulation (or "control," in Bennet and Chorley's terms) as merely a bilateral optimization between economy and ecology with the help of new systems technologies. These guidelines are based on the recognition that in planning and designing our human systems we should rely on laws of nature. These have been successfully governing, for many thousands of years, those complicated processes with which we have had to cope only recently, such as interlaced energy consumption, production, waste, transport, congested regions, diseases, stress, and civilization. In Vester's (1980) words, "A new comprehension of reality guided by the regulating cycles of nature has to take over from our technocratic view, which is guided by linear blinker thinking."

This new biocybernetic system thinking by Vester and his interdisciplinary study group is lucidly described in a great number of publications, exhibitions, films, and books (some of which have become best sellers in Germany and have been translated into many languages, but not yet into English). His biocybernetic system studies are not an attempt to create a new discipline, but a starting point for an orientation on the only viable system that has "not gone bankrupt"—the biosphere, of which we are a living entity (or holon). Therefore, contrary to regular cybernetic technology, we do not influence from outside but are, as regulators, always part of the system, which again may affect us in the very sense spoken of by Jantsch.

These studies recently culminated in the preparation of a comprehensive sensitivity model on the integrated planning of a congested landscape in southwest Germany, as part of the UNESCO-MAB program of regional planning, which is discussed in more detail in Chapter 3 (Vester and von Hesler, 1980). In the preliminary study of this project, Vester (1976) formulated eight basic biocybernetic rules. These can be used as a checklist for a reliable evaluation of the long-term stability and viability of our systems, a checklist derived not from any isolated economic or sociological theory but from the economic behavior of the biosphere itself, the most viable system ever tested:

1. Negative feedback couplings dominate over positive ones in interlaced regulation cycles.
2. Function is independent of quantitative growth.
3. Orientation is toward functions, instead of production, through multiple functions.
4. The principle used in the Asian method of self-defense (jujitsu)—steering and diverting of available forces—is utilized in energy cascades, chains, and couplings.

There is:

5. multiple use of products, processes, and organizational units;
6. recycling with couplings of input and output processes;
7. synthesis with utilization of small-scale diversity; and
8. basic biological design through compatability of technical and biological structures, feedback planning, and development.

These eight biocybernetic rules, which reinforce each other, are sufficient in Vester's view to guarantee the vital self-regulation of any system by minimum input of energy and material consumption. This self-regulation is demonstrated in Vester's writings, especially in his most recent book (Vester, 1980), by many examples from those open landscapes threatened by urban–industrial and agro-industrial expansion and pollution. They differ not only from the technocratic view, but have also advanced far beyond the technosystemic and cybernetic approaches discussed above, in his reliance on the guiding values of the biosphere itself and its evolutionary strategies as open, nondeterministic systems.

Vester (1980, p. 49) warns us that the biosphere—this most subtle, delicate, but at the same time most tenacious, membrane embracing our planet—can rid itself of any disturbing subsystems that do not obey its rules by either freezing to death or exploding in accelerating positive runaway feedbacks through its overruling self-stabilizing negative feedback couplings.

> Human civilization, as one of the most complex subsystems, is also one of the most vulnerable and has the greatest dependencies. We might go on for some time eliminating more and more (other) species and desertifying more and more landscapes, poisoning our air and changing climates. But we shall be the first to be hit by the first greater perturbation of the total system, long before the living world as such will suffer serious damage. In the atomic testing regions of Nevada, where nobody can stay because of their deadly pollution, new colonies of microbes and insects have already established young viable ecosystems and have very soon overcome these disruptions, mortal for mankind, in the same way as they have also rapidly developed resistance against man-made pesticides.

Vester (1976) used the concept of the "critical space"—the minimum area required by each individual of a species for its existence—and the "critical situation," which develops by increasing the population density and violating the critical space. He showed, by examples from social organisms [including termite colonies, which were also used by Prigogine (1976) as examples of self-organizing dissipative ecosystem structures—see above], that this density stress can either lead to the destruction of a large part of the population, and thereby again to a lower density, or can force the population to alter its behavior by "jumping to a higher level of organization." Our industrial civilization and its artificial systems (which we shall describe later in more specific terms as technoecosystems), after having reached this critical situation, must therefore undergo such a radical change in their behavior and jump to a higher level of organization through integration with the biosphere, as the only chance for survival.

This biocybernetic symbiosis should not be interpreted as a romantic call "back to nature"—to the Stone Age or primitive agriculture. On the contrary, it implies progress to even more advanced and sophisticated, but chiefly small-scale and intermediate, technologies. In these technologies, "bionics" (the combination of biology and technology) and the application of biological structure and organization, the principles of recycling and multiuse of the biosphere will play important roles. Vester (1974, 1980) rightly compares the biosphere to a most efficient "superfactory" which, in spite of an annual turnover of 200 billion tons of carbon and organic matter, the production of 100 billion tons of oxygen, and the conversion of many millions of tons of light and heavy metals, has never suffered from any shortages in raw material or waste disposal problems and has never gone bankrupt. Similar ideas of combining "soft technologies" with high-powered biotechnologies and electronics also play an important role in the postindustrial ecological and cultural revival of the "third wave," as envisaged by Toffler (1980) and to be described later in this chapter.

Vester has succeeded in communicating his ideas and concepts to the broader public in most comprehensive and convincing ways, not only writing popular books but also using other communications media—exhibitions, films, and audiovisual educational programs. He has recently gone a step further in developing practical tools for our "survival techniques," translating this biocybernetic symbiosis into quantitative parameters of a sensitivity model for decision making in land-use planning. This was prepared in cooperation with Dr. A. von Hesler, a leading West German land planner, along with a team of landscape ecologists and computer scientists, within the framework of the West German MAB project on ecology and planning in congested regions. This model will be discussed in Chapter 3 as an example of a holistic land-use planning tool, but its underlying principles are of great epistemological importance for landscape ecology as a total human ecosystem science. According to Jantsch (1975) they represent operative planning in the realm of purposeful systems by inventing new operational targets.

The main target of Vester and von Hesler (1980) in this model is the determination of the local systems' actual sensitivity by "tapping" its dynamics with the help of a planning instrument for complex systems. For this purpose, the actual data of the lower main (Untermain) region were chosen and system analyses were carried out enabling the diagnosis of the system structure and dynamics in regard to stability and viability and, as mentioned earlier, oriented toward the evolutionary survival strategies of the biosphere. The basic assumption was that a macroscopic system inspection is sufficient in order to ascertain the cybernetic interlacing from which the system dynamics can be determined, and that this can be evaluated in a sound way only in reference to the basic biocybernetic rules.

In each matrix of cybernetic criteria the following seven most relevant system compartments were evaluated: economy, population, land use, human ecology, natural resources, infrastructure, and the communal and public sectors. At the same time, this matrix also covered the basic physical system criteria, namely, energy, material, and information, as well as the dynamic criteria of structure and flow and other important systems criteria, such as degree of couplings, input–output relations, diversity, and irreversibility. These variables were reduced, with the help of a screening matrix, into 14 key indicators, which were then evaluated from different points of view of data processing, history, planning expertise, and scientific relevance, and from questionnaires of the local populations involved. In these, not only linear but also nonlinear relations of higher order were computed and established in table functions.

In Figure 2-10 an example of the feedback couplings between the populations, open landscapes, and their land use as part of the system compartments within the interaction system is presented. Here, population and open and built-up landscapes interact with ecological burdens as a tension field. Growing populations with greater needs for housing increase the demands for building

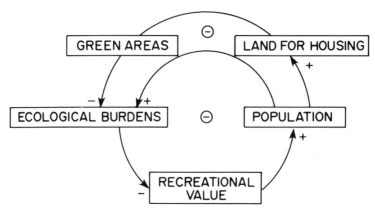

Figure 2-10. Feedback loops between population and open and built-up land-scapes (from Vester and von Hesler, 1980).

space, which leads to a reduction of green areas. Their size and ecological structure affect the ecological burdens and their capacity for air regeneration and filtration influence the groundwater level and water purification, so that their abating effects are reduced. Via the recreational value of the open landscape a negative influence is exerted on population immigration, thereby acting as a negative feedback loop in the regulation of the influx of population. A similar relationship is also true for the second negative feedback loop of populations, acting as positive feedback in the increasing ecological burdens that, in turn, is closed again by the negative feedback of the recreational value. Similar feedback couplings were also constructed for many other interactions between population, economy, communal life, human ecology, land clearing, and the like, and all these individual scenarios were then linked together in a coherent model of the total interaction system. Such a relatively simple hologram allowed the recognition of the crude overall relationships and influences from which the general behavioral trends could be derived with the aid of a suitable interpretive model. However, for further illumination of these numerous behavioral system patterns, quantitive simulation models were developed, which will be referred to in more detail in Chapter 3 when we present this simulation model.

Here it is worthwhile to stress that this model differs from other regional and global simulation models, such as those prepared by Forrester (1971) and others. In contrast to these, the dynamic through-playing of the interaction system is not handled as a "closed machine," letting the model predict the next 50 years as an interpolation of different starting values. Instead, the model is treated as an open system. It is operated by mixed manual and computerized compartment models (by HP 85 or ST table computers) with a stepwise hierarchical structure. This enables constant rechecking by the operator, who himself acts as an internal feedback regulation, simultaneously controlling all steps of the simulation process and thereby himself becoming an integral part of the investigated interaction system.

From the early interpretation model until the final cybernetic evaluation of the sensitivity model, the planner is confronted with an entirely new set of relationships that he has never encountered in the conventional planning process, for example, whether the enlargement of a feeder highway is a control element or a feedback loop or an exchange factor for other variables of this regulation cycle; whether planting of green areas introduces or releases negative feedback loops; whether the construction of a new rapid transport system is an active, passive, or buffering element of the interaction system. The latter terms were introduced by Vester (1976) as mutual influences in a "paper computer model" and an environmental simulation game, and they have also been adopted by one of the authors for multiple-benefit evaluation of mediterranean uplands (Naveh, 1979), to which we shall refer in Chapter 4.

It is obvious that these models cannot be used for extrapolation of the air pollution or water pollution levels in a certain community five years after construction of a chemical plant; but they will predict general trends within the region as a whole and whether flexibility toward perturbation will be lowered or strengthened thereby, whether self-regulation will be disrupted or even dangerous, or whether accelerated runaway coupling will be introduced.

This planning instrument cannot, therefore, replace the conventional planning steps, but can be used as a supplemental aid to the elementary planning process. It is, however, of great didactic value for comprehending the behavior of complex systems and for holistic thinking in general, not only for the planning team but also for decision makers and users, even if not professionally trained, because it enables a continuous dialogue with the model. Its mathematical functions serve merely as a simplifying aid and not as an indisputable oracle, and the strategic instructions are meant as indications for problem solving.

Toward a New Paradigm of Human Ecosystemology and Landscape Ecology

The Ecological Hierarchy

Having explored some of the basic premises and recent developments of GST and biocybernetics, we are now reaching the final building stone of our theory of landscape ecology as a human ecosystem science. This is human ecosystemology, with a transdisciplinary concept of the THE.

For this purpose we shall first deal with the ecological hierarchy, as defined by Egler, and then reevaluate the ecosystem term in the light of above-mentioned premises, thereby setting the scene for our final considerations and conclusions.

In discussing Miller's living system theory (D-20), we have already emphasized the fundamental differences between the biological–physiological integrative levels of living systems and the above-organismic ecological levels. We have also referred several times to the pioneering contribution of Egler in introducing the holistic point of view and the "sister idea" of *emergent evolution* into ecological sciences in the English-speaking world (Egler, 1942). His views were summarized more recently in his appeal for a new science (Egler, 1970). Here, he outlined the development of modern biology from classical organismic biology into two opposite directions: downward to the suborganic cellular and molecular level and upward to the supraorganismic population, biocommunity, ecosystem, and human ecosystem level. Together with subatomic particles and elements, he thus arrived at nine levels of integration, or subject matters. Egler (1970, p. 127) emphasized that "the nine Levels of Integration are not water-tight compartments, either in nature, or should they be in our minds. The idea is simply a "model" that helps us to understand nature. Any one of the nine wholes is but a part of the next larger whole, and is itself composed of parts which themselves are smaller wholes. Thus we have a boxes-within-boxes philosophy that gives us a totally different understanding from the divided apple pie which is exemplified by most university administrations. Such apple pies, or rather whole shelves of apple pies each to be cut into pieces, are the obsolete "models" that our present understanding of science is based upon." These views have now been accepted by most ecologists and are reflected in modern textbooks, such as Odum's (1971a) *Fundamentals of Ecology,* which has been most instrumental in spreading them.

Egler, as a truly intellectual pioneer and philosopher (he is at the same time also a highly professional practicing ecologist), has again set the pace for a further, most significant, epistemological evolution of the ecosystem concept by adding a ninth and last level, that of the human ecosystem (Egler, 1970, p. 126).

> By human ecosystem, we certainly do not mean a virgin, climax, primeval wilderness, which man has utilized, exploited, raped or ruined and which would return to its 'balance of nature' if only man would 'preserve' it. This is the archaic view of those scientists who are pegged at the Eighth (Ecosystem) Level, and have hit their intellectual ceiling. On the contrary, the idea of the Total Human Ecosystem [THE] is that *man-and-his-total-environment* form one single whole in nature that can be, should be and will be studied in its totality.

At the same time, in regarding the THE as the apex of the ecological hierarchy and the organism as its bottom level, we should realize that, in Koestler's (1978, p. 55) words, "The grand holarchies of existence—whether social, biological, or cosmological—tend to be 'open-ended' in one or both directions." Thus, our terrestrial "holarchies" can be considered as a microhierarchy of the macrohierarchy of the cosmos, which ranges from the space–time field, energy continuum, and quarks, up to the entities of astronomy—planets, stars, star clusters, galaxies, and galaxy clusters.

In presenting this ecological hierarchy in Figure 2-11, we have followed Koestler's (1969) example in depicting it as a "living tree, a multi-leveled stratified out-branching pattern of hierarchical organization and subject matters." By making a horizontal cross section across the tree and drawing a schematized bird's eye view which shows how the twigs stem from branches, branches from

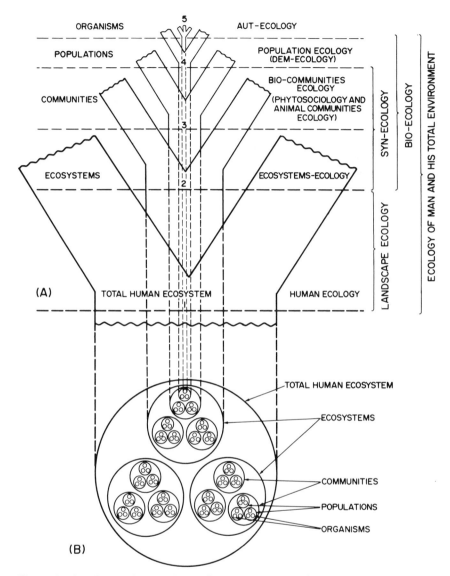

Figure 2-11. The ecological hierarchy and its scientific disciplines. The hierarchy of five levels as a combination of (A) the three; (B) the Chinese box, derived from a cross section through level 5 of the hierarchy tree (adapted from Koestler, 1969).

limbs, and so on, the tree diagram is amplified by a Chinese box diagram. On the right side of the diagram are some of the major ecological disciplines studying these branches. These are linked by integrative sciences, such as synecology and bioecology, dealing with bioecosystems (this term will be discussed later in more detail) and their lower integrative levels. Landscape ecology is depicted as another integrative science and as a branch of THE science, bridging between bioecology and human ecology and dealing with Egler's eighth and ninth integration levels of ecosystems and human ecosystems. According to Egler

> The chief goal of Human Ecosystem Science is the knowledge of, and humanitarian-oriented technology towards, a permanent balance between man and his total environment—both operating as part of a single whole that will afford a life of highest quality . . . . In adding man as a component of a larger whole, we bring into our orbit, as part of one and the same "system," all man-centered fields of knowledge, not only medicine and law, political sciences and economics, psychology and sociology but even the arts and humanities. Human Ecosystem Science aims at bringing into relationship, as parts of one whole, not only the natural sciences, but also knowledge of man himself. It is at this point that quality, judgment, value, art and humanitarianism again enter into this one unity, as real parameters which—if we cannot intelligently manage—we can at least describe. (Egler, 1970, pp. 127–128)

One of the first successful attempts at such a description of an urban human ecosystem was carried out recently by an interdisciplinary team of the Human Ecology Group of the Australian National University, under the leadership of N. Boyden. They did a study of the ecology of the city of Hong Kong and its people (Boyden et al., 1981), as part of the MAB Project 11 of integrated ecological studies of human settlements. Although they did not use the term "THE" and did not apply the system-theoretical and biocybernetic approaches outlined here, their concepts of human ecology as an interplay between the total environment and the human living conditions and biophysical state have much in common with our approach, and the practical implications of their important study lend strong support to our basic assumptions.

The Ecotope as the Concrete Above-Organismic Holon

In our attempt to arrive at a more rigorous conceptualization of human ecosystemology and the Total Human Ecosystem, we should reassess the ecological hierarchy in the light of the above-described epistemological developments of systems concepts, especially that of the holon concept.

If we view these ecological boxes as holons of a biological–ecological continuum and Janus-faced spectrum, embracing the below-organismic and above-organismic integration levels, then we must ask ourselves the following questions: Can plant communities and ecosystems truly be regarded as "real" above-organismic phenomena presented to us by the biological object? Or are they not just conceptual tools and therefore "fictions of a speculative mind," continuing the paraphrase of Weiss (1969). Are they real, natural, concrete systems or just abstracted organizational systems, designed by the ecologist to explore operational dynamics? Or are they both?

Egler (1942), in his pioneering paper on vegetation as an object of study, discussed five emerging levels of increasing complexity, with vegetation above plant and biotic communities and below ecosystems. In his recent challenging book on the use and misuse of vegetation (Egler, 1977), he treats vegetation as an entity in its own right, to be studied, like organisms, from physiological, morphological, chorological, and geographic points of view. Vegetation thus becomes the actual space–time-defined totality of the plant cover of a certain landscape and geographic region and therefore a concrete system through which one can walk and that one can measure, cut, and weigh. In recent years, the term "vegetation science," proposed by Egler in 1942, is coming to replace the European "phytosociology" as the science that focuses on quantitative aspects in the study of plant communities as concrete vegetation units. We shall refrain here from discussing the numerous and diverging definitions and meanings of "plant community," as both concrete and abstracted units. But we shall cite one of the most recent authoritative experts of contemporary vegetation science, the late R. O. Whittaker (1975), in his excellent textbook, *Communities and Ecosystems.* He did not bother much about semantics or epistemological problems and applied an empirical approach. He defined a "natural community" as a living system with distinct composition, structure, environmental relations, development, and functions, and the ecosystem as a functional system with complementary relationships of transfer and circulation of energy and matter. Thus, in his view, an oak forest or patch of desert can be approached either as a community in which productivity, diversity, and the like can be determined, or as an ecosystem in which energy/material cycling is studied.

In California, A. M. Schultz (1967) has been concerned with system-theoretical and epistemological aspects and has expressed a different view in a penetrating paper that can be regarded as an important landmark in the formulation of a transdisciplinary ecosystemological paradigm. For this purpose he restated the laws of integrative levels, as formulated by Feiblemann (1954) from an ecosystem point of view, as follows:

1. Each level organizes the level below it plus one emergent quality.
2. Complexity of the levels increases upward.
3. In any organization, the higher level depends on the lower.
4. For an organization at any level, its mechanism lies at the level below and its purpose at the level above.
5. The higher the level, the smaller the population of instances.
6. It is impossible to reduce the higher level to the lower.

Schultz followed Rowe (1951), who argued that the ecosystem, integrating living systems and their environment, should be considered as the next higher level of integration above organisms, as a space–time unit (or, as defined by us in D-18, as a concrete system). According to Schultz (1967, p. 144): "It cannot be species, population, vegetation or community. None of these are 'environment' for individual plants nor are they any specific volumetric functional part of the ecosystem. Community is an abstracted category which brings together spatially separated parts that have no direct functional relationship. Only by way of the operational environment of the parts (the organism) is there any functional relation between community members." Therefore, continuing the volumetric criteria also for the higher levels of integration, he added a further, seventh, law to express these volumetric relationships of spatial inclusiveness:

7. "The object of study, at any level whatever, must contain, in the volume sense, the objects of the lower level and must itself be a volumetric part of the levels above. Each object in any given level constitutes the immediate environment (in the sense of impinging surroundings) of objects on the level below. Each object is a specific structural functional part of the object at the level above."

Since the object at each level is heterogeneous, the ecosystem consists of the community plus "whatever is spatially interwoven." Its emergent qualities are, therefore, those of the system, namely, productivity, stability, cyclicity, diversity, and the like. In Schultz's opinion, therefore, when ecologists study attributes of communities, such as pattern, abundance, and vegetation dynamics, their objects of study, which are perceptible, remain at the individual organism level.

In order to resolve these contradictory views, we should first distinguish between the above-described levels of the general conceptual ecological hierarchies, as outlined by Egler, and the much more rigorously defined ecological holons of concrete systems at the above-organismic level. For the latter we should accept Schultz's "law of spatial inclusion" as our guideline and therefore rule out population, community, and vegetation, as these supraorganismic ecological holons. But the ecosystem, in its presently used ambiguous and vague connotations, is also inappropriate and needs further elucidation. Thus, for instance, it is considered by Schultz (1967) to be both a conceptual tool—enabling us to look at "big chunks of nature" as an integrated system—and the actual "chunk of nature" studied in the field as the concrete system of a given landscape, and as such, an ecological, topological, and geographic space–time entity.

This confusion goes even further if the same single terminus is applied both to the "ecosystems" of the naturalist and those of the planner, including human social systems, as was pointed out in our first chapter. What Schultz called a "chunk of nature," as the concrete volumetric above-organismic level, is in fact the above-organismic holon, and as such, a uniform, definable (as well as mappable) *landscape unit*. It consists of viable pieces of the biosphere, with its living soil substrate of the pedosphere and its physical environment of the geosphere, namely, the rock parent material of the lithosphere and the water table of the hydrosphere below and the air from the atmosphere above. (In the case of limnological landscape units, the place of the pedosphere and lithosphere is, of course, taken directly by the hydrosphere.)

As mentioned in Chapter 1, for this smallest, above-organismic, homogeneous, and mappable landscape unit, the "ecotope" term has been introduced by landscape ecologists. Thus Schmithüsen (1959) defined the ecotope as a "spatially well-defined and, in its internal interaction system, uniform landscape unit." This ecotope definition has also been recently adopted by European vegetation scientists concerned with the delineation of ecological uniform vegetation complexes as building blocks of the landscape (Pignatti, 1978). In environmental geochemistry, these volume units are called "landscape prisms" (Fortescue, 1980), and in pedology, Jenny's (1958) "tessera" comes closest to our "ecotope holon": "The entire landscape can be visualized as being composed of such small landscape elements. This picture is comparable to the elaborate mosaic designs on the walls of Byzantine churches, which are made up of little cubes, dice or prisms called *tesseras*. We shall use the same term, tessera, for a small landscape element."

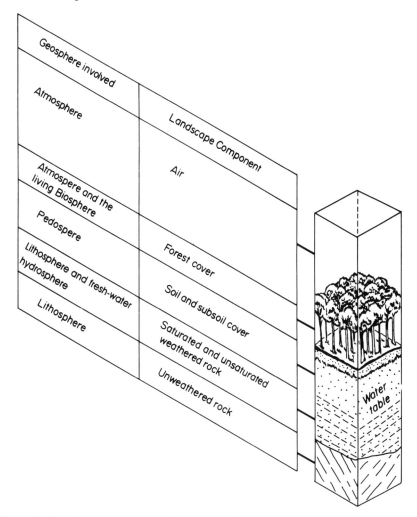

Figure 2-12. A landscape prism (or ecotope) drawn to show relationships between the geosphere and landscape components in a typical forest ecosystem (from Fortescue, 1980).

In a later paper, however, Jenny (1961), while actually referring to this "tessera" ecotope as a portion of the landscape with arbitrary boundaries, applied the "ecosystem" term to both the observed concrete and the abstracted "larger system." He used the following general state factor equation to express the functional/factoral relationships between dependent variables, or internal properties, and independent state factors:

$$l, s, v, a = f(L_0, P_x, t)$$

where the internal properties are l, the ecosystem property; s, soil; v, vegetation; and a, animals, as a function of three state factors: L_0, the initial state of the "large system" at time zero of genesis or observation as soil parent material, topography–geosphere components; P_x, the external flux potential of energy, nutrients, and organisms in and/or from and out and/or to adjacent systems; and t, the age of system. In this way the "environment" outside the ecotope is defined clearly by the position of the observer "inside" the system. P inside and P outside are defined as concentration potentials and intensity factors outside and inside the boundary, and their difference per distance x constitutes the gradients:

$$\text{flux} = -\left(\frac{P_{\text{outside}} - P_{\text{inside}}}{\Delta x}\right) m$$

In this way rainfall, as a flux potential, might be measured just above the vegetation canopy but anything happening to the rain above this boundary is of no concern here (although relevant, of course, to the meteorologist studying the larger systems). Seasonal rainfall is considered as an independent variable.

If we study the effect of burning on a forest, the deer population creates a "deer pressure" at the ecotope boundary and the actual number of deer (or the number of newly invading plants) will depend on deer and plant gradients and the m coefficient as a multivariate function of the above-mentioned "independent" state factors. The same is true, of course, in the study of the effect of recreation in a national park or reserve, with a "human pressure" gradient depending not only on these natural factors, but also on cultural factors, as will be shown later.

It is also obvious that the genecologist studying ecotypic variations of plant populations, the phytosociologist studying plant communities, and the pedologist studying the soil of these ecotopes will apply different boundaries, and their level of discrimination will be very different from that of the landscape ecologist, who studies the ecotope as part of the whole landscape. An outstanding example of such a study of a watershed landscape as a functional unit is the recently summarized one on the Hubbard Brook ecosystem (Likens et al., 1977), in which the hydrological cycles were observed by a large team of ecologists, hydrologists, and geochemists in order to determine the nutrient budget. Here, as in other studies of this kind, "ecosystem" was indiscriminately used for the actual watershed ecotope studies, for the experimental forest in which this study was carried out (sometimes replaced by "landscape"), for "forested ecosystems," and for temperate or tropical forests in general.

A Functional Classification of Ecosystems—The Ecological Holarchy

Having provided a clearer and more formal definition of the ecotope as the concrete, above-organismic, ecological holon, we shall now attempt to further clarify and classify the ecosystem term and its relation to natural and cultural landscapes and the ecological holarchy in general. For this purpose the functional ecosystem classification proposed by Ellenberg (1973) is of great value and will be used as a starting point. In Chapter 1 we mentioned the important contribution of Ellenberg to the development of ecosystem ecology in West and Central Europe, and in Figure 1-3 we presented his hologram model of an ecosystem in which man either fits into the food chain as an omnivore or acts as a "supernatural" factor, influencing all parts of the ecosystem and thereby demonstrating his dichotomic position. Ellenberg (1973) defined the ecosystem as an interacting system formed by living organisms and their abiotic environment and regulating itself to a great degree. Therefore, he subdivided the biosphere, as the largest and all-encompassing global ecosystem, as a functional unit, doing so in a deductive way from the most extensive and complex ecosystem downward, according to six main functional criteria in a five-level hierarchy that could eventually be enlarged if more information on smaller units becomes available. These criteria included the dominant life media, primary producers and their limiting factors for production, material gains and losses, the relative roles of micro- and macroconsumers, and the role of man in the creation of the ecosystem and its energy/material cycling, especially in regard to additional energy sources. Thus, he distinguishes, in a major subdivision of the largest megaecosystems, between the natural and close-to-natural (marine, limnic, semiterrestrial, and terrestrial) ecosystems, which are more or less dependent on solar radiation as their energy source; and artificial urban–industrial ecosystems, which depend on human supply of fossil energy, and recently also on nuclear energy. Ellenberg subdivided only the first group according to size and function, but we shall follow up this basic subdivision in the light of our earlier findings.

As stated by Ellenberg and shown in the previous section, the term "ecosystem" is rankless and has been applied to both the abstracted system type (e.g., "limnic ecosystem") and the concrete space–time system (e.g., "Sea of Galilee"). For the latter, and for any actually mapped, observed, studied, and managed landscape unit, we have proposed the term "ecotope," which can now be specified in this classification as the *smallest concrete ecosystem*. However, the biosphere can be regarded only as the concrete global ecosystem of the first group of Ellenberg's natural and close-to-natural ecosystems. This is also true for other, man-modified agroecosystems in which autotrophic organisms are the primary basis for biological productivity. All these ecosystems are maintained by input of solar energy and by natural biotic and abiotic resource materials, and are regulated chiefly by biophysical information as "adaptive biological systems" (sensu Jantsch, 1975). They can therefore be called biological or, for short, *bioecosystems*. However, the second group, Ellenberg's "artificial ecosystems," lack these biological functions of photosynthetic energy conversion. They depend, instead, on the technological conversion of man's and his domestic animals' muscle energy; and of sun, wind, water and (since the Industrial Revolution) chiefly on fossil energy. Therefore, they can be called technological ecosystems or *technoecosystems*. They are constructed and maintained by man's inputs of energy and material derived directly or converted from the geosphere and biosphere, and are regulated by noospheric cultural, scientific, technological, political, and spiritual information processed by human civilizations or "inventive human (or human action) systems," sensu Jantsch. In these technoecosystems man has replaced the primary autotrophic producers by becoming himself a primary heterotrophic producer, whereas all other organisms fulfill only secondary functions. In accordance with the rankless definition of ecosystems, they form the largest concrete, global technoecosystem of the *technosphere*.

As mentioned earlier, the biosphere should be defined as the largest functional ecological unit and global interaction system, and not merely as the thin layer of the earth's crust and geographic living space. The recent "Gaia hypothesis" (Gaia is the earth mother in Greek mythology), proposed by Margulis and Lovelock (1974), widens this dynamic view by regarding the biosphere together with the atmosphere as a worldwide self-regulating, self-renewing, and self-maintaining system that evolved after the oxidation of the surface of the earth and the oxygen enrichment of the atmosphere by the prokaryotes (the nucleus-free cells), the first life forms, some 2000 million years ago.

The technosphere, however, is the recent creation of inventive human systems of the noosphere, but not of the adaptive biological systems of the biosphere. Its technoecotopes occupy, together with these biological systems, the physical living space on earth; but from the functional and structural point of view, they are not only not integrated into, but are even antagonistic toward, the Gaia system as a whole and its self-regulating functions. This is clearly shown by the adverse global effects of the technosphere on the biogeochemical cycles. Therefore, a specific new term must be given to this concrete, global space-time entity, embracing all biological and technological ecosystems together with relevant parts of the geosphere, namely, the *ecosphere*. (Man, with his technological skill, has overstepped, in the technosphere, the ecological and spatial limits set to life for the biosphere, and thus the ecosphere of "Homo industrialis" includes all those parts of the geosphere—the atmosphere and hydrosphere—that are directly or indirectly influenced by him. Actually, he has already overstepped the limits of the geosphere in the stratosphere, with far-reaching ecological implications for life on earth.)

If the ecotope is the smallest landscape unit of concrete bio- and technoecosystems, it follows that the ecosphere is the largest global landscape and concrete system of the THE, integrating, at least visually and spatially through the geosphere, the biosphere and the technosphere, as shown in Figure 2-13. Thus

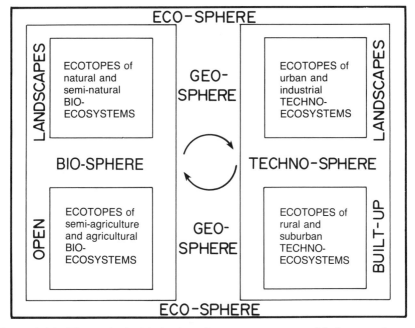

Figure 2-13. The ecological holarchy of concrete systems with the ecosphere as the largest system and global landscape of the total human ecosystem, integrating ecotopes of the biosphere and technosphere through the geosphere (after Naveh, 1980).

the functional classification, based on concrete ecological systems as holons above the organismic level, has yielded a multistructured holon hierarchy, or holarchy, not to be confused with the conceptual ecological hierarchy presented earlier. With the help of this classification, the ecosystem term has lost its vague meaning, for three reasons: (1) A distinction is made between abstracted, functional interaction systems (for which the general "ecosystem" term has been retained) and concrete voluminous "chunks of nature" as the concrete above-organismic holons and landscape units, for which the term "ecotope" has been adopted by landscape ecologists and planners. These landscape ecotopes can of course be specified according to the bio- or technoecosystems that they present, plus their energy/matter and information inputs. (For more detailed explanation, see below and Figure 2-14.) (2) These ecosystems are subdivided into two different classes of bio- and technoecosystems. (3) In addition to these natural ecosystems [in the broad meaning of Laszlo (1972)], a higher level of integration and organization of the THE is created. We prefer to use the word "total" in order to emphasize this integration of man with his total environment (which has been defined by Egler but is now specified more concisely by the integration of the biosphere and technosphere) and in order to distinguish it from the vaguely used "human ecosystem" term, existing *besides* natural ecosystems and not as a higher integration level.

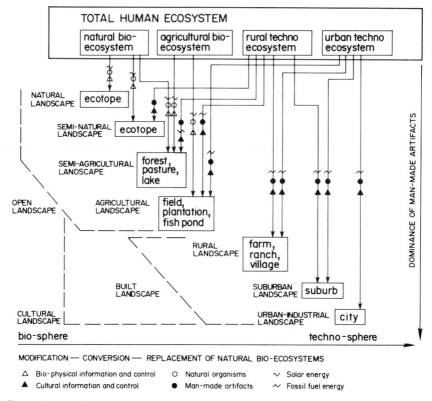

Figure 2-14. Ordination of landscape ecotopes according to energy, matter, and information inputs from bio- and technoecosystems. The achievement of a new balance between the left and right poles of this ordination is the major goal of landscape ecology (from Naveh, 1980).

This also brings into focus the epistemology and semantics of the "environment." By definition, the environment is everything outside the system. But for man, as part of the THE, whatever has been called "natural environment" or "physical environment" or "urban environment," and so on, is actually a part of his system. For this reason, the term "environmental system," if not dealing with the extraplanetary, astronomical, cosmic, and geological systems beyond the ecosphere (see page 82) is an unfortunate and misleading misnomer. This is true to a great extent in the areas of "environmental" education, protection, management, conservation, and the like. However, it will be very difficult to change these widely used terms. Whatever we want to educate, protect, manage, and conserve must be part of the ecosystem in which we live—otherwise we would not be able to affect it in one way or the other, and as "environment" it would be out of our reach. The THE concept has made it even more implicit that whatever is effected by us and whatever we are affected by in the physical world surrounding us is part of the ecosphere and its concrete landscape units, belonging, with us, to the THE.

The Leap to the Third Wave of Change
of Postindustrial Civilizations

It should be emphasized that neither the "ecosphere" nor the "technosphere" term was invented by us; both have been used before with slightly different connotations. "Ecosphere" has become well known through the issue of a collection of essays from *Scientific American* (Ehrlich et al., 1971) on man and the ecosphere, introduced by an article by Cole (1971) in which he defined the ecosphere as a "combination of the biosphere and ecosystems."

"Technosphere" was used by Toffler (1980), the author of the popular *Future Shock*, in a very challenging book in which he (among other things) made a penetrating exploration of the technosphere and its closely interwoven cultural, sociopsychological, economic, and ecological implications. His thought-provoking futuristic outlook is basically guided by an enlightened holistic view that has much in common with our biocybernetic system approach, and it provides a suitable background for our conclusions. He has called this book *The Third Wave* (of change), meaning the present Postindustrial Revolution, following the "second wave" of the Industrial Revolution, which was preceded by the "first wave" of the Agricultural Revolution. Toffler (1980, p. 41) states: "What we see, therefore, if we take these changes together, is a transformation of what might be called the 'technosphere'." All societies—primitive, agricultural, or industrial—use energy; they make things; they distribute things. In all societies, the energy system, the production system, and the distribution system are interrelated parts of something larger. This larger system is the technosphere, and it has a characteristic form at each stage of social development.

As the second wave swept across the planet, the agricultural technosphere was replaced by an industrial technosphere: Nonrenewable energies were directly plugged into a mass production system, which in turn spewed goods into a highly developed mass distribution system.

In describing the threat to the biosphere and the planet as a whole by the urban–industrial complexes and their side effects, Toffler (1980, p. 32) comments:

> Among these was the rampant, perhaps irreparable damage done to the earth's fragile biosphere. Because of its indust-real bias against nature, because of its expanding population, its brute technology and its incessant need for expansion, it wreaked more environmental havoc than any preceding age . . . . Industrial society raised the problems of ecological pollution and resource use to a radically new level, making the present and past incommensurable. Never before did any civilization create the means for literally destroying not a city but a planet.

But two changes, by themselves, make the "normal" continuation of industrial civilization no longer possible. "We have reached a turning point in the 'war against nature'." The biosphere will simply no longer tolerate the industrial assault. Second, we can no longer rely indefinitely on nonrenewable energy, until now the main subsidy of industrial development. "This means that all

future technological advances will be shaped by new environmental constraints."
(Toffler, 1980, p. 134)

But a civilization is more than simply a technosphere.

> Every civilization operates in and on the biosphere, and reflects or
> alters the mix of population and resources. Every civilization has a
> characteristic technosphere—an energy base linked to a production sys-
> tem which in turn is linked to a distribution system. Every civilization
> has a socio-sphere consisting of interrelated social institutions. Every
> civilization has an info-sphere—channels of communication [for pro-
> ducing and distributing information] through which necessary informa-
> tion flows. (Toffler, 1980, p. 359)

Every civilization also has its psychosphere, and Toffler (1980, p. 136) describes
very vividly the crisis of the second wave in all these spheres as a closely interwoven event:

> We can ignore the connections between the energy crisis and the per-
> sonal crisis, between new technologies and new sexual roles and other
> such hidden interrelationships. But we do so at our peril. For what is
> happening is larger than any of these. Once we think in terms of suc-
> cessive waves of interrelated change, of the collision of these waves, we
> grasp the essential fact of our generation—that industrialism is dying
> away—and we can begin searching among signs of change for what is
> truly new, what is no longer industrial. We can identify the Third Wave.

The main part of Toffler's book is devoted to the identification of these
changes and their recognition as creative innovations in the different spheres and
their close interrelations:

> It is now possible to see in relationship to one another a number of
> Third Wave changes usually examined in isolation. We see a transfor-
> mation of our energy system and our energy base into a *new techno-
> sphere*. This is occurring at the same time that we are de-massifying the
> mass media and building an intelligent environment, thus *revolutioni-
> zing the info-sphere* as well. In turn, these two giant currents flow
> together to change the deep structure of our production system, alter-
> ing the nature of work in factory and office and, ultimately, carrying us
> towards the transfer of work back into the house. (Toffler, 1980,
> p. 246)
>
> By themselves, such massive historical shifts would easily justify the
> claim that we are on the edge of a new civilization. But we are simul-
> taneously restructuring our social life as well, from our family ties and
> friendships to our schools and corporations. We are about to create,
> alongside the Third Wave techno-sphere and info-sphere, a Third Wave
> *socio-sphere* as well. (Toffler, 1980, p. 217)

Among some of the changes most relevant for our discussion he visualizes
those induced by the pressures on the biosphere resulting from our escalating
demands and the consequential alarm signals of pollution, desertification, signs
of toxification in the oceans, and subtle changes in climate. These force the

corporations, as the main organizers of economic production and the "key producers" of environmental impacts—the managers of tomorrow—to assume
responsibility for converting their environmental impacts from negative to
positive.

> They will assume this added responsibility voluntarily or they will be
> compelled to do so, for the changed conditions of the biosphere make
> it necessary. The corporation is being transformed into an environmen-
> tal, as well as economic, institution—not by do-gooders, radicals, ecolo-
> gists or government bureaucrats, but by a material change in the rela-
> tionship of production to the biosphere. (Toffler, 1980, p. 246)

Taking up the traits of the influential ecologically minded British economist
Schumacher (1971), Toffler sees another encouraging sign in the growing
recognition that "small within big is beautiful" and that striving for appropriate
scale is gradually replacing the second-wave obsessive emphasis on maximiza-
tion, together with standardization, synchronization, centralization, specializa-
tion, and concentration. Toffler devotes an important chapter to the deep
changes in our view of nature, evolution, progress, time, and space. In a vein
similar to the trends described by us above, he sees a major shift from a second-
wave culture that emphasized the study of things in isolation from one another
to the third-wave scientific culture that emphasizes contexts, relationships, and
wholes, culminating in a cybernetic system approach and in a philosophical
breakthrough initiated by Prigogine and his group. Toffler points out, rightly,
that Prigogine's work not only combined chance and necessity but actually
stipulated their relationships to one another. When dissipative structures
mutate from one stage of reorganization to a higher one and leap to a new stage
of complexity, thereby creating order through fluctuation, it is impossible to
predict which of many forms this order will take. But once a pathway has been
chosen and a new structure comes into being, determinism again dominates.
Toffler compares this with the leap of the second wave to the third wave of
change that will shape the rest of our lives and that bears hope for us and our
children if turned into a wave of reconstruction. In Toffler's words:

> I am under no Pollyannish illusions. It is scarcely necessary today to
> elaborate on the real dangers facing us—from nuclear annihilation and
> ecological disaster to racial fanaticism or regional violence .... Never-
> theless, as we explore the many relationships springing up—between
> energy patterns and the new forms of family life, or between advanced
> manufacturing and the self-help movement, to mention only a few—we
> suddenly discover that many of the very same conditions that produce
> today's greatest perils also open fascinating new potentials.
> The Third Wave shows us these new potentials. It argues that, in the
> very midst of destruction and decay, we can now find striking evidence
> of birth and life. It shows clearly and I think indisputably, that—with
> intelligence and a modicum of luck—the emergent civilization can be
> made more sane, sensible and sustainable, more decent and more demo-
> cratic than we have ever known. If the main argument of this book is

correct, there are powerful reasons for long-range optimism, even if the transitional years immediately ahead are likely to be stormy and crisis-ridden. (Toffler, 1980, pp. 16–17)

In concluding, the basic hope for the future of mankind, expressed by Prigogine in thermodynamic terms as a mutation of dissipative structures toward a new order, and by Jantsch in evolutionary terms as noogenesis of inventive human systems, or by Vester in biocybernetic terms as a jump to a higher system level of organization, has been described in Toffler's sociocultural terms as the leap from the second wave to the third wave of change.

The Total Human Ecosystem Ecosphere Landscapes:
A New Symbiosis between Human Society and Nature

The final two sections will be devoted to some of the cultural, philosophical, and educational ramifications of these perspectives and changes, as viewed within the context of human ecosystemology and landscape ecology.

For this purpose, it is important to stress again, as done already in defining space (D-6) and in the previous sections, that we are living in both a physical-geographic–ecological landscape space and a conceptual space with its noospheric components of the info-, socio-, and psychospheres. This development of conceptual space is a result of the cultural evolution of homo sapiens during millions of years throughout which primeval man evolved as an independent ecosystem state factor. It added unique psychosocial technoeconomic and spiritual dimensions to his biophysical nature which released him from his total dependence on the biosphere and its bioecosystems and made him their dominating state factor by modifying all other ecosystem variables. As described earlier, it enabled him to add an entirely new class of concrete ecological systems, the technoecosystems of his rural, urban, and finally, urban–industrial landscapes. In this way, the cultural evolution of homo sapiens—his noogenesis—yielded the agricultural landscapes of Toffler's first wave and the industrial landscapes of the second wave, and it will, we hope, lead to creation of the postindustrial landscapes of the third wave.

As explained earlier, cultural landscapes are Gestalt systems, and as such they reflect not only the economic, but also the spiritual, ethical, and aesthetic values of those who shaped them throughout history. According to the above-mentioned laws of integrative levels, the question of purpose and values ("What for?") has to be answered at the higher level of integration and is closely related to the crucial question of industrial man's relation to nature. This "nature" is represented in the open landscape, as its last physical and biotic remnant. The norms applied in land-use decisions are crucial for its fate, and therefore also crucial for landscape ecology as a total human ecosystem science.

Presently, our technoecosystem–industrial landscapes are being driven by positive feedback loops of energy/matter and cultural information resulting from our human inventive systems with their exponentially rising demands and materialistic expectations. They are therefore growing rapidly into increasingly larger urban–industrial complexes and are thus threatening the natural and semi-natural ecotopes by direct replacement, pollution, and other syndromes of "neotechnological landscape degradation" (Naveh, 1973), for which the recently published comprehensive "Global 2000 Report to the President" (of the United States; Barney, 1981) has provided a very substantial factual basis.

As can be seen in the ordination model of the relationships between the major bio- and technoecosystem holons of the THE and their concrete landscape ecotopes (Figure 2-14), there is an alarming tendency toward the lower right-hand corner of ordination with increasing human impact of accelerated, exponential urban–industrial expansion and increasing dominance of man-made artifacts, leading to more and more monotonous urban–industrial cultural landscapes. Increasing inputs of fossil energy, man-made waste material, and cultural information are unfortunately accompanied by increasing losses of natural and spontaneously occurring organisms and of the negative natural feedback loops ensuring environmental stability and resilience.

Man, as a biological creature, is dependent for his existence on the viability of natural and agricultural bioecosystems. Therefore, his technoecosystems cannot survive without the biosphere, and their exponential growth may also endanger the technosphere itself. We can thus view man in a cybernetic way as occupying a position of mutual causality—as a receiver of vital inputs from the biosphere and geosphere, but at the same time as a modifier of the biosphere and geosphere through the outputs of his technosphere. He is thus affecting and being affected by these modifications, as shown in Figure 2-15. This dichotomy of dependence and independence of man's position in nature, resulting from his closely interwoven biogenesis and noogenesis, is one of the reasons for the already mentioned confusion about the semantic and epistemological relations between natural and human ecosystems. It is also reflected in the many converging philosophical, historical, and ecological interpretations of man's responsibility to nature described by Passmore (1974), and more recently in a much more critical way in an annotated bibliography of "ecophilosophy" by Sessions (1979).

Figure 2-15. Mutual causality among industrial man, the biosphere, and the geosphere. Waste products and stressors from the technosphere are causing adverse changes in man's health, well-being, and behavior. This conflict can be resolved only by a new balance between man's self-transcending and self-asserting holon properties toward the biosphere and geosphere as part of the total human ecosystem–ecosphere (from Naveh, 1980).

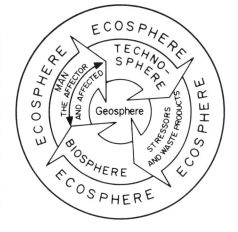

In our opinion, however, it can be resolved by recognizing the holon properties of man and the biosphere operating as autonomous wholes toward their subordinates and as dependent parts of a higher controlling whole, the total human ecosystem. As Koestler (1978) has pointed out, this dichotomy of opposing tendencies of integration or self-transcendence, functioning as part of the larger whole, and of self-assertion to preserve its autonomy is causing a basic polarity in every type of hierarchical structure. Under favorable conditions the self-asserting and integrative tendencies are more or less equally balanced, and a kind of dynamic equilibrium is created in which the two Janus faces complement each other. But unfavorable conditions distorting this equilibrium may have dire consequences. This is most obvious in social hierarchies, but it is also true in our present ecological hierarchy because of the overwhelming role man's self-assertive tendencies are playing in his relation to the biosphere and ecosphere. In primitive societies these self-transcendent tendencies lead to a symbiotic I-thou relation of man to nature sanctified in a process of man's cybernetic adaptation. But the development of modern technology has disrupted this cybernetic adaptation and has replaced man's respect for natural ecosystems with the ecological hubris of his punishable arrogance and ignorance and has created an I-it exploitative relation. Unfortunately, present reductionistic trends in biology, and even in ecology under the darwinistic influence of viewing natural selection solely as competition and struggle for survival of the fittest, has given much scientific sanctity to these self-assertive tendencies. At the same time, however, important integrative tendencies of cooperation and symbiosis in evolution have been overlooked and much neglected.

We cannot deny the evolution of our neocortex and the noospheric cultural revolution, which has given us the present superior status in the ecological hierarchy of nature, as human inventive systems. We cannot return to the landscapes of the Garden of Eden, nor to the primeval symbiotic status of man with nature. Nor can we return to the landscapes of the first agricultural wave, although—as will be explained in Chapter 4—we would like to conserve as much as possible of these (especially in the Mediterranean Basin). At the same time, we cannot continue to expand those of the second industrial wave with the arrogant, exploitative, anthropocentric attitudes, rightly criticized by "deep" ecologists and environmentalists. If we regard human evolution as a dynamic process of self-realization, then through self-transcendence the new cybernetic symbiosis to be attained between man and nature can be envisaged as a reconciliation between our self-asserting and self-transcendent holon behavior toward our landscapes. In this further step of noogenesis, self-organizing positive and self-stabilizing negative feedback loops, steered by pragmatic, cultural, and biophysical information from natural–adaptive and cultural information from human–inventive biosystems will ensure the functional and structural integration of the biosphere, the technosphere, and the geosphere in the THE ecosphere. In this way, the new I-thou relationship between man and nature will find its expression in man's attitudes toward concrete natural and cultural ecotopes. The jump to a higher level of complexity and the mutation of biosystems and technosystems as dissipative structures will yield a higher level of ecosphere organization with emerging qualities of diversity, stability, productivity, beauty, and utility in the new third wave landscapes.

This new "balance of nature" is, therefore, not a stationary, homeostatic equilibrium to be achieved either by technocratic control models of "intervention" or by linear, deterministic, global models of "synthesis" (Bennet and Chorley, 1978), and it goes far beyond "interfacing of socio-economic and physio-ecological systems," as proposed by these authors. Its goal should be to ensure a dynamic flow equilibrium through our integral participation as holons and regulators in the systems to be regulated. This implies "our moral engagement in regulating human systems, including ourselves" (Jantsch, 1975), achievable neither by anthropocentric attitudes of "shallow" ecology nor by biocentric egalitarian attitudes of "deep" ecology, but by creative human ecosystemology.

There are encouraging signs that this reconciliation of man with nature has already started. The distinguished biologist and environmental educator and policymaker, Lord Ashby (1978), has shown in a very illuminating way the evolution of new social norms and ethics arising in our dealing with the intrinsic values of nature as reflected in the actual decision–making process in land-use planning and management. However, in view of the critical situation of our ecosphere, these trends very urgently need to be strengthened and broadened. Landscape ecology, as an interdisciplinary human ecosystem science, must play an important role in this "Third Wave cultural evolution" by supplying scientific, technological, and educational information as regulative feedbacks for the structural and functional integration of the biosphere and technosphere. Of great importance in this respect is its task-solving-oriented interdisciplinary outlook, which should be based on the paradigms of transdisciplinary THE—as outlined in this chapter. We shall conclude here by discussing briefly these educational aspects of landscape ecology.

Landscape Ecology and Ecosystem Education

As was explained in Chapter 1, by overstepping the purely natural realm of classical bioecological sciences and entering the realm of human-centered fields of knowledge and their sociological, psychological, economic, historical, cultural, and other aspects involved in modern land uses, landscape ecology has itself become an interdisciplinary human ecosystem science. In this chapter an attempt has been made to deepen and broaden this interdisciplinarity with the help of General Systems Theory and its recent conceptual, epistemological, and methodological developments. Of special relevance for the interdisciplinary educational aspect is the contribution of *structuralism,* as presented by its most important proponent, Piaget, whose influence on modern education cannot be overstressed.

According to Steiner (1978), the editor of a recent comprehensive volume of more than 60 articles appraising Piaget's work and its influence on contemporary psychology, he is not only a "Giant of Developmental Psychology" but also a system builder; he has succeeded in designing, in his genetic epistemology, an all-embracing system on the epistemological foundations of biology, sociology, and psychology. This system is now affecting again, in a mutual causal way, the disciplines that served as its starting points.

Piaget (1972) has given a special interpretation of systems in relation to structure which differs from the use made here of this term. He defined structure as a transformation system presenting laws independently of the properties of its parts and capable of self-regulation. He therefore follows very much the biological General Systems Theory approach developed by Bertalanffy, and that of the Gestalt-oriented psychological view. But according to Piaget (1968), the notion of structure as a self-regulating system should be carried beyond the individual organism, beyond even population, to encompass the complex of milieu, phenotype, and genetic pool. Further, the interpretation of self-regulation as a cybernetic loop and other advances in the systems approach in biology are of first importance for evolutionary theory and to structuralism. Joined to ethology (the comparative study of animal behavior), they furnish the basis for psychogenetic structuralism.

But structuralism is a method whose scope also includes all human, social, and psychological phenomena. In these fields structure is not used as an analytical concept to break down sets into their constituent elements, but is used with a holistic approach. Giving logical priority to the whole over its parts, the structuralism method attempts to study not the elements of a whole, but the complex network of relationships that link and unite these elements.

Piaget (1968) coined the term "operational structuralism" to denote an approach that he believes surpasses the schemes of atomist association on the one hand and emergent totalities on the other. It adopts, from the start, a relationship perspective according to which it is neither the elements nor a whole that comes about in some manner (one knows not how), but the relations among elements that count. He applied these principles to the study of genetic epistemology (the intellectual development of the human mind) and in his work on cognitive psychology, which was most instrumental in dispelling the myth of reductionistic and mechanistic trends of behavioral psychology as represented by the influential psychologist B. F. Skinner. As will be shown later, these views are also most important for our perception of systems as concrete landscape entities.

As summarized by Bakuzis (1974, p. 485),

> structuralism is a method and not a philosophy—a method consisting of identifying the characteristics of a whole, which is regarded as something more than a simple aggregation of antecedent elements, and arriving at these characteristics by processes which may be formal or partly experimental, as the case requires, but which are always verifiable. As such, structuralism is necessarily interdisciplinary. In the domains of logic and mathematics, which are wholly self-sufficient disciplines, the fact remains that, to establish the full epistemology of the

structures and not to stop short at their technical aspects, they must be compared with the structures whose psychogenetic construction can be followed. However, the study of this construction sooner or later calls into play our knowledge of physical, biological and social structure, and reciprocally, comprehension of these three sorts of structures is a required resource, as much to psychogenetic analysis as to formal methods of construction. In a word, any structure is always located at the intersection of a multiplicity of disparate disciplines, so that no general theory of structures can possibly escape the requirement that it be not simply multidisciplinary but authentically interdisciplinary.

Within the scope of this book we cannot go into greater detail [see Piaget's writings and a special review article on his structuralism by Turner (1973)]; further references can be found in the excellent summary of this subject by Bakuzis (1974).

In concluding, we use Piaget's (1970) own words to show that we no longer have to divide reality into watertight compartments or mere superimposed stages corresponding to the apparent boundaries of our scientific disciplines. On the contrary, we are compelled to look for interactions and common mechanisms. Interdisciplinarity thus becomes the prerequisite of progress in research, instead of being a luxury or bargain article.

Of further great relevance to our discussion are several important lectures presented at a symposium on environmental education at the Second International Congress of Ecology in Jerusalem in 1978 (Bakshi and Naveh, 1980). At this symposium, Schultz (1980) suggested replacing the poorly defined "environmental education" with "ecosystem education" and regarding as its primary goal educating ourselves about our ecosystems by developing tools for proper whole-system thinking, seeing, and even talking. He cited Jantsch's (1975) stream analogy for this ecosystem education. In this you are not just an objective observer on the bank of the stream ("the rational approach") or going downstream in a boat and feeling it ("the mythological approach") but you are the stream, in tune with the whole system ("the evolutionary approach"). Schultz coined the term "holistic paralysis" for those unable to cope with the sheer complexity of the whole system, usually connected with the multidisciplinary programs without real purposeful interdisciplinary integration and interaction and without the transdisciplinary concepts outlined above. He also pleaded for the education and training of unique ecosystem experts, instead of continuing to use traditional professionals. These should not be called "generalists" because they really would be specialists, knowing how to manage a particular ecosystem after thorough study of its interrelationships, cutting across the lines of conventional disciplines. Schultz himself, without doubt, is contributing much to such an ecosystem education by his inspiring lectures on ecosystemology on the Berkeley campus, which are frequented year after year by many students from all disciplines of natural sciences and humanities.

In a similar vein, G. Schaefer (1980), who has developed valuable biocybernetic high school teaching programs, together with his team from the Institute for Science Education (IPN) at Kiel, West Germany, regards environmental education as a new philosophy of teaching the total ecosystem. In this education, exclusive thinking should be supplemented by inclusive thinking. The following is a summary of comparisons from this and another important lecture by Schaefer (1978):

Exclusive thinking:	Inclusive thinking:
1. Disciplinary thinking within fixed systems with rigid borders	1. Integrated (inter- and transdisciplinary) thinking with flexible borders
2. Thinking in models, laws, and "securities"	2. Thinking in realities, regressions, and probabilities
3. Purely logical and typological thinking	3. Logical thinking accompanied by free associations (with the freedom of fantasy) and by variability
4. "Economic" thinking toward maximum progress in a specific field with decisions of either/or and first this/then that	4. Ecological thinking, with progress in one field taking into consideration also others and coexistence and compromises with this/as well as that; here this/there that
5. "A-historical" thinking in terms of identities, reproducible objects, and great exactness. Postulates of contemporary science: full reproducibility, fullest security now, deterministic thinking	5. Historical thinking, with autonomous growth, individuality, and discrimination; fuzziness of single answer and low reproducibility, probabilistic and evolutionary thinking

An eminent science educator from Cornell University stressed the importance of applying principles from cognitive learning theory (Novak, 1977, 1980) to environmental education. He suggested proceeding by teaching a sequence of concepts jointly shared by the teacher and learner, from the most general and inclusive concepts to the more specific ones. He pointed, however, to the difficulty of using such concepts in environmental education, because these notions tend to be unusually dependent on the meaning of many related concepts. This is true if we use the vaguely defined and abstracted "ecosystem" term or—even worse—the ill-defined "environment" as such a broadly inclusive concept. But, as explained in more detail at the above-mentioned symposium (Naveh, 1980), for such conceptual and practical building stones in the learning process, the concrete space-defined entities of natural, agricultural, and urban landscape ecotopes can be very useful. These ecotope holons also provide a logical sequence in the biological continuum from the subcellular, cellular, and organismic levels to which the student has been exposed already, both in the classroom and in everyday life. From these, the THE can then be derived, as the broadest inclusive concept.

Actual natural, agricultural, and urban landscapes can serve as the smallest geographic and ecological units for instruction on their tangible structural and functional reality in which their patterns and processes can be traced, studied, and demonstrated. Thus the changes in time and space resulting from the combined action of abiotic forces, together with biotic and anthropogenic forces (including historical, cultural, and economic) as related to past, present, and predicted future land uses, are available for study, examination, and analysis in the field and for illustration and discussion in the classroom.

The efficient use of these instructional media for field-oriented environmental education might help to break down the artificial barriers between the humanistic, realistic, and biological sciences, and between their branches, which have been erected in our minds through rigid disciplinary teaching methods. It might also provide a rare opportunity for a more comprehensive and balanced transdisciplinary outlook on man's impact on the earth (Naveh, 1975).

At the above-mentioned symposium, the development and application of teaching programs around the THE and landscape-ecological concepts in Israel were reported in relation to both urban and rural environments. In the city of Haifa, this is part of a junior high school curriculum on man and nature initiated by M. Ben-Perez (1980) with the help of a team of experienced biology, geography, and history teachers. In this curriculum the concepts of system and cybernetic feedback loops are introduced in dealing with familiar and concrete issues concerning the interrelationships between human society and its natural and modified environments. Pupils are involved in active inquiry into these principles in their own landscapes. In the original version, a historical–cultural approach was employed in order to stimulate the interaction between teachers of biology, history, geography, the Bible, and so on (Naveh, 1975), but in the rigid discipline-structured schooling system of Israel (as elsewhere) it is very difficult to carry out such programs. This was, however, achieved successfully in a special project on environmental education in the Environmental High School established by the Desert Research Institute of the Beer Sheba University at Sde Boker (Naveh, 1980). Here the core curriculum on environmental education is implemented in field workshops called "eco-shops." Each eco-shop is built on integration of the actual team studies in different natural, agricultural, and urban landscapes, lasting for several days in each landscape ecotope, with relevant subjects covered in classroom teaching before and after the workshops (Bar-Lavie, 1980). They are planned on a hierarchy of increasing complexity, culminating shortly before matriculation in the 12th grade with a model of the Negev THE. In this way a field-oriented transdisciplinary teaching program has been conceived, integrating the different disciplines relevant for the study of these landscapes in the eco-shops and coordinated through curriculum development and the in-service teacher training in systems concepts, biocybernetics, and landscape ecology, toward the THE, as shown in Figure 2-16.

The final part of the above-mentioned symposium (Bakshi and Naveh, 1980) was conducted as a workshop in Sde Boker, and for more details the reader is referred to this source.

In academic education, the conceptualization and formalization of the THE and the ecological holarchy, as described in this chapter, should be carried fur-

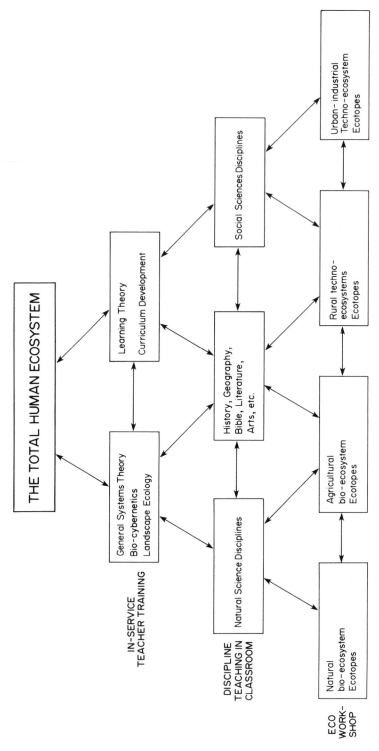

Figure 2-16. The eco-shop version of ecosystems education at Sde Boker: A transdisciplinary multilevel, multigoal teaching project, coordinated toward the common purpose of the total human ecosystem.

ther. Accordingly, landscapes should not be approached as mere physical entities and "environmental systems," requiring solutions by technological intervention with the help of deterministic computerized models. They should be taught and studied as complex Gestalt systems with self-transcendent openness and as cybernetic interaction systems with the insights gained recently into self-organization of nonequilibrium biosystems. For this purpose, novelty-seeking, creative information should be optimized with confirmation-seeking scientific information to achieve maximum pragmatic information.

In such a way, landscape ecology can serve as an urgently needed counterbalance to the above-mentioned trends toward one-sided reductionistic biological education, presenting man as detached from nature and as the almighty manipulator of life who knows more and more about that part of nature which can be taken apart, isolated, and analyzed but less and less about real nature and its life-supporting systems in action.

In the education of engineers, also, this approach can fulfill the much-needed function of an antidote to the hubris created by the illusion of the scientific and technological supremacy of man and its disastrous results. Landscape ecology can serve also as an important link between the major bioecological, socioecological, technoeconomic, and engineering branches of "environmental sciences." It may even help to prevent the erection of the rapidly rising ivory towers of biological, technological, and mathematical "cultures" and "cults" within ecology itself.

A typical example of such a cult is the "energy-model" one, presented at the above-mentioned symposium by E. C. and H. T. Odum (1980) as a basis for environmental education. Such models, promoted by the Odums with great zeal and success, can be very useful in demonstrating energy flows in different bio- and technoecosystems. But they are not sufficient to serve as a cognitive basis, as contended by these authors, to help the student "see complexity and component relationships at the same time." They can even be misleading in their "energy–materialistic" interpretation of economic systems as an energy flow in one direction and money flow in the opposite one, and in their neglect of qualitative aspects of energy steering through pragmatic biological and cultural information, as explained earlier in this chapter.

That the ecological reality of complex urban human ecosystems can be presented and used as an educational tool in a comprehensive way, not by resorting to such energy flow models but by describing the interplay patterns of the "city metabolism" and the total material, biosocial, and cultural aspects of human experience and behavior, has been shown in the above-mentioned study of the ecology of Hong Kong and its people (Boyden et al., 1981).

This new conceptual and educational framework of landscape ecology as a THE science, outlined here, is also of special significance for the Man and Biosphere (MAB) program. Initiated in 1971 by UNESCO, this international research and training program aims at providing knowledge and trained personnel for the management of natural resources in a rational and sustained manner. It already embraces over 1000 field projects in 100 countries in 14 major areas, including the establishment of networks of focal regional pilot projects and 177 Biosphere Reserves in 46 countries.

According to M. Batisse (1980), one of the main conceivers and leaders of this program, the pilot projects should be conducted on an interdisciplinary basis and in association with planners, decision makers, and the population concerned, and with constant interaction between disciplines in the natural sciences and the social and human sciences. "The Programme is focused on human-use systems and on problems of land and resources management. Land and resources are managed for people and by people." In addition to actual studies on the impact of man on his natural and cultural landscapes, much emphasis is given to environmental education at the levels of land managers, decision makers, and the public.

A good opportunity to critically assess some of these projects in the light of these statements was provided by the recent international MAB–UNESCO scientific conference, "Ecology in Practice–Establishing a Scientific Basis for Land Management" (September 1981, UNESCO, Paris), marking the tenth anniversary of the MAB program. Without doubt, MAB, as presented at the conference, in its publications, and in the illuminating exhibition, can be considered the most outstanding achievement on an international and intergovernmental level in the education toward new man–land relations. At the same time, however, it became very apparent during the discussions that the communication gaps between the scientist, the land planner, manager, decision maker, and the public, as well as between the scientists from the different disciplines taking part in these projects, are far from being bridged. As has been explained in this chapter, effective communication as a process of information transfer requires the bridge of commonly shared signals and symbols and their proper decoding. In our case, this means common concepts, premises, and paradigms, translated into the language of research, management, and training. It seems that the common platform of "ecology" in the conventional sense, with its vague "ecosystem" and "environmental" concepts and equally vaguely defined common goal of "rational sustained use of natural resources," is not sufficient for this purpose. We hope, therefore, that landscape ecology, as presented in this book, can provide such a transdisciplinary platform and communication bridge for the MAB research and training program at all levels.

In the following chapters, some examples of attempts at the solution of these problems in land-use planning, landscape evaluation, and conservation management practices will be provided. One of the conceptual tools that can be applied in order to study and teach the complex interactions between natural and man-induced forces operating in seminatural and cultural mediterranean landscapes will be described in the last chapter, as a semiformal, multivariate state factor equation corresponding to the major phases in the history of anthropogenic functions of landscape modifications, and as a further development of the above-described ecotope state factor equation.

In the near future we shall undoubtedly witness further important developments, based on biocybernetic modeling, catastrophe theory, fuzzy sets, and other advanced tools for meaningful quantitative expressions of these complex bio- and socioecological interactions, that are applicable to interdisciplinary ecosystem education.

Conclusion

In this chapter an attempt has been made to outline the theoretical, conceptual, and epistemological framework of landscape ecology as an emerging branch of THE science. This could best be defined as a General Biosystems Theory, a holistic scientific theory of a hierarchical order of holons as open systems with biocybernetic self-regulation. A central paradigm in this thesis is a recognition of the THE as the highest level of ecological integration, with the ecosphere as its largest concrete global landscape entity and the ecotope its smallest one. This visual and spatial integration of the biotechnogeosphere must be complemented by their functional and structural integration through the creation of a new dynamic flow equilibrium at a higher level of organization and complexity as the chief goal of THE science. Landscape ecology contributes to this goal by supplying scientific, ecological, and educational information for these negative and regulatory feedback couplings in the critical interphase of land use. As such, it has to define new operational targets of a transdisciplinary landscape-ecological determinism, requiring new tools for holistic analysis, diagnosis, planning, designing, and management, some of which are described in the following chapters.

Without doubt, one of the greatest challenges facing landscape ecology as an ecosystem education science is to help in the creation of the cognitive basis for the transformation of the "soft" intangible values of nature and its noneconomic richnesses into workable parameters in educational and decision-making processes and at the same time to create the affective basis for not only understanding but also loving nature, as represented in its concrete landscape ecotopes.

References

Ashby, W. R. 1964. Introduction to Cybernetics. Chapman & Hall, London.

Ashby, E. 1978. Reconciling Man with the Environment. Stanford University Press, Stanford, California.

Bakshi, T. S., and Z. Naveh (Eds.). 1980. Environmental Education: Principles, Methods and Applications. Plenum Press, New York and London.

Bakuzis, E. V. 1974. Foundations of Forest Ecosystems. Lecture and Research Notes. College of Forestry, University of Minnesota, St. Paul.

Bar-Lavie, B. 1980. Eco-shop development at an environmental high school in Israel. In: T. S. Bakshi and Z. Naveh (Eds.), Environmental Education: Principles, Methods and Applications. Plenum Press, New York and London, pp. 255–262.

Barney, G. O. (Ed.). 1981. The Global 2000 Report to the President. A report prepared by the U.S. Council on Environmental Quality and the Department of State, Washington, D.C.

Batisse, M. 1980. The relevance of MAB. Environ. Conserv. 7:179–184.

Beddington, J. R., C. A. Free, and J. H. Lawton. 1976. Concepts of stability and resilience in predatory/prey models. J. Anim. Ecol. 45:791–816.

Bennet, R. J., and R. J. Chorley. 1978. Environmental Systems: Philosophy, Analysis and Control. Methuen, London.

Ben-Perez, M. 1980. Application of some guiding principles in the development of a curriculum for teaching the Total Human Ecosystems: A case study. *In:* T. S. Bakshi and Z. Naveh (Eds.), Environmental Education: Principles, Methods and Applications. Plenum Press, New York and London, pp. 175–182.

Bernstein, B. B., and M. Jung. 1979. Selective pressures and coevolution in a kelp canopy community in Southern California. Ecol. Monogr. 49:335–355.

Bertalanffy, L. von. 1968. General System Theory, Foundations, Development and Applications. George Braziller, New York.

Bertalanffy, L. von. 1969. Chance or law. *In:* A. Koestler and J. R. Smithies (Eds.), Beyond Reductionism: New Perspectives in the Life Sciences. Hutchinson of London, pp. 56–84.

Boulding, K. E. 1980. Universal physiology. Behav. Sci. 25:35–39.

Boyden, S., S. Miller, K. Newcombe, and B. O'Neill. 1981. The Ecology of a City and Its People. The Case of Hong Kong. Australian Natl. University Press, Canberra, London, Miami.

Bunge, M. 1969. The metaphysics, epistemology and methodology levels. *In:* L. L. Whyte, A. G. Wilson, and D. Wilson (Eds.), Hierarchical Structures. American Elsevier, New York, pp. 17–28.

Cole, L. C. 1971. The ecosphere. *In:* P. R. Ehrlich, J. P. Holdren, and R. W. Holm (Eds.), Man and the Ecosphere: Readings from Scientific American. W. H. Freeman, San Francisco, pp. 11–16.

Dörner, D. 1976. Problemlösen als Informationsverarbeitung. Verlag W. Kohlhammer, Stuttgart, Berlin, Köln, Mainz.

Egler, F. E. 1942. Vegetation as an object of study. Philos. Sci. 9:245–260.

Egler, F. E. 1970. The Way of Science: A Philosophy of Ecology for the Layman. Hafner, New York.

Egler, F. E. 1977. The Nature of Vegetation, Its Management and Mismanagement. Aton Forest, Norfolk, Connecticut.

Ehrlich, P. R., J. P. Holdren, and R. W. Holm (Eds.). 1971. Man and the Ecosphere: Readings from Scientific American. W. H. Freeman, San Francisco.

Eigen, M., and P. Schuster, 1979. The Hypercycle: A Principle of Natural Self-Organization. Springer-Verlag, New York, Berlin, Heidelberg.

Ellenberg, H. (Ed.). 1973. Ökosystemforschung. Springer-Verlag, Berlin, Heidelberg, New York.

Feiblemann, J. K. 1954. Theory of integrative levels. Brit. J. Philos. Sci., Vol. 5. pp. 559–566.

Forrester, J. W. 1971. World Dynamics. Wright-Allen, Cambridge, Massachusetts.

Fortescue, J. A. C. 1980. Environmental Geochemistry: A Holistic Approach. Ecological Studies 35. Springer-Verlag, New York, Heidelberg, Berlin.

Frankl, V. E. 1969. Reductionism and nihilism. *In:* A. Koestler and J. R.

Smithies (Eds.), Beyond Reductionism: New Perspectives in the Life Sciences. Hutchinson of London, pp. 396-408.

Gallopin, G. C. 1972. Structural properties of food webs. *In:* B. C. Patten (Ed.), System Analysis and Simulation in Ecology, Vol. 2. Academic Press, London and New York, pp. 241-282.

Grinker, R. R. 1976. In memory of Ludwig von Bertalanffy's contribution to psychiatry. Behav. Sci. 21:207-217.

Halfon, E. 1978. Theoretical Systems Ecology: Advances and Case Studies. Academic Press, New York.

Harmon, L. D. 1973. The recognition of faces. Sci. Am. 229:75.

Holling, C. S. 1973. Resilience and stability of ecological systems. Ann. Rev. Ecol. Syst. 4:1-23.

Holling, C. S. 1976. Resilience and stability in ecosystems. *In:* E. Jantsch and C. H. Waddington (Eds.), Evolution and Consciousness: Human Systems in Transition. Addison-Wesley, Reading, Massachusetts, pp. 73-92.

Jantsch, E. 1975. Design for Evolution: Self-Organization and Planning in the Life of Human Systems. George Braziller, New York.

Jantsch, E. 1976. Self-transcendence: New light on the evolutionary paradigm. *In:* E. Jantsch and C. H. Waddington (Eds.), Evolution and Consciousness: Human Systems in Transition. Addison-Wesley, Reading, Massachusetts, pp. 9-10.

Jantsch, E. 1980. The Self-Organizing Universe. Pergamon Press, Oxford and New York.

Jantsch, E., and C. H. Waddington (Eds.). 1976. Evolution and Consciousness: Human Systems in Transition. Addison-Wesley, Reading, Massachusetts.

Jenny, H. 1958. Role of plant factor in the pedogenic function. Ecology 39: 5-16.·

Jenny, H. 1961. Derivation of state factor equations of soils and ecosystems. Soil Sci. Soc. Am. Proc. 25:385-388.

Jones, D. D. 1975. The application of catastrophe theory to ecological systems. *In:* G. S. Innis (Ed.), New Directions in the Analysis of Ecological Systems, Vol. 5. Simulation Councils Proc. Ser. Society for Computer Simulation, La Jolla, California, pp. 133-148.

Klaus, G. 1969. Wörterbuch der Kybernetik. Fischer-Bücherei, Frankfurt am Main.

Koestler, A. 1969. Beyond atomism and holism—the concept of the holon. *In:* A. Koestler and J. R. Smithies (Eds.), Beyond Reductionism: New Perspectives in the Life Sciences. Hutchinson of London, pp. 192-216.

Koestler, A. 1978. Janus: A Summing Up. Picador, Pan Books, London.

Koestler, A. 1982. Bricks to Babel. Picador, Pan Books, London.

Koestler, A., and J. R. Smithies (Eds.). 1969. Beyond Reductionism: New Perspectives in the Life Sciences. Hutchinson of London.

Kuhn, T. S. 1970. The Structure of Scientific Revolutions. University of Chicago Press, Chicago, Illinois.

Laszlo, E. 1972. Introduction to Systems Philosophy: Toward a New Paradigm of Contemporary Thought. Harper Torchbooks, New York.

Lehninger, A. L. 1965. Bioenergetics. W. A. Benjamin, New York and Amsterdam.

Likens, G. E., F. H. Bormann, R. S. Pierce, J. S. Eaton, and N. M. Johnson. 1977. Biogeochemistry of a Forested Ecosystem. Springer-Verlag, New York, Heidelberg, Berlin.

Lindeman, R. L. 1942. The trophic-dynamic aspect of ecology. Ecology 23: 399-418.

Margalef, R. 1968. Perspectives in Ecological Theory. University of Chicago Press, Chicago, Illinois.

Margulis, L., and J. E. Lovelock. 1974. Biological modulation of the earth's atmosphere. Icarus 21:471-489.

Maruyama, M. 1963. The second cybernetics: Deviation-amplifying mutual causal processes. Cybernetics 1:1-8.

Maruyama, M. 1976. Toward cultural symbiosis. In: E. Jantsch and C. R. Waddington (Eds.), Evolution and Consciousness: Human Systems in Transition. Addison-Wesley, Reading, Massachusetts.

Maturana, H. R., and F. Varela. 1975. Autopoietic Systems. Report BCL 9,4. Biological Computer Laboratory, University of Illinois, Urbana, Illinois.

May, R. M. 1973. Model Ecosystems. Princeton University Press, Princeton, New Jersey.

Milsum, J. H. (Ed.). 1968. Positive Feedback. Pergamon Press, Oxford and New York.

Miller, J. G. 1975. The nature of living systems. Behav. Sci. 20:343-365.

Miller, J. G. 1978. Living Systems. McGraw-Hill, New York.

Mitchell, R. R., A. Mayr, and J. Downhower. 1976. An evaluation of the biome programmes. Science 192:859-865.

Monod, J. 1970. Le Hasard et la Nécessité. Editions de Science, Paris.

Naveh, Z. 1973. The neo-technological landscape degradation and its ecological restoration. In: E. S. Barrekette (Ed.), Pollution: Engineering and Scientific Solutions. Plenum Press, New York and London, pp. 168-181.

Naveh, Z. 1975. Landscape ecology as an educational and scientific tool in environmental education. Proc. Int. Conf. Environ. Education (Hebrew University, Jerusalem), pp. 97-104.

Naveh, Z. 1979. A model of multiple-use management strategies of marginal and untillable mediterranean upland ecosystems. In: J. Cairns, G. P. Patil, and W. E. Waters (Eds.), Environmental Biomonitoring, Assessment, Prediction and Management: Certain Case Studies and Related Quantitative Issues. Int. Coop. Publ. House, Fairland, Maryland, pp. 269-286.

Naveh, Z. 1980. Landscape ecology as a scientific and educational tool for teaching the total human ecosystem. In: T. S. Bakshi and Z. Naveh (Eds.), Environmental Education: Principles, Methods and Applications. Plenum Press, New York and London, pp. 149-163.

Naveh, Z. 1982. The dependence of the productivity of a semi-arid Mediterranean hill pasture ecosystem on climatic fluctuations. Agriculture and Environment 7:47-61.

Naveh, Z., and J. Kinski. 1975. The effect of climate and management on species diversity of Tabor oak savanna pastures. In: Proc. 6th Sci. Conf. Israel Ecol. Soc., Tel Aviv University, Tel Aviv, Israel, pp. 284-296.

Naveh, Z., and R. H. Whittaker. 1979. Structural and floristic diversity of shrublands and woodlands in northern Israel and other mediterranean areas. Vegetatio 41:171-190.

Nicholis, G., and I. Prigogine. 1977. Self-Organization in Non-equilibrium Systems. Wiley, New York.

Novak, J. D. 1977. A Theory of Education. Cornell University Press, Ithaca, New York.

Novak, J. D. 1980. A theory of education as a basis for environmental education. *In:* T. S. Bakshi and Z. Naveh (Eds.), Environmental Education: Principles, Methods and Applications. Plenum Press, New York and London, pp. 129-148.

Odum, E. P. 1971a. Fundamentals of Ecology, 3rd edition. Saunders, Philadelphia, Pennsylvania.

Odum, H. T. 1971b. Environment, Power and Society. Wiley, New York.

Odum, E. C., and H. T. Odum. 1980. Energy systems and environmental education. *In:* T. S. Bakshi and Z. Naveh (Eds.), Environmental Education: Principles, Methods and Applications. Plenum Press, New York and London, pp. 213-231.

Pankow, W. 1976. Openness as self-transcendence. *In:* E. Jantsch and C. H. Waddington (Eds.), Evolution and Consciousness: Human Systems in Transition. Addison-Wesley, Reading, Massachusetts, pp. 16-36.

Passmore, J. 1974. Man's Responsibility for Nature: Ecological Problems and Western Traditions. Duckworth, London.

Pattee, H. H. (Ed.). 1973. Hierarchy Theory: The Challenge of Complex Systems. George Braziller, New York.

Patten, B. C. (Ed.). 1972. Systems Analysis and Simulation in Ecology, Vol. 2. Academic Press, New York.

Piaget, J. 1968. Structuralism. Basic Books, New York.

Piaget, J. 1970. General problems of interdisciplinary research and common mechanisms. *In:* Main Trends of Research in the Social and Human Sciences, Pt. 1. Social Sci., Mouton, UNESCO, Paris, pp. 467-528.

Piaget, J. 1972. The concept of structure. *In:* Scientific Thought. UNESCO, Division of Philosophy, Mouton, Paris and The Hague, pp. 35-56.

Pignatti, S. 1978. Zur Methodik der Aufnahme Gesellschaftskomplexen. *In:* R. Tüxen (Ed.), Assoziationskomplexe (Sigmenen) und ihre praktiche Anwendung. Rinteln 4.-7.4, 1977. J. Cramer, Vaduz, pp. 27-41.

Platt, J. R. 1974. Guiding the wagon train of history. *In:* Biology in Human Affairs. Voice of America, Forum Series, Washington, D.C., pp. 293-305.

Porter, A. 1969. Cybernetics Simplified. The English Universities Press, London.

Prigogine, I. 1973. Can thermodynamics explain biological order? Impact of Sci. on Soc. 23:159-179.

Prigogine, I. 1976. Order through fluctuation: Self-organization and social system. *In:* E. Jantsch and C. W. Waddington (Eds.), Evolution and Consciousness: Human Systems in Transition. Addison-Wesley, Reading, Massachusetts, pp. 93-130.

Rapoport, A. 1976. General systems theory: A bridge between two cultures (3rd Ann. Ludwig von Bertalanffy Mem. Lec.), Behav. Sci. 21:228-239.

Rowe, J. S. 1951. The level of integration concept and ecology. Ecology 42: 420-427.

Sachsse, H. 1971. Einführung in die Kybernetik. Vieweg & Sohn, Braunschweig.

Schaefer, G. 1972. Kybernetik und Biologie. J. B. Metzlersche Verlagsbuchandlung, Stuttgart.

Schaefer, G. 1978. Inklusives Denken-Leitlinie für den Unterricht. *In*: G. Trommer and K. Wenk (Eds.), Leben in Ökosystemen. Leitthemen 1/78, Westerman, Braunschweig.

Schaefer, G. 1980. Environmental education—a new word or a new philosophy of teaching? *In:* T. S. Bakshi and S. Naveh (Eds.), Environmental Education: Principles, Methods and Applications. Plenum Press, New York and London, pp. 3–7.

Schultz, A. M. 1967. The ecosystem as a conceptual tool in the management of natural resources. *In:* S. V. Cirancy-Wantrup and J. J. Parsons (Eds.), Natural Resources: Quality and Quantity. University of California Press, Berkeley, pp. 139–161.

Schultz, A. M. 1969. A study of an ecosystem: The arctic tundra. *In:* G. M. van Dine (Ed.), The Ecosystem Concept in Natural Resource Management. Academic Press, New York, pp. 77–93.

Schultz, A. M. 1980. Systems theory and environmental education. *In:* T. S. Bakshi and Z. Naveh (Eds.), Environmental Education: Principles, Methods and Applications. Plenum Press, New York and London, pp. 139–148.

Schumacher, E. F. 1971. Small is Beautiful! Earth Island. Blond & Briggs, London.

Schmithüsen, J. 1959. Allgemeine Vegetationsgeographie. Gruyter, Berlin.

Sessions, G. 1979. Ecophilosophy (No. 2). Philosophy Department, Sierra College, Rocklin, California.

Shannon, C. E., and W. Weaver. 1949. The Mathematical Theory of Communications. University of Illinois Press, Urbana, Illinois.

Shannon, C. E., and W. Weaver. 1963. The Mathematical Theory of Communication. University of Illinois Press, Urbana, Illinois.

Simon, H. A. 1962. The architecture of complexity. Proc. Am. Philos. Soc. 106: 467–482.

Steiner, G. (Ed.). 1978. Die Psychologie des 20 Jahrhunderts, VII: Piaget und die Folgen. Kindler, Zurich.

Smuts, J. C. 1926. Holism and Evolution (2nd printing, 1971). Viking Press, New York.

Tansley, A. G. 1935. The use and abuse of vegetational concepts and terms. Ecology 16:284–307.

Thom, R. 1970. Topological models in biology. *In:* C. H. Waddington (Ed.), Towards a Theoretical Biology, Vol. 3. Edinburgh University Press, Edinburgh, pp. 89–116.

Toffler, A. 1980. The Third Wave. William Collins Sons, London.

Turner, T. 1973. Piaget's structuralism. Am. Anthropologist 75:351–373.

van Dobben, W. H., and R. H. Lowe-McConnell (Eds.). 1975. Unifying Concepts in Ecology. Report of the Plenary Sessions of the First International Congress of Ecology, The Hague, September 1974. Dr. W. Junk, The Hague.

Vester, F. 1974. Das Kybernetische Zeitalter: Neue Dimensionen des Denkens. S. Fischer, Frankfurt am Main.

Vester, F. 1976. Urban Systems in Crisis. Deutsche Verlags-Anstalt GmbH, Stuttgart.

Vester, F. 1978. Dënken, Lernen, Vergessen. Deutscher-Taschenbuch Verlag, München.

Vester, F. 1980. Neuland des Denkens. Vom technokratischen Zeitalter zum kybernetischen Zeitalter. Deutsche Verlags-Anstalt Gmbh, Stuttgart.

Vester, F., and A. von Hesler. 1980. Sensitivity Model. Ecology and Planning in Metropolitan Areas. Regionale Planungsgemeinschaft Untermain. 6000 Frankfurt am Main 1. Am Hauptbahnhof 18.

Waddington, C. H. 1975. A catastrophe theory of evolution. The Evolution of an Evolutionist. Cornell University Press, Ithaca, New York, pp. 253–266.

Waddington, C. H. 1977. Tools for Thought. Paladin, Granada Publ., Frogmore, England.

Weiss, P. A. 1969. The living system: determinism stratified. *In:* A. Koestler and J. R. Smithies (Eds.), Beyond Reductionism: New Perspectives in the Life Sciences. Hutchinson of London, pp. 3–55.

Weizsäcker, E. von (Ed.). 1974. Offene Systeme I: Beiträge zur Zeitstruktur von Information, Entropie und Evolution. St. Klett, Stuttgart.

Westmann, W. E. 1978. Measuring the inertia and resilience of ecosystems. BioSci. 28:705–710.

Whittaker, R. H. 1975. Communities and Ecosystems, 2nd edition. Macmillan, New York.

Whyte, L. A., A. G. Wilson, and D. Wilson (Eds.). 1969. Hierarchical Structures (Proc. Symp. Douglas Adv. Res. Lab., Huntington Beach, California). American Elsevier, New York.

Young, G. L. 1974. Human ecology as an interdisciplinary concept: A critical inquiry. Adv. Ecol. Res., Vol. 8, pp. 1–105.

Woodcock, A., and M. Davis. 1978. Catastrophe Theory. Penguin Books, New York.

Zadeh, L. A. 1965. Fuzzy sets and systems. Proc. Symp. Systems Theory. Polytechnic Press, New York.

Zeeman, E. C. 1976. Catastrophe theory. Sci. Am. 334:65–83.

Conceptual and Theoretical Basis of Landscape Ecology as a Human Ecosystem Science

Introduction

Chapter 2 introduced some innovative approaches to studying organized ecological complexity. These system-theoretical and biocybernetic concepts contrast with conventional thermodynamic and deterministic paradigms of equilibrium and stability. As shown in a very lucid way by Jorgensen (1992), these concepts have been recognized as integral parts of contemporary ecosystem theories and have become the foundations for a more generalized theory of complexity and change. The organization of complexity in landscape structure and function has become a central issue in contemporary landscape ecology. Principles of hierarchical organization, self-organization, order and disorder, homeorhesis, information theory, and catastrophe theory are being adopted by landscape ecologists (O'Neill, 1991; O'Neill et al. 1986, 1989). The theory of dissipative structures "creating order out of chaos," and the creative process of self-organizing systems with autopoietic ("self-making") capacities have been further developed by Prigogine and Stengers (1984) and summarized by Nicolis (1991). The relationship of these theories to ecosystem evolution has been discussed by di Castri (1991). Ulanowicz (1991) has pointed out to the importance of positive feedback loops in the development of ecological networks in a more general way, and DeAngelis et al. (1986) have shown the importance of positive-feedback loops in biological and ecological systems. These serve as driving forces in coevolution, in heterogenous landscapes, and in the colonization process of islands and isolated land patches. Their findings are inconsistent with the equilibrium-based classical Theory of Island Biogeography (McArthur and Wilson, 1963), to which much attention has been given by landscape ecologists. A comprehensive and critical discussion of the relationship between island biogeography and conservation management was provided by Shafer (1990).

New Transdisciplinary Notions of Order

In a groundbreaking book, Bohm and Peat (1987) opened new vistas for understanding order in relation to science and creativity. They laid the foundations for a new metatheory of order of which landscape ecologists should be aware. They noted that relaxing our rigid commitments to familiar notions of order, will enable us to perceive new hidden orders behind simple regularity and randomness: "It is possible for categories to become so fixed a part of the intellect that the mind finally becomes engaged in playing false to support them. Clearly as context changes, so do categories" (Bohm and Peat, 1987, p. 115).

Order is neither subjective nor objective, for when a new context is revealed, a different notion of order will appear. No one order will fully cover the human experience, and as contexts change, orders must be constantly created and modified. This is true also for the Cartesian grid of coordinates which has dominated the basic order of physical and geographical reality for the last three hundred years, including that of landscape ecology. Its general appropriateness is questioned by Bohm and Peat (1987). They propose the concept of different degrees of order: "The flowing river gives a good image of how a simple order of low degree can gradually change to a chaotic order and eventually to a random order. In this process, complex whirlpools may develop and the water may break up into foam, bubbles, and spray. The origin of this behavior lies in the relationships between the elements of the flowing water. Each element would, if left to itelf, follow an order of low degree. But in fact, each is affected by all others, which are, for it, external influences, that change its motion."

The flowing river provides a good image of how a simple order of low degree can gradually change to a chaotic order of high degree and eventually to random order. Bohm and Peat show that between the two extremes of simple regular order and chaos there is a rich new field for creativity, a state of high energy conducive to fresh perception. In their view, full scientific creativity requires free communication.

Bohm and Peat (1987, p. 151), following an important earlier publication by Bohm (1980) on wholeness and order, arrive at a new kind of order, generative order: "This order is primarily concerned with a deeper and more inward order out of which the manifest form of things can emerge creatively. Indeed this order is fundamentally relevant both in nature and consciousness."

Mandelbrot (1982) has formalized such a generative order through fractals, a generation of forms which proceed by repeated application of a similar shape on a decreasing scale. As will be explained below, this new kind of order has been applied widely by landscape ecologists.

The idea of generative order is not restricted to mathematics and to spatial dimensions of landscape patterns and processes. As shown by

Bohm and Peat, it is of relevance to all areas of experience. Modern physics and especially quantum field theory have opened the way for the realization of higher levels of order, which organize the lower ones in this hierarchy. They have thereby extended the notion .of generative order beyond what can be done by fractals and introduced the concept of implicate or enfolding order. In this order, the process of unfolding is related to the whole, and not (as in the order of fractals) to a local order of space. To such a new, higher holistic order belongs also the hologram systems perception, discussed on page 54.

Bohm and Peat (1987) claim that an overall common generative order will bring together science, nature, society, and consciousness. By helping landscape ecology bridge the "two cultures" of science and the humanities (Naveh, 1990a), and foster a synthesis between science and art (Caldwell, 1990), it could have far-reaching implications.

Fractal Geometry and Chaos

Fractal geometry is the generative order which underlies the geometric regularity of self-similarity or scale invariance of repeated patterns. As an innovative method for the study of organized landscape complexity, it enables quantification of the shape and texture of landscape features and the prediction of the multiscale dynamics of landscape processes. It has brought a new perspective to landscape ecology.

Burrough (1981) was the first to refer to the usefulness of fractals in landscape ecology. He described fractal dimensions as a helpful indicator of the complexity of autocorrelation over many scales of natural phenomena. With the aid of fractal dimensions, Krummel et al. (1987) could distinguish between small-scale, human-controlled and large scale, topographically controlled scales in landscape patterns of a decideous forests. They showed that by examining the fractal dimensions of forest patches they could formulate hypotheses concerning the spatial scale of process-pattern interactions. Since then, the use of fractal dimension has become an integral part of many quantitative landscape ecology studies. In a comprehensive review of the role of fractal models in landscape patterns, Milne (1991) described fractal geometry as a continuously developing calculus of heterogeneity which reveals novel features of landscape patterns. He showed examples of their use for the quantification of natural and human-caused irregularities of landscape patchiness and fragmentations.

A good example of different ways of applying fractals in landscape-ecological study has been developed by Loehle's (1990) new method of fractal analysis to the numerical (as opposed to pictorial) description of animal home range, size, shape, and concentration. This could prove very beneficial for the conservation management of endangered animal species, especially such raptors as hawks. He also supplied a simple algorithm in the form of FORTRAN and Pascal subroutines for the calculation of fractal dimensions.

Burel (1992) applied fractal dimensions for the description of hedgerow networks as related to the dynamics of carabid species assemblages and biodiversity. Many more examples could be cited which show how the application of fractal geometry is enhancing our comprehension of the complex interactions between geomorphological, biotic, and anthropogenic factors operating at different space-time scales.

According to Juergens et al. (1990), fractal geometry can be considered as the *geometry of chaos*. Scale invariance has a noteworthy parallel in contemporary chaos theory. Such chaotic events as atmospheric turbulence or fluctuating populations exhibit an underlying long-term order that is neither wholly deterministic nor wholly random. We do not yet have sufficient experimental evidence from homeorhetic landscapes (see p. 61), but it is likely that such highly fluctuating systems may reveal similar chaotic behavior. They all show similar patterns at different time scales, much as scale-invariant objects show similar structural patterns on different spatial scales.

A clear, nonformal description of fractal landscape dimensions is presented by Allen and Hoekstra (1992). A more formal description is given by Milne (1988). Gleick (1988) has presented chaos and fractals in a very popular and well-illustrated book. Li (1993a) has recently updated developments in analysis and description of patch patterns and patch dynamics by fractal geometry and its limitations.

Information Theory

Another promising approach to the study of organized complexity in natural and cultural landscapes is the application of the mathematical principles of information theory (discussed in Chapter 2, pp. 31–33). Using the ecotope and its ecological features as the holistic unit of information transfer, we assume that entropy and disorder will be at their maximum if all ecotope types are randomly distributed. But the "uncertainty" of the occurrence of a given ecotope type will be reduced by a neighboring ecotope type with redundant information. Order in the landscape can therefore be determined by the degree of predictability of an ecotope type as a neighbor of another ecotope type as spatial "mutual information" (Kwakernaak, 1984; Phipps, 1984). As described in the supplement to Chapter 4, this method has also been used in a comprehensive study of vanishing Mediterranean mountain landscapes in Tuscany by Vos and Stortelder (1992).

A widely used application of information theory is the transformation of the nonspatial Shannon-Weaver and Simpson diversity indices into spatially explicit measures for the analysis of landscape structure, and as indices of landscape patterns, heterogeneity, and diversity. These include "ecotope (type) diversity" (Haber, 1990), "biotic landscape diversity" (Romme, 1982; Hoover and Parker, 1991), and the change of spatial grain resolution (Gardner and O'Neill, 1991). Most of these indices have been reviewed by Turner (1989).

For systematic measurement of multidimensional landscape ecodiversity (see supplement to Chapter 1), Naveh (1991, 1993) suggested conversion of Whittaker's (1972) species diversity indices into spatial indices of (1) "alpha ecotope diversity" of biotic and cultural components within each ecotope; (2) "beta ecotope diversity" between ecotopes of each landscape unit; (3) "gamma land unit diversity" within each regional land system, and eventually also (4) "delta land system diversity" within each regional landscape.

In an important paper based on system theory and information theory, Norton and Ulanowicz (1992) suggested a hierarchical approach for the protection of biodiversity at the landscape level of ecological organization. They introduced the concept of sustained "ecosystem health" which is supported by the autopoietic, self-organizing capacities of the ecosystems. For its measurement they computed the average mutual information of all the trophic connections and temporal sequences, as the overall diversity of flows, and called this functional diversity index of the network topology "ascendancy."

It would be very worthwhile to unify these structural and functional landscape diversity indices into a practical quantitative parameter for the evaluation of overall landscape health and diversity.

The application of information theory and other systems-theoretical approaches to landscape ecology have been described in detail in the first French book on landscape ecology, edited by Berdoulay and Phipps (1985).

Fuzzy Set Theory and Expert Systems

The application of fuzzy set theory and its combination with expert systems should be regarded among the most significant advances in dealing with large, heterogeneous and complex systems. This is especially the case with landscapes, which have many measurable variables with low precision and "soft" variables which are unmeasurable by conventional quantitative methods. As mentioned briefly in Chapter 2 (p. 45), the mathematical foundations for fuzzy set theory were laid by Zadeh (1965) as states of fuzzy systems. He generated it to deal with "approximate reasoning" in decision-making problems involving complex systems (Zadeh, 1973). In the foreword to a recent textbook introducing fuzzy set theory and its applications (Zimmermann, 1992), Zadeh stated:

As its name implies, the theory of fuzzy set is, basically, a theory of grade concepts—

a theory in which everything is a matter of degree or, to put it figuratively, everything has elasticity. Since its inception, the theory has matured into a wide-ranging collection of concepts and techniques for dealing with complex phenomena which do not lend themselves to analysis by classical methods based on probability theory and bivalent logic.

Because this new type of "fuzzy logic" is contrary to the generally accepted Aristotelean bivalent logic of "either true or false," of crisp sets, it has been opposed by many mathematicians and engineers. In spite of this, it has found wide application in many field—including ecology. As shown first in Japan, it has proved itself superior to traditional tools for which complex control systems are required such as washing machines, motor cars, and even subways (Kosko, 1993).

Here we outline some of the features of fuzzy set theory most relevant to landscape ecology, and refer the reader to the above-mentioned introductory text book by Zimmermann and to an excellent introduction to expert systems and fuzzy systems by Negoita (1985). Both require only a basis knowledge of mathematics and set theory.

To appreciate fully the potential of fuzzy sets it is important to realize that "fuzziness" is meant in the sense of imprecision rather than lack of information about the value or parameter. Fuzzy set theory provides a strict mathematical framework ("There is nothing fuzzy about fuzzy set theory"), in which imprecise conceptual phenomena can be precisely and rigorously studied.

Fuzzy set theory can be thought of as a generalization of classical set theory, using instead of the binary choice of two elements, weighted membership with more than two values. This membership function allows a continuum of possible choices and can be used to describe imprecise terms. For example "old" can be defined according to the universe of ages of humans (or forest trees, hedgerows, buildings, etc.). Clearly, a man over seventy is old, so the degree of membership of an age seventy or greater is 1.0. But for a sixty- year-old one can say that he is "partially old" and assign an age value of, say, 0.7. In this way the vagueness of the term "old" can be captured mathematically and dealt with in an algorithmic fashion (Negoita, 1985).

This concept enables us to deal with qualities rather than strictly with quantities. This has been shown, for instance, by Bosserman and Ragade (1982) for wildlife habitat quality determination in Kentucky. It is true in a very general way for all qualitative and "soft" landscape parameters to which we can assign relative value indices, with a goal of characterizing the objects of a set as the basis for classification ("beautiful," "vital," "unique," "worthwhile for protection," etc.). But the degree of membership in each subset is a value judgment by the observer.

In this way, fuzzy mathematics makes it possible to use verbal models exactly as mathematical values are used in physics. As a result, linguistic models and heuristic knowledge can be handled by computers. This has also removed the barrier between formal, scientific languages and natural language, in which the qualities of landscapes as self-transcendent Gestalt systems are expressed (see Chapter 2, pp. 56–58).

The application of fuzzy set theory for hierarchical decision-making in land-use quality classifications of ecosystems and landscapes can result in more realistic solutions than probabilistic or deterministic classifications determined by Boolean Algebra models. This has been shown most recently by Burrough et al. (1992) in their study of land suitability for agriculture from multivariate soil profiles and topographically controlled site drainage conditions. Their studies revealed that fuzzy methods allowed the matching of individuals to be determined on a continuous scale instead of the conventional crisp boundaries, which rejected many sites.

Applications of fuzzy systems to community ecology for the study of vegetation/environment relations and vegetation dynamics have been discussed by Roberts (1989). Allen and Hoekstra (1992) have discussed in a very original way the value of "fuzzy criteria" as one of the ways to broaden "the breadth of vision" for basic ecological research from single to multiple-use criteria. Their hierarchical approach will be discussed in more detail below. For this purpose they used the example of removing trees by selective cutting. Cutting affects the plant community, the ecosystem, and the landscape to varying degrees along a trajectory of fuzzy influences, according to the scale of cutting. Allen and Hoekstra state that "the scheme for the fuzzy assignment of major ecological criteria to management actions offers one of the few intellectual frames for the unified treatment of the wealth of information" (Allen and Hoekstra, 1992, p. 299).

Some of recent advancements achieved in the application of fuzzy logic to ecological modeling were presented in a recent symposium of the International Society for Ecological Modeling at Honolulu, Hawaii, August 1992. Of special interest for landscape ecology were the lectures by D. W. Roberts (1993) from Utah State University, on fuzzy modeling for the analysis of the dynamics of vegetation on landscape scales and the lecture by Bai-Lian-Li from Texas A&M University College Station on its value for facing the uncertainty challenges in ecology.

In our opinion, recent developments in knowledge engineering are of great relevance for future developments in landscape ecology. *Expert systems* are interactive computer programs that can directly help managers make decisions by posing and answering questions which are related to knowledge and information derived from human experts. As explained above, fuzzy set theory enables the combination of disparate landscape variables and subjective evaluations, including intangible values and functions, which are more effectively described in words than numerically. As shown by Negoita (1985), such verbal models can be used in fuzzy linear programming for the automatic transfer of knowledge from linguistic values in optimization functions. This opens the way for efficient integration of complex ecological, cultural, and socioeconomic information (quantitative as well as qualitative) in landscape development masterplans. "Hard" and "soft" values and their marketable and noneconomic richness may be expressed as functional parameters of production, regulation, protection, carrier, and information functions. Instead of monetary values, relative values of fuzzy membership functions may be connected in a decision-making matrix. The most recent promising development for this purpose is the combination of fuzzy logic with neural networks in expert fuzzy systems. In these, a neural system learns the rules from data or from human experts and each input to the fuzzy system fires all the rules to some degree as in a massive associative memory. Fuzzy chips perform this associative mapping from input into an averaged output thousands or millions of times per second. Each map from input to output defines one fuzzy logical inference per second (Kosko, 1993).

Scaling and Hierarchy Theory

The choice of an adequate space-time and perceptional framework for the consideration of multiple scales is of the outmost importance and has become a major issue in contemporary landscape ecology (Golley, 1989). Landscape ecologists have to scale up and down from the smallest ecotope in-situ field studies to the larger regional and even global landscape units with the help of remote sensing and satellites. As discussed in detail by Dale et al. (1989) in a special symposium, attempts are made in theoretical studies to extrapolate and to predict from one landscape scale to another with the help of "neutral" models. Neutral models lack effects due to topography, contagion, disturbance, and other ecological processes, and are used as standards for the simulation of the relationships between patterns and processes (such as animal movements), thereby incorporating the effects of spatial heterogeneity (Gardner and O'Neill, 1991). In practic, however such extrapolations may be justified only if the scale of observations does not change the basic landscape patterns. This is the case, for instance, in certain mountain landscapes where the scale invariance and fractal structural features of hillslopes and ecotones of rivers have not yet been distorted by intensive human land uses (Lathrop and Peterson, 1992). On the other hand, scale transformations across a hierarchical level are much more complex. It is therefore important to identify criteria thresholds in which slight changes in landscape patterns produce sudden changes in the response of the process. Therefore, the extrapolation of information across a critical threshold requires information on the behavior of the system on both sides of the threshold.

In the above-mentioned symposium, O'Neill et al. (1989) presented a hierarchical framework for the analysis of scale, based on nonequilibrium thermodynamics and catastrophe theory. Rosen (1989) discussed the same problems from a system-theoretical point of view with the help of cup catastrophe theory (see p. 60 in Chapter 2), in which bifurcation points separate two dissimilar regimes of behavior.

It is therefore obvious that scales and scaling have to be considered within the context of hierarchical organization, presented in Chapter 2 as part of general systems theory. Allen and Star (1982) introduced it to ecologists as "hierarchy theory" and O'Neill et al. (1986) attempted to use this hierarchy theory as the basis for a new conceptualization of ecosystems. Hierarchical organization is conceived not in the context of increasing complexity, but only as as a result of different process rates. This isolates lower levels with faster rates from higher ones with lower rates. An excellent overview of hierarchy theory has been provided by O'Neill (1991).

Contrary to Koestler's original definitions of holons as the units of the *vertical* hierarchical structure (Chapter 2, pp. 51–52), Allen and Star (1982), followed by O'Neill et al. (1986), are using holons as subsystems of the *horizontal* structure of each hierarchical level.

In their recent book, Allen and Hoekstra (1992) used a similar contextual and rate-dependent view of hierarchical organization. In this, the *upper level*, which displays lower-frequency behavior and is spatially larger or more constant over time, *becomes the context of the lower level*. The conventional biological hierarchical levels from the organism to the biosphere are treated in an original way as criteria for observation, in which each criterion has its own scale-defined hierarchical levels. Accordingly, the context of community is not necessarily the ecosystem, nor should the ecosystem by regarded as a key set for larger-scale landscapes, as suggested by Forman and Godron (1986). Allen and Hoekstra's ecosystem criteria are conceived as a set of interlinked, different-scale properties. These may be diffuse in space but easily defined in turnover times. Accordingly, ecosystems should be regarded as "intangible," in contrast to landscapes, as the most tangible of the ecological criteria. These are concrete systems in their own right with different scales, serving *as the spatial matrix for organisms, populations, and ecosystems* (Allen and Hoekstra, 1992).

These definitions have much in common with those presented in Chapter 2. However, Allen and Hoekstra (1992) differ from us in their narrower view of the role of landscape ecology. In their representation the goal of landscape ecology is the spatial contiguity of landscapes, as their only ordering principle. Thereby it becomes merely the study of the spatial ramifications of the other ecological criteria. The unique role played by European landscape ecology as a problem-solving-oriented science of landscape planning, management, conservation, and restoration is mentioned in only one short sentence in relation to ecological management.

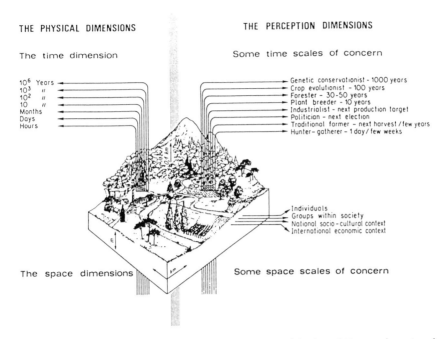

THE PHYSICAL DIMENSIONS   THE PERCEPTION DIMENSIONS

The time dimension   Some time scales of concern

10^6 Years
10^3 "
10^2 "
10 "
Months
Days
Hours

Genetic conservationist - 1000 years
Crop evolutionist - 100 years
Forester - 30-50 years
Plant breeder - 10 years
Industrialist - next production target
Politician - next election
Traditional farmer - next harvest / few years
Hunter - gatherer - 1 day / few weeks

Individuals
Groups within society
National socio-cultural context
International economic context

The space dimensions   Some space scales of concern

Figure S2-1. Physical and Perceptual Dimensions and Scales of Research on Land Use (after di Castri and Hadley 1986).

Hierarchical Concepts of Interdisciplinary Land-Use Studies

In an important paper summarizing their rich experience as leaders of the UNESCO-MAB projects in interdisciplinary land-use research (see Chapters 2, p. 98 and Chapter 4, pp. 327–330), di Castri and Hadley (1986) adopted the hierarchical model of Jantsch (1970) with the highest level of a common transdisciplinary goal to solve a central ecological problem. As a way to link the ideas of biological systems and social systems they proposed the concept of "human use systems" as organizational systems. For this purpose, they included the perceptual dimension together with the physical dimension in their land-use research model (Figure S2-1).

As pointed out in the context of fire ecology (Naveh, 1990b), the transdisciplinary challenge of the landscape ecologist lies in his capacity not only to realize these different fragmentized hierarchical perceptual views and their role in conceiving and shaping the landscape. He should also be able to apply a more integrative and multidimensional approach at the level of the Total Human Ecosystem.

Such a broader, transdisciplinary conception of hierarchical organiza-

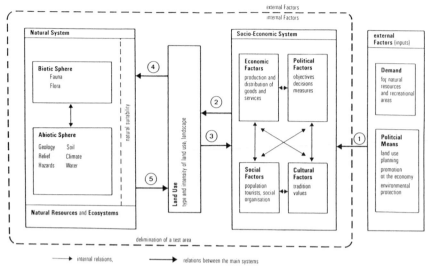

① Influences of the external factors; ② Socio-economic development and connected change of land use; ③ Feedback of change in land use on population, economy and society; ④ Effects of change in land use on the natural system; ⑤ Feedback of the change in the natural system on the (human) conditions of living. (Figure published in GEOGRAPHICA HELVETICA, 1978, volume 4)

Figure S2-2. Schematic Presentation of a Regional Socio-Economic Ecological System (Messerli and Messerli 1978).

tion, based on systems theory and biocybernetics, has been applied by leading Swiss and German landscape ecologists and geographers in the Man and Biosphere (MAB 6) studies of human impacts on mountainous ecosystems.

The project in the Swiss Alps on "Socioeconomic development and ecological burden capacity in the mountain regions" was carried out between 1979 and 1985 under the leadership of B. Messerli and P. Messerli (1978), from the Geography Department of the University of Bern, in four different locations as an extensive, interdisciplinary project with fifty academic participants and contributors from the natural sciences and the humanities. For the study of these regional alpine systems, a holistic human ecosystem concept was applied, integrating natural and socio-economic subsystems, as shown in Figure S2-2.

Scaling of these human ecosystems ranged from the individual farmer, his fields, and agro-silvo-pastoral units, to the humanity and biosphere.

The rapid transformation of the Swiss Alps was described by Messerli and Brugger (1984) as a process of oscillations between self-reliance and dependence in the economic and ecological realms. This is a good illustration of the dichotomic holon properties of human society being both wholes and parts, described in Chapter 2, pp. 88–90. The final results and the practical implications of this "unique interdisciplinary field experiment"—focusing on the triangle of economy, ecology, and culture—were presented by Messerli (1989). One of the main conclusions was that the ecological stability, sustainable productivity, natural diversity, and scenic beauty of these endangered mountain landscapes can be ensured only by creating a new balance between man and nature and between agriculture and tourism. This balance should be based on qualitative rather than quantitative development and transformation. Thereby nature and landscape can be protected as a whole and not pushed into reservations. In other words, the dilemma of human holons' being both "wholes" of their technosphere systems and "parts" of their natural and agricultural biosphere systems can be solved only by a new symbiosis between man and nature at the higher hierarchical level of the Swiss ALP Total Human Ecosystem.

The Swiss MAB model was further developed by W. D. Grossmann (1983). As described in more detail by Haber et al. (1984) and by Haber (1990), this model was applied in the MAB 6 project in Berchtesgaden on "Interactions between human ecosystems and high mountainous ecosystems," and in the West German Forest Die-Off Project. Both studies, carried out by W. Haber and colleagues (Haber, 1990), continued and enlarged considerably the research and planning projects mentioned in Chapter 1, pp. 15–16.

In order to obtain a complete typology of the relationships between biosystems of different scales and human systems of different scales, Grossmann (1983) combined the two different ekistic logarithmic scales of Doxiadis (1977) for populations and space, starting with the individual farmer and his biosystem of 10^{-2} km^2, and continuing with the family and its biosystem of 5.1^{-2} km^2 (afterwards expanding by a factor of seven for both). Grossman offered a hierarchical control stucture for these complex human ecosystems according to the priority of action. Human perception was included as an important ordering principle. On the lowest level, all subprocesses of the process are controlled, such as metabolism and regeneration in ecosystems. These are all amenable to measurement, mostly simple, linear and therefore predictable and can be documented and mapped with the help of GIS.

On the intermediate levels, processes are dynamic and are regulated by complex feedback-control mechanisms with lower certainty and predictability, such as fire, seasonal nutrients, and the adaptation of an ecosystem to seasonal water availability and external disturbance. The number of these units is smaller, but their connectivity is much higher than that of those at the bottom level. Haber (1990) has shown that indicator organisms and processes which can be derived from the lower-level studies may serve as keys for investigations of middle-level dynamics.

On the highest level, typical objectives concern viability and resilience of the whole system, e.g, "strategic risks and opportunities"; long-term, cyclic succession; and adaptations. External influences are very strong and unpredictable, and neither feedback responses nor experimental approaches can be applied. Here we have to deal with scenarios for which

Problems	The Hierarchy	Characteristics	Methods
Preserve viability. To do so: decide structure of all lower levels subject to general principles. Decrease risks. Explore, recognize and exploit opportunities. Prepare system so that it can better cope with "whatever may happen" (Strategic risks)	Highest layers	Uncertainties in structure and data. High influence of the outside environment.	Strategic management. R&D. Evolution, succession. Bio-cybernetic approach (Vester 1976, 1981). Principle of viability and resilience. Importance of subjectivity and experience. Scenarios.
Within the defined structure: Preserve the structure, keep the system going. Problem solving such that interdependencies and feedback reactions are taken into account.	Intermediate layers	Uncertainty in data and to a lesser degree in structure. Considerable influence of the outside environment. Many interdependencies and competition as well as cooperation.	Holistic approaches. Considerations must include (feedback) reactions. Aggregated dynamic models. Preserving of balance (e.g., liquidity)
Solve the many routine jobs quickly and precisely	Lowest layers	Preciseness in data and structure. Many usually not interacting decision systems. Low influence of the outside environment. Nearly no uncertainty in data or structure.	Optimization both heuristic and exact. Automatic control. Real time process control.

Figure S2-3. A Hierarchical Description of a Complex Human Ecosystem, According to Grossmann (1983).

the biocybernetic sensitivity models of Vester (1976, 1980) and Vester and von Hesler (1980) are most appropriate (see Chapter 2). In Figure S2-3, the characteristics, data structure, and problems and methods of such a hierarchy are presented.

In these complex (total) human ecosystems, uncertainties increase from the lowest to the highest layers owing to increasing outside influence and additional subsystems with unpredictable behavior.

Conclusions

Following the MacArthur school of theoretical ecology, there are a considerable number of landscape ecologists who still maintain that landscape ecology will achieve "scientific maturity" only when it can make exact predictions in a mechanistic sense, as in physics. The studies described above show clearly the severe limitations for predicting across different hierarchical levels. This may be possible, as long as landscapes are regarded in a reductionistic and mechanistic way as nothing more than spatially heterogeneous areas of repeated patterns of natural ecosystems. If, however, we treat landscapes as complex hierarchically ordered holons, with unique natural and cultural properties and nondeterministic and chaotic behavior, then we have to be aware of misleading extrapolations and must satisfy ourselves with fuzziness, probabilities and uncertainties. *We cannot predict precisely the fate of Total Human Ecosystem land-*

scapes, but we are able to offer different scenarios of their future dynamics under different land-use strategies and conservation policies.

Our present methods for categorization of ordered landscape complexity are based chiefly on the simple regularities of Euclidean geometry for the description of formal structures, and on randomness for the mechanistic interpretation of information theory. As stated by Bohm and Peat (1987), in order to perceive hidden order behind these regularities, we have to free ourselves of rigid commitments to these familiar notions, and if we change the context, we must also change these categories. Therefore, when we perceive landscapes as self-transcendent natural and cultural Gestalt systems, the change in context demands entirely new categories. The generative order of fractals and contextual scaling in hierarchical levels are important steps in this direction, and transcend these regularities. However, in their present use by landscape ecologists, they remain almost exclusively within the categories of the natural sciences. Therefore, one of the greatest challenges for landscape ecology as a transdisciplinary science is the inclusion of further enfolding orders, with new categories for intrinsic natural values, health, and self-organization, as well as for human consciousness and creativity. Of great importance for this purpose are recent developments in systems theory and fuzzy sets. These provide a unifying framework and innovative tool to integrate numerical data, linguistic statements and expert experience to deal with vague and qualitative variables, relationships or events in landscape ecology.

References

Allen, T. F. H., and T. Starr. 1982. Hierarchy Perspectives for Ecological Complexity. Chicago University Press, Chicago.

Allen, T. F. H., and T. W. Hoekstra. 1992. Toward a Unified Ecology. Columbia University Press, New York.

Berdoulay, M., and M. Phipps. (Eds.). 1985. Paysage et Systeme Ecologique. Presse de'l Universite' d'Ottawa, Ottawa, Canada.

Bohm, D. 1980. Wholeness and the Implicate Order. ARK Paperback, London and New York.

Bohm, D., and F. Peat. 1987. Science, Order, and Creativity. A Dramatic New Look at the Roots of Science and Life. Bantam Books, New York.

Bosserman, R. W., and T. K. Ragade. 1982. Ecosystem analysis using fuzzy set theory. Ecological Modeling 16:191–208.

Burrel, F. 1992. Effect of landscape structure and dynamics on specics diversity in Hedgrow networks, Landscape Ecology 6:161–174.

Burrough, P. A. 1981. Fractal dimension of landscapes and other environmental data. Nature 294:243.

Burrough, P. A. 1983. Multiscale sources of spatial variation in soil. I. Application of fractal concepts to nested levels of soil variations. J. Soil Sci. 34:577–597.

Burrough, P. A. MacMillan, R. A., and W. Van Deursen. 1992. Fuzzy classification methods for determining land suitability from soil profile observations and topography. J. Soil Sci. 43:193–210.

Caldwell, L. K. 1990. Landscape, law and public policy: conditions for an ecological perspective. Landscape Ecology 5:3–8.

Dale, V. H., R. H. Gardner, and M. G. Turner. 1989. Predicting across scales—comments of the guest editors of Landscape Ecology. Landscape Ecology 3:147–152.

DeAngelis, D. K., W. M. Post, and C. C. Travis. 1986. Positive Feedback in Natural Systems. Springer-Verlag.

Di Castri, F. 1991. Ecosystem evolution and global change. In: O. T. Solbrig and G. Nicolis (Eds.), Perspectives on Biological Complexity. IUBS, Paris, pp. 189–217.

Di Castri, F., and Hadley, M. 1986. Enhancing the credibility of ecology: Is interdisciplinary research for land use planning useful? Geojournal 13(4):299–325.

Doxiadis, C. A. 1977. Ecology and Ekistics, Elek, London.

Forman, R. T. T., and M. Godron. 1986. Landscape Ecology. Wiley and Sons, New York.

Gardner, R. H., and R. V. O'Neill. 1991. Pattern, process, and predictability: The use of neutral models for landscape analysis. In: M. G. Turner and R. H. Gardner (Eds.), Quantitative Methods in Landscape Ecology. The Analysis and Interpretation of Landscape Heterogeneity. Springer-Verlag, New York, pp. 289–308.

Gleick, J. 1988. Chaos: Making a New Science. Penguin Books, New York.

Golley, F. B. 1989. A proper scale. Editor's comment. Landscape Ecology 2:71–72.

Grossmann, W. D. 1983. Systems approaches toward complex systems. In: P. Messerli and E. Stucki (Eds.), Colloque International MAB 6 Les Alpes Modele et Synthese, Pays -D'Enhaut, 1–3 Juin 1983. Fachbeitrage zur Schweizerischen MAB-Information Nr. 19. 1983, pp. 25–57.

Haber, W. 1990. Basic concepts of landscape ecology and their application in land management. In: H. Kawanabe, T. Ohgushi, and M. Higashi (Eds.), Ecology for Tomorrow. Physiology and Ecology in Japan, 27:131–146.

Haber, W., W. D. Grossmann, and J. Schaller. 1984. Integrated evaluation and synthesis of data by connection of dynamic feedback models with geographic information systems. In: J. Brandt and A. P. Agger (Eds.), Methodology in Landscape Ecological Research and Planning. Proceedings of the First International Seminar of IALE, Roskilde University Center, October 15–19, 1984, Vol. IV. Roskilde Universitetsforlag, Roskilde, Denmark, pp. 147–161.

Hoover, S. R., and A. J. Parker. 1991. Spatial components of biotic diversity in landscapes of Georgia, USA. Landscape Ecology 5:125–136.

Jantsch, E. 1970. Inter- and transdisciplinary university: A systems approach to education and innovation. Policy Sciences 1:403–428.

Juergens, H., H. O. Peitgen, and D. Saupe. 1990. The language of fractals. Scientific American, August 1990:60–67.

Kosko, B. 1993. Fuzzy Thinking. The New Science of Fuzzy Logic. Hyperion, New York.

Kwakernaak, C. 1984. Information applied in ecological land classification. In: J. Brandt and A. P. Agger (Eds.), Methodology in Landscape Ecological Research and Planning. Proceedings of the First International Seminar IALE,

Roskilde University Center, October 15–19, Vol. III. Roskilde Universitetsfor-
larg, Roskilde, Denmark, pp. 5–15.

Lathrop, R. G., and D. L. Peterson. 1992. Identifying structural self-similarity in
mountain landscapes. Landscape Ecology 6:233–338.

Li, B. L. 1993a. Fractal geometry applications in analysis and description of patch
patterns and patch dynamics. Ecological Modelling (in press).

Li, B. L. 1993b. Facing the uncertainty challenge in ecology. Ecological Modelling (in press).

Loehle, C. 1990. Home range: A fractal approach. Landscape Ecology 5:39–52.

Mandelbrot, B. B. 1982. The Fractal Geometry of Nature. Freeman and Co.,
New York.

Messerli, B., and P. Messerli, 1978. MAB Schweitz. Geographica Helvetica No.
4.

Messerli, P., and E. A. Brugger, 1984. Theoretical guidelines for the analysis of
problems in mountain areas. In: E. A. Brugger, G. Furrer, B. Messerli, and P.
Messerli (Eds.), The Transformation of Swiss Mountain Regions. Verlag Paul
Haupt, Bern, Stuttgart, pp. 43–66.

Messerli, P. 1989. Mensch und Natur in alpinen Lehensraum: Risiken, Chancen,
Perpesktiven. Haupt, Stuttgart.

Milne, B. T. 1988. Measuring the fractal geometry of landscapes. Appl. Math.
Comput. 27:67–79.

Milne, B. T. 1991. Lessons from applying fractal models for landscape ecology: A
review and prognosis. In: M. G. Turner and R. H. Gardner (Eds.), Quantitative Methods in Landscape Ecology. The Analysis and Interpretation of
Landscape Heterogeneity. Springer-Verlag, New York, pp. 199–239.

Naveh, Z. 1990a. Landscape ecology as a bridge between bio-ecology and human
ecology. In: H. Svoboda (Ed.), Cultural Aspects of Landscapes. Pudoc,
Wageningen, pp. 42–55.

Naveh, Z. 1990b. Fire in the Mediterranean—A landscape ecological perspective.
In: J. G. Goldammer and M. J. Jenkins (Eds.), Fire in Ecosystem Dynamics.
SPB Academic Publishing The Hague, pp. 1–20.

Naveh, Z. 1991. Biodiversity and ecological heterogeneity of Mediterranean
uplands. Linea Ecologica 12:47–61.

Naveh, Z. 1993. Biodiversity and landscape management. In: K. C. Kim (Ed.),
Biodiversity and Landscapes: A Paradox of Humanity. Cambridge University
Press, New York (in press).

Negoita, C. V. 1985. Expert Systems and Fuzzy Systems. Benjamin/Cummings
Publishing Co., Menlo Park, California.

Nicolis, G. 1991. Non-linear dynamics, self-organization and biological complexity. In: O. T. Solbrig and G. Nicolis (Eds.), Perspectives on Biological Complexity. IUBS, Paris, pp. 7–50.

Norton, B. G., and R. E. Ulanowicz. 1992. Scale and biodiversity policy: A
hierarchical approach. Ambio 21:244–249.

O'Neill, R. V. 1991. Perspectives in hierarchy theory. In: R. M. May and J.
Roughgarten (Eds.), Perspectives in Theoretical Ecology. University Press,
Princeton.

O'Neill, R. V., A. R. Johnson, and A. W. King. 1989. A hierarchical framework
for the analysis of scale. Landscape Ecology 3:193–205.

O'Neill, R. V., D. L. DeAngelis, J. B. Waide, and T. F. H. Allen. 1986. A

Hierarchical Concept of the Ecosystem. Princeton Press, Princeton, New Jersey.

Phipps, M. 1984. Rural landscape dynamic: The illustration of some key concepts. *In*: J. Brandt and A. P. Agger (Eds.), Methodology in Landscape Ecological Research and Planning. Proceedings of the First International Seminar of IALE, Roskilde University Center, 2 October, 15–19, 1984. Vol. I. Roskilde Universitetsforlag, Roskilde, Denmark, pp. 17–27.

Prigogine, I., and I Stengers. 1984. Order Out of Chaos. Man's New Dialogue with Nature. New Science Library. Shambhala, Boulder, London.

Roberts, D. W. 1989. Fuzzy systems vegetation theory. Vegetatio 83:71–80.

Roberts, D. W. 1993. Landscape vegetation modelling with vital attributes and fuzzy systems theory. Ecological Modelling (in press).

Romme, W. H. 1982. Fire and landscape diversity in subalphine forests of Yellowstone National Park. Ecol. Monogr. 52:199–221.

Rosen, R. 1989. Similitude, similarity, and scaling. Landscape Ecology 3:207–216.

Shafer, C. L. 1990. Nature Reserves—Island Theory and Conservation Practice. Smithsonian Institution Press, Washington, London.

Turner, M. G. 1989. Landscape ecology: the effect of pattern and process. Ann. Rev. Ecol. Systematics 20:171–197.

Turner, M. G. and R. H. Gardner (Eds.). 1990. Quantitative Methods in Landscape Ecology. Springer-Verlag, New York.

Vester, F. 1976. Urban Systems in Crisis. Deutsche Verlags-Anstalt GmbH, Stuttgart.

Vester, F. 1980. Neuland des Denkens. DVA, Stuttgart.

Vester, F., and A. von Hesler. 1980. Sensitivity Model. Ecology and Planning in Metropolitan Areas. Regionale Planungsgemeinschaft Untermain, Frankfurt am Main.

Vos, W., and A. H. F. Stortelder. 1992. Vanishing Tuscan Landscapes. Landscape Ecology of a Submediterranean-montane Area (Sonalo Basin), Tuscany, Italy. PUDOC, Wageningen.

Whittaker, R. H. 1972. Evolution and measurement of species diversity. Taxon 21:213–251.

Zadeh, L. A. 1965. Fuzzy sets and systems. *In*: J. Fox (ED.), System Theory. Microwave Research Institute Symposia Series IV. Polytechnic Press, New York, pp. 29–37.

Zadeh, L. A. 1973. Outline of a new approach to the analysis of complex systems and decision processes. IEEE Trans. Syst. Man Cybern. 2:28–44.

Zimmermann, H. J. 1992. Fuzzy Set Theory and its Applications. Kluwer Academic Publishers, Boston, Dodrecht, London.

SECTION II

APPLICATIONS
OF LANDSCAPE ECOLOGY

Introduction to Section II

The first section of this book has emphasized the epistemology, evolution, and conceptual and theoretical bases of landscape ecology. Section II will explore the applicative dimensions of landscape ecology, including the use of such important tools as remote sensing in its several forms and selected landscape analysis and planning methodologies that have recently been developed. Underlying the applications of landscape ecology—as a part of total human ecosystem science—to various fields, including that of comprehensive land-use planning, is the recognition that holistic and transdisciplinary approaches are called for in the implementation of landscape evaluation, planning, design, management, conservation, and reclamation. Thus, the second half of this book will focus on the applications of landscape ecology.

In making the transition from theory to application, it is imperative to remember the comparative infancy of landscape ecology, as well as the evident fact that the terminology employed by those involved in the practical applications of landscape ecology (landscape planners, landscape managers, and engineers, among others) often does not coincide with that developed in the first part of this book. In short, the science and terminology of landscape ecology are relatively new, and those who apply landscape ecology often use language that is much different from that employed by landscape ecology researchers. Despite this gap, which the authors hope they have helped to bridge through the publication of this book, there is a discernible increase in the application of landscape-ecological knowledge in decision making in several parts of the world. Theory and application are coming together; more uniform terminology accompanies this progress. In the meantime, it is well to remember the words of Zonneveld (1979): "Any comprehensive, physiographic or integrated approach to the survey of separate land elements is, in fact, making use of landscape ecology, even when the user has never heard of the actual term."

The goal of Chapter 3 is to sharpen the focus on tools and methodologies being utilized as part of, and to make possible the contributions of, landscape ecology. Examples will be drawn from activities in different climatic regions and countries, including the United States, Canada, Holland, West Germany, Australia, and Israel.

3

Some Major Contributions
of Landscape Ecology:
Examples of Tools, Methods,
and Applications

Recent decades have been witness to highly productive activities in the evolving fields of land-use capability analysis, regional landscape evaluation, ecologically based land-use planning and design, ecological management, and landscape reclamation. Major new capabilities and approaches have been developed that make possible intensive utilization of ecological insights and principles in addressing, in an integrative way, the relationship of human communities with their environment in the landscape of particular areas. The activities of leaders in several disciplines are discernible in this period of rapid evolution (Cocks and Austin, 1978; Steinitz et al., 1977; Vester, 1980; Whittaker, 1975; Walter, 1979; Whyte, 1976; Zonneveld, 1979; McHarg, 1969; Hills, 1976). Strides in developing methods and procedures that reflect a holistic viewpoint are evident in their work and writings. The interdisciplinary human ecosystem science of landscape ecology has increasingly contributed to research into, and planning and management of, our environment.

In announcing the International Congress on Perspectives in Landscape Ecology for 1981, The Netherlands Society for Landscape Ecology has noted that we face an increasing demand for a thorough theoretical basis for landscape ecology, and for application of ecological principles to the planning and management of urban, rural, and natural landscapes. The main themes of the congress reflect the focuses and contexts of landscape ecology in relation to research, planning, and management of the environment.

1. *Theoretical concepts.* The following theoretical subjects were discussed: the landscape as a holistic entity; integration of land attributes; stability (resistance/resilience) in landscapes; regulation/selection problems; horizontal

relations between landscapes; segregation or integration of functions; research methodology in landscape ecology.

2. *Rural areas.* Participants discussed the following problems of research, planning, and management specific to rural areas: segregation or integration of functions; possibilities and constraints of combination (integration) of agricultural use with recreation or other forms of land use; methodology of research (inventory and mapping techniques, etc.); methodology of planning (procedures, environmental impact assessment, etc.); management aspects specific to rural areas (intensity of land use, land amelioration).

3. *Urban–rural relationships.* Such rural–urban problems as flows of energy and matter, horizontal relations, changes in the landscape, and the influence of urban human behavior on rural landscapes were discussed. (In this context, problems restricted to the urban environment were not subject to discussion.)

4. *Natural areas.* Problems specific to research, planning, and management of natural landscapes were discussed at a landscape-ecological level. Subjects included internal and external management; output and input problems; habitat construction or creation; management of semi-natural ecosystems (or landscapes); border areas; buffer zones; integration/segregation problems.

The emphasis represented in these themes conveys a sense of the breadth of the science of landscape ecology and its "consideration of the landscape as a holistic entity, made up of different elements all influencing each other. This means that land is studied as the 'total character of a region,' and not in terms of the separate aspects of its component elements" (Zonneveld, 1979). Thus a holistic land survey is a survey that takes into account or utilizes the principle of the landscape's being more than merely the total of its separate observable attributes and measurable parameters. The term is in fact almost synonymous with "integrated" when it is defined as a combination of components such that the result is more than the sum of its parts.

The stress in Chapter 2 on the nature of holistic considerations provides a meaningful background for the application of such thinking presented in Chapters 3 and 4. The reader is reminded of the ultimate link between holistic theory and practice afforded by the emergence of landscape ecology and its adherents and practitioners.

There are, of course, intermediate approaches as well, ranging from physiographic soil surveys, which study much more than just "soil," to the "land" or "landscape" concept. Vegetation surveyors display similar orientations, studying the ecology as a whole rather than just its vegetational features.

Detailed treatment of integrated Australian approaches appears in the latter part of this chapter under methodologies for landscape evaluation and landscape planning and design. Here, however, we want to take further note of the term "integrated" as applied to surveys, for integrated inventories are becoming more common. Multiple-resource, multiple-product, and multiple-variant inventories, as they relate to natural resource management, notably draw from advances in biological sciences, land-use planning, and related fields. In the work of the

United States Bureau of Land Management an evolving pattern of multiple-resource inventories is discernible (Lund and McNutt, 1979). The integrated surveys developed in Australia use the word "land" to denote the complex combined effect and interactions of climate, topography, soil, and vegetation (Christian and Stewart, 1947, 1953). "Land" is used to refer to land surface, and includes all characteristics important to man's existence and success. Thus, one finds, in addition to plant and animal populations, attention paid to past and present human activity. The many features of the total profile and their many combinations and interactions are noted as resulting in a wide array of land types. Each of these has its own potential and limitations for agriculture or forestry, "each presenting its own specific barriers to the achievement of maximum plant or animal production" (Christian, 1963).

Comprehensive land planning can make good use of the enlarged orientation that holistic, integrated thinking provides. Several large-scale physical planning and landscape management endeavors undertaken in both developed and less-developed countries in recent years have benefited from such thinking. Regrettably, more limited, restricted (and one might add, eventually self-defeating) orientations remain the rule. This is especially disturbing because scientific planning breakthroughs make enlarged approaches possible. Indeed, new tools, exemplified by remote sensing, and new planning capabilities, as represented by recently developed methodologies, place landscape-ecological approaches within easier reach. In this chapter we consider tools and methodologies with the objective of furthering the use of landscape-ecological insights. Selection, therefore, has been made from among those reflecting holistic orientations. References are cited that encompass several additional techniques and modes.

Remote Sensing:
An Important Tool for Holistic Landscape Evaluation

Introduction

"Within the ecotopes, the integrated and dynamic viewpoint provided by aerial photo interpretation enables us to outline the ecosystem and to evaluate productive capacity or establish capability classes."—Herman Joseph Bauer from "Ecological Aerial Photo Interpretation for Revegetation in the Cologne Lignite District" in *Ecology and Reclamation of Devastated Land,* p. 471.

Troll, who introduced the term "landscape ecology," and who regarded landscape as a fully integrated holistic entity, appreciated the great potential value of aerial photographic interpretation of landscapes.

While the conceptual basis of landscape ecology was being developed, tools, techniques, and methodologies to enhance the quality of landscape evaluation also evolved. One of the major relatively recent tools for a holistic approach to landscape evaluation is *remote sensing* in its several forms, including aerial photography and satellite imagery.

A useful publication by the Dames and Moore Company (Alexander et al., 1974) treats the state of the art of remote sensing and its environmental and geotechnical applications. It covers such applications as soils mapping, exploration geology, site analysis, land-use planning, and hydrology (groundwater and surface-water mapping, flood monitoring, and flood-damage assessment). In addition, it looks at physical, chemical, and thermal pollution studies, snow and ice studies, vegetation mapping, forestry, wetlands inventory, and biome analysis, and identifies elements of remote sensing and its application.

Remote sensing is simply defined as the acquisition of knowledge by unattached means. Remote sensing includes, in addition to aerial photography and satellite imagery, radar, thermal imagery, and side-looking radar. Cameras and films, multispectral scanners, and microwave sensors are thus part of the array of remote sensing systems, with aerial photography still the most important of all. Aerial photography is regarded by many as the most adaptable system because of the wide range of cameras and film types now available.

The technology of remote sensing and its application to site planning and design, land-use planning at urban and regional levels, and the exploration of resources have made rapid progress in the last decade. At the same time, advanced sensor systems and platforms to carry them have been developed.

The technology of remote sensing comprises several elements:

1. *Imagery and other forms of data.* Aircraft or orbiting platforms may be employed for the acquisition of imagery and other data.
2. *Information processing techniques.* These range from labor-intensive processes and conventional visual interpretation to machine-intensive, computer-interactive (man–machine) systems. As will be noted in this chapter, low-cost photographic processing may well suffice for providing an information base for decision (land-use and resource management) making.
3. *Reduction of information to employable formats.* Included here are base-map overlays, computer-generated maps, and tabular data, as well as electronically displayed color-coded thematic maps. The needs of the user, relative to data requirements and information needs, are taken into account in considering appropriate formats.

As a term, "remote sensing" is comparatively new. Photogeology, soils mapping, aerial photographic intelligence, surveying for agriculture and forestry, and topographic mapping by aerial photogrammetry are all art and science arenas in which remote sensing has its roots. Aerial photographs are the most common form of remote sensing product; however, other pictorial and nonpictorial forms of data collection are also produced by some remote sensing techniques.

The technology, international problems, and prospects of remote sensing were treated in a 1977 publication of the United States National Academy of Sciences, *Resource Sensing from Space: Prospects for Developing Countries.*

Technical detail in depth on the concepts and foundations of remote sensing, photographic systems, aerial photo interpretation, photogrammetry, aerial thermography, multispectral scanning, and spectral pattern recognition, among other topics, can be found in a major work, *Remote Sensing and Image Interpretation,* by Lillesand and Kiefer (1979).

These references should be consulted for further insights into specific technical and policy aspects of remote sensing. Because it is felt that they serve as sufficiently informative technical background sources, the remainder of this chapter will focus mainly on relevant remote sensing approaches for application in the practice of landscape ecology, and thus will cover the intricate technical elements of remote sensing in only a very general way.

Glossary of Selected Remote Sensing Terms. As was done with other terminology in Chapter 2, we explain here some of the major terms connected with remote sensing. Unless otherwise indicated, most of the definitions have been derived from three sources: Lillesand and Kiefer (1979), Schwarz et al. (1976), and Alexander et al. (1974). The terms are listed in alphabetical order.

Density slicing. The continuum of grey tones in an original image is sliced into a series of discrete grey scale ranges, each assigned a unique color. The grey level at each point in the photograph is then converted to the appropriate color on a color TV monitor. The resulting color-coded image can often be more readily interpreted than a simple black and white original because interpreters can differentiate far more colors than grey tones.

Instruments designed to measure density by shining light through film transparencies are called *transmission densitometers.* Density measurements may also be made from paper prints with a reflectance densitometer, but more precise measurements can be made on the original film material.

Electromagnetic spectrum. All objects reflect, absorb, transmit, or radiate electromagnetic energy in the form of electromagnetic waves or radiation. Radiated or reflected energy can be characterized by its wavelength. The human eye is sensitive to radiation in only a small portion (the visible region) of the electromagnetic spectrum (National Academy of Sciences, 1977); see Figure 3-1.

Infrared, color. Color infrared film is manufactured to record green, red, and infrared (to about 0.9 μm) scene energy. "False color" film is film in which blue images result from objects reflecting primarily green energy and green images from objects reflecting primarily red energy. Red images result from objects reflecting primarily in the photographic infrared portion of the spectrum, 0.7–0.9 μm (see Figure 3-2).

Multiband photography. Photographs taken simultaneously from the same geometric vantage point, but with different film–filter combinations, are multiband photos.

Multispectral scanners (MSS). Multispectral scanning devices are electro-optical instruments that provide data in a multiband mode similar to that obtained from multiband camera systems (see Figure 3-3).

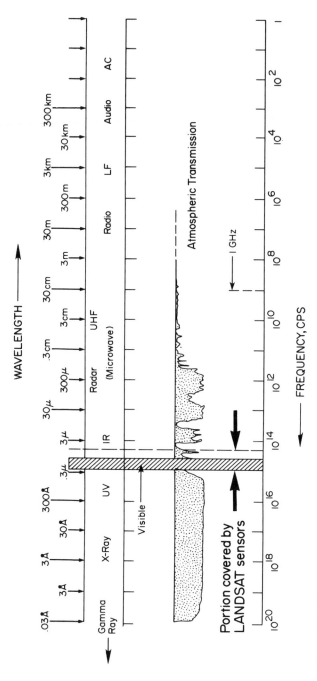

Figure 3-1. Electromagnetic spectrum. Aboard Landsats 1 and 2, bands in sensors covered visible (shaded area) and a part of infrared (IR) portions of the magnetic spectrum. Reproduced from *Resource Sensing from Space*, National Academy Press, Washington, D.C., 1977.

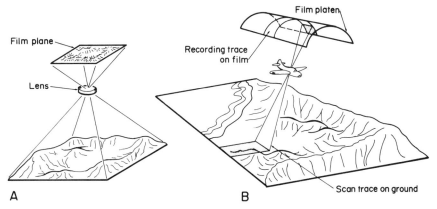

Figure 3-2. Basic aerial camera types: framing cameras (A) expose the scene instantaneously; panoramic cameras, (B) expose the scene as a swath, "painting it" on the film as the camera platform moves the flight path. Courtesy of Dames and Moore.

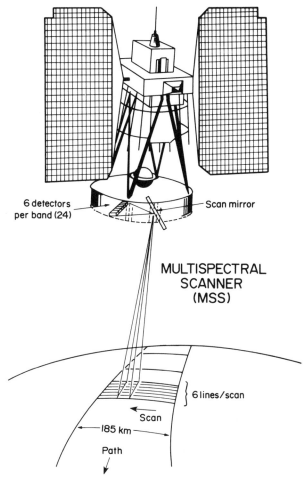

Figure 3-3. Multispectral scanner system pattern. Reproduced from *Resource Sensing from Space,* National Academy Press, Washington, D.C., 1977.

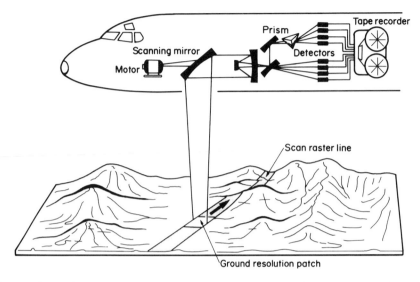

Figure 3-4. Multispectral scanners. Used in both aircraft and spacecraft, the scanner senses reflected and emitted energy of the ground scene line-by-line. Through optics of the system, it separates the energy beam into discrete wavelength bands. Signals from each band are recorded simultaneously on magnetic tape. (Separate images for each band may be produced or the tape input may be computer processed to determine the spectral characteristics of the different features in the ground scene for separation into thematic maps.) Courtesy of Dames and Moore.

Microwave sensors. Sensors that operate in the microwave portion of the electromagnetic spectrum are classified by their mode of operation and in terms of operating wavelengths or frequency bands. Active systems transmit as well as receive microwave signals, whereas passive systems only receive incoming radiation.

Microwaves are capable of penetrating the atmosphere under virtually all conditions. Depending on the wavelengths involved, microwave energy can "see through" haze, light rain and snow, clouds, smoke, and the like.

Remote sensing. As already noted, remote sensing is simply the acquisition of knowledge by unattached means; for example, any data or information acquisition method that uses airborne techniques and/or equipment to determine the characteristics of an area (see Figures 3-2–3-5).

Return-beam vidicon (RBV) system. The RBV system consists of three television-like cameras aimed to view the same 185 km × 185 km ground area simultaneously. Spectral sensitivity of each camera is essentially akin to that of a single layer of color infrared film: 0.475–0.575 μm (green); 0.580–0.680 (red); and 0.690–0.830 μm (reflected infrared). Nominal ground resolution of the cameras is about 80 m (see Figure 3-5). They are also used in aircraft with much higher resolution than 80 m. Also,

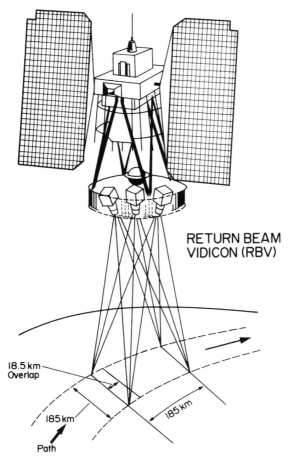

RETURN BEAM
VIDICON (RBV)

18.5 km
Overlap

185 km

185 km

Path

Figure 3-5. Return beam vidicon scan pattern. Reproduced from *Resource Sensing from Space,* National Academy Press, Washington, D.C., 1977.

they do not have to be related to color IR; each camera can be set to select a wave length, although not through a very wide range

Side-looking airborne radar (SLAR). The SLAR system carries its own "illumination source," which transmits a radar beam off to the side of the aircraft, normal to the flight path (whence the term "side-looking"), and detects reflections from ground objects. In a single system, the beam is directed off toward one side only; in a double system, beams are transmitted from both sides of the aircraft.

Thermal scanner system. Thermography is that branch of remote sensing concerned with measuring the radiant temperature of earth surface features from a distance. Thermal scanners produce images. A thermal scanner builds up a two-dimensional record of radiant temperature data for a swath beneath the aircraft. Thermal scanner data are normally recorded

Table 3-1. Guidelines for Aerial Surveys

Description of task	Film type	Season	Scale
Forest-mapping: conifers	Pan	Fall, winter	1:12,000–1:20,000
Forest mapping: mixed stands	IR	Late spring, fall	1:10,000–1:12,000
Timber volume estimates	Pan or IR	Spring, fall	1:5,000–1:20,000
Location of property boundaries	Pan	Late fall, winter	1:10,000–1:25,000
Measurement of areas	Pan	Late fall, winter	All scales
Topographic mapping; highway surveys	Pan	Late fall, winter	1:5,000–1:10,000
Urban planning	Pan	Late fall, winter	1:5,000–1:10,000
Automobile traffic studies	Pan	All seasons	1:2,500–1:6,000
Surveys of wetlands or tidal regions	IR	All seasons–low tide	1:5,000–1:30,000
Archeological explorations	IR	Fall, winter	1:2,500–1:20,000
Identification of tree species	Color	Spring, summer	1:600–1:5,000
Assessment of insect damages	IR color	Spring, summer	1:600–1:5,000
Assessment of plant diseases	IR color	Spring, summer	1:1,000–1:8,000
Assessment of water resources and pollution	Multispectral	All seasons	1:5,000–1:8,000
Agricultural soil surveys	Color	Spring or fall, after plowing	1:4,000–1:8,000
Mapping of range vegetation	Color	Summer	1:600–1:2,500
Real estate assessment	Color-negative	Late fall, winter	1:5,000–1:12,000
Assessment of industrial stockpile inventories	Color-negative	All seasons	1:1,000–1:5,000
Recreational surveys	Color-negative	Late fall, winter	1:5,000–1:12,000
Land-cover mapping	Color-negative	Spring, summer	1:20,000–1:100,000

Source: Adapted from Avery (1970). Reproduced from *Photo Interpretation for Land Managers*, Eastman Kodak Company, Rochester, New York, 1970.

in flight on magnetic tape. Tape data are in a form that can be computer processed.

Some Guidelines for Aerial Photography

The matter of guidelines that can be used for aerial surveys, involving description of the task to be accomplished, film type, season, and scale, is addressed in Table 3-1. The tasks included fall into such diverse categories as forestry mapping and timber volume estimates, urban planning, archeological explorations, assessment of water resources and pollution, agricultural soil surveys, real estate assessment, and land-cover mapping. [See the reference by Belcher et al. (1967) for additional listings of tasks.]

In addressing altitudes for aerial photography flights, Avery (1977) notes that for most nonmilitary aerial photography, upper altitudinal limits might be regarded as around 20,000 m (65,000 ft) above Mean Sea Level (MSL), which can be achieved by such planes as the U-2. The lower limit for fixed-wing aircraft might be arbitrarily given as about 300 m (1000 ft) above ground, depending on topography, special hazards, and air safety regulations. The generalized relationships of platform, altitudes, and scale of conventional intermediate- and high-altitude aerial photography information gathering for urban planning, with a particular focus on metropolitan growth pressures for fringe areas and open space, are alluded to by Grey et al. (1973) and found in Table 3-2.

Advantages of Remote Sensors. One of the advantages of remote sensors is the synoptic overview they provide. Ground survey methods are obviously unable to achieve such an overview. A total scene is recorded as an image, not just the group of data points that ground methods can collect. In addition, remote sensing images have a high information density compared with graphic textual or

Table 3-2. Aerial Photography–Platform, Altitudes, Scale

---

1. *Conventional aerial photography*
   (mainly black and white panchromatic)
   scale range: 1:2000–1:30,000
   platform: low-altitude piston aircraft; altitude, 1000–15,000 ft

2. *Intermediate-altitude photography*
   (black and white, color, color IR, multispectral; also thermal IR and radar)
   platform: scale range: 1:30,000–1:60,000 and smaller pressurized turboprop
   aircraft (typically); altitude 15,000–30,000 ft

3. *High-altitude aerial photography*
   (color, color IR, multispectral)
   platform: pressurized jet aircraft; altitudes 30,000–75,000 ft

---

Source: Reproduced with permission from Proceedings of a Symposium on Remote Sensing, held at Sioux Falls, South Dakota, October 20–November 1, 1973. Copyright 1973, by the American Society of Photogrammetry.

electronic storage media. The cost–benefit ratio between overhead coverage for a given area greatly favors the remote sensing approach, except in cases where investigation of only a very small area is required (Alexander et al., 1974). It is cheap, stores easily, and provides an image at a known point in time.

This leads us to a brief consideration of various remote sensing vehicles, specifically Landsat-1, -2, -3, and -4 satellite sensors, a low-cost method for processing satellite imagery, and to uses of remote sensing in regional-scale inventories and in natural resource change analysis, as well as to its applications in land-use planning. This will provide a backdrop for considering the role now being played by remote sensing in holistic land-use and natural resource analysis approaches and methodologies.

Remote Sensing Vehicles: Aircraft, Satellites

Much remotely sensed data is obtained by aircraft. This includes inexpensive light aircraft, in which hand-held 35-mm cameras or mapping cameras are operated over open sections in the floor of the plane. More sophisticated systems for broad regional coverage use multiengined prop and jet aircraft for moderate- and high-altitude missions.

A program in the late 1960s of the United States National Aeronautics and Space Administration (NASA) developed a series of Earth Resources Technology Satellites (ERTS), the first (ERTS-1) being launched on July 23, 1972, and operating until January 6, 1978. The ERTS system can image the entire globe from its nominal orbital altitude of about 500 nautical miles in 100 × 100 nautical-mile increments. Sensing is always conducted at a constant sun angle.

It should be noted that in 1975 the ERTS program was renamed "Landsat" to distinguish it from the planned oceanographic satellite program known as "Seasat." At that time ERTS-1 was renamed, retroactively, Landsat-1; Landsat-2 was launched in January 1975 and Landsat-3 in March 1978, Landsat-4, 1982.

Each satellite has the capability of repeating its coverage of any given swath of earth every 18 days, or about 20 times a year. The orbit of the satellite progresses westward each day, and yields images that overlap those of the previous day (Figures 3-6 and 3-7).

Landsat Sensors. Landsat-1 and Landsat-2 had two remote systems on board when launched. One was the three-channel return-beam vidicon (RBV) system and the other a four-channel multispectral scanner (MSS) system. In the RBV system, television cameras are used, with each camera filtered for a different visual band. The bands are designated as channels 1, 2, and 3. The RBVs have more cartographic fidelity than the MSS aboard Landsat, because they image an entire scene (185 km × 185 km) at once. Moreover, RBVs do not contain film; RBV images are exposed by a shutter device and stored on a photosensitive surface within each camera. This surface is scanned by an internal electron beam to produce a video signal. The RBV on Landsat-1 produced scenes between July 23 and August 5, 1972, when a tape recorder switching problem forced a system

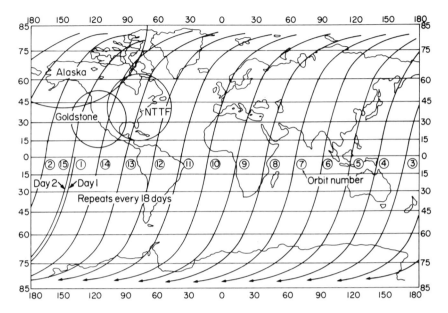

Figure 3-6. Typical daylight ground trace of Landsat satellite. Reproduced from *Resource Sensing from Space,* National Academy Press, Washington, D.C., 1977.

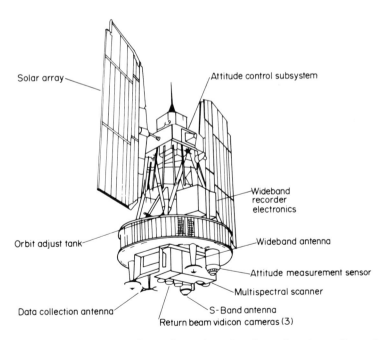

Figure 3-7. Landsat spacecraft configuration showing subsystems. Reproduced from *Resource Sensing from Space,* National Academy Press, Washington, D.C., 1977.

shutdown. "The RBV on Landsat-2 was primarily for engineering evaluation purposes. Consequently, the MSS became the primary data source on board Landsat-1 and -2" (Lillesand and Kiefer, 1979).

The MSS system generates individual scenes covering a 185-km swath in four wavelength bands. Each scene is scanned to produce four similar images simultaneously: Band 4 records in the 0.5- to 0.6-μm range (green); band 5 in the 0.6- to 0.7-μm range (red); band 6 in the 0.7- to 0.8-μm range; and band 7 in the 0.8- to 1.1-μm range.

As the vehicle travels along its orbit, the MSS system's image recording and information processing follows this pattern: Each frame overlaps the previous one by approximately 10%, thus giving continuous coverage. The adjacent parallel orbits are such that side-lap coverage is about 10% at the equator; greater side-lap occurs nearer the poles. The scanner output signals are recorded on magnetic tape, and these are held in storage for line-of-sight transmission to one of three United States ground stations. Those images which are taken within line of site of a ground station are transmitted in real time. The central ground processing facility at NASA's Goddard Space Flight Center in Beltsville, Maryland, is sent the data, and conversion is made to 70-mm film masters. Black and white renditions of each scene, one for each of the different filter bands, are generated, giving data for several applications (i.e., water–land interface, vegetation analysis, etc.). Three bands may be combined to produce a single color-infrared presentation; usually bands 4, 5, and 7 are used (Alexander et al., 1974).

In addition to the ground stations in the United States, Landsat data are directly received in a number of other nations, including Canada (at two locations), Brazil, Italy, and Sweden.

On Landsat-3, a thermal channel (10.4–12.6 μm) was added to the MSS, and the RBV was improved. The thermal channel failed shortly after launch, after developing operating problems. Nevertheless, the other four channels (identical to bands 4–7 of the previous Landsats) continue to provide data .(See Note Added in Proof, p. 255.)

It should be noted here that in 1973, Skylab, America's first space workshop, was launched. Astronauts on Skylab took more than 35,000 images of the earth. An Earth Resources Experiment Package (EREP) on board included a six-camera multispectral array, a long focal length "earth terrain" camera, a 13-channel multispectral scanner, and two microwave systems. "These EREP experiments were the first to demonstrate the complementary nature of photography and electronic imaging from space" (Lillesand and Kiefer, 1979). Skylab could be regarded as almost entirely experimental.

Development of a Low-Cost Photographic Method
for Processing Satellite Imagery

During the period 1972–1975, many research projects were conducted on applications of data from Landsat. One such project, carried out by the Resource Information Laboratory (RIL) at Cornell University, had the objective of developing a low-cost method of enhancing satellite data, one that would preclude the need to purchase expensive equipment for image processing. The satellite imagery presents data in a 70-mm format, with each 70-mm image representing a ground area of 10,000 square miles at a scale of 1:3,300,000. While other groups were developing computerized methods for enhancing the imagery, the RIL researchers at Cornell were developing an inexpensive process that depended on photographic techniques. The RIL has continued to maintain its capability to produce low-cost satellite data.

The photographic process developed by RIL has gained international attention. Many of those who work in international land-use planning and resource management and who focus on the needs of developing nations view the process as an important tool. It allows developing nations to benefit from satellite imagery and yet not become burdened with the acquisition of costly sophisticated equipment, The RIL process will undoubtedly be viewed with disdain by those with a "technological fix." However, it represents an opportunity for developing nations (as well as those areas in developed nations with limited economies and lack of skilled personnel) to utilize satellite observations.

An example of the application and apparent cost effectiveness of this RIL method is found in work on spatial characteristics (Senykoff, 1978). Low-cost photographic enhancement of the Landsat imagery was considered in relation to the costs of aerial photography. Based on per-unit-area figures, a projection is made of the cost of complete Landsat coverage for that country, whose area is 582,646 km^2 (224,960 square miles). To prepare Landsat imagery would cost about $9200. High-altitude photography would cost $682,000. It is estimated that a 20–30% discount would occur because of the scale of such an endeavor, bringing the cost figure for aerial photography down to between $545,600 and $477,400. (Such a discount would of course be related to flight requirements and time allotted for the photography.) In any event, in comparison with low-cost photographic enhancement of Landsat imagery, aerial photography cannot compete if the same data can be derived from satellite remote sensing. Aerial photography will have to be used to some extent because satellite data is far lower in spatial resolution. Multilevel data should most likely prove successful in developing countries.

Although satellite data have limited use in developing highly detailed natural resource inventories, it has application in the planning and presentation phases of natural resource inventories. Satellite imagery provides a single synoptic view of an area. It shows large-scale associations and interrelationships and places the study area in a regional context. Satellite data provide a dramatic impact and degree of credibility that mapped information often lacks. They can be used to produce an inventory of regions where the smallest mapping unit is 10 acres or more.

Nearly every use of satellite data requires that they first be enlarged. The photographic techniques used by the RIL enlarge the image to 1:1,000,000, producing an image size of approximately 7 in. × 7 in. Further enlargement can be developed at this point. The most common size now developed for projection is 1:250,000.

Remote Sensing and Landscape-Ecological Analysis
in Developing Countries

As part of an assessment of the prospects for the use of remote sensing in developing countries (National Academy of Sciences, 1977), the existing state of remote sensing activities in many countries of Africa, Latin America, South and East Asia and the Near East is tabulated (see Tables 3-3–3-5). Included are agency/institutional involvement, identifiable remote sensing programs and pro-

Table 3-3. Remote Sensing Activities in South and East Asia and the Near East

	Source of Information	Remote Sensing Infrastructure	Governmental Agencies	Military	Individual Scientists	Commercial/Private Sector	Academic Sector	Identified?	Cartographic	Aerial Photographic	Landsat	Other Spaceborne	Other Sensors	Automatic Data Processing
Bangladesh	•	1	•		•		•	•	•	•	•	•		•
Burma		4			•	■		•	?	•				
India	•	1	•		•		•	•	•	•	•		•	
Indonesia	•	4	•	•	•	•	•	•	•	•	•			
Korea	•	3	•		•			•	•	•	•			
Malaysia	•	2	•		•				•	•	•	•		
Mekong Area	•	1	•				•		•	•	•			5
Pakistan	•	2	•		•			•	•	•	•			
Philippines	•	1	•		•	•	•		•	•	•			
Sri Lanka	•	4	•							•	◉			
Taiwan (Rep. of China)	•	3	•		•			•	•	•				
Thailand	•	1	•	•	•		•	•	•	•	•		•	6
Aden		4						•	■	•				
Iraq	•	2	•		•			•	•	•	•			
Iran	•	2	•		•	•	•	•	•	•	•	•	•	•
Jordan	•	4							•	•	•			
Kuwait	•	4			•			•	•	•				5
Lebanon	•	4			•				•	•				
Muscat	•	4						•	■	•	•			
Oman	•	4						•	■	•	•			
Saudi Arabia	•	3	•				•	•	•	•	•			
Turkey	•	3	•	•	•		•	•	•	•	◉			
Yemen	•	3	•							•	•			

Reproduced from *Resource Sensing from Space*, National Academy Press, Washington, D.C., 1977.
See Tables 3-4 and 3-5 for listing of symbols.

jects, facilities and personnel available, education and training, and activities of external organizations within the country. The variability of available equipment and skilled personnel frequently is cited as a factor that limits the ability to interpret and utilize satellite imagery. The lack of trained photointerpretation personnel and/or basic analytical instruments and cartographic facilities hinders effective use of such imagery.

In instances where there is some equipment, however, the potential exists for garnering valuable resource information, in reconnaissance surveys at a scale of

Facilities and Personnel Available							Education and Training			Activities of External Organizations Within the Country			
Basic Capability		Data Collection			Data Interpretation								
Photographic Reproduction	Cartographic Capability	Aircraft Capability	Sensor Capability	Data Collection Military Only	Manual Interpretation	Automatic Data Processing	Academic Programs Offered?	Users Agencies Conduct Training Programs	External Seminar/Workshop Participation	International Agencies	Bilateral Assistance	Commercial/Private Sector Activities	Notes
●	●	●			7	●	■			●	●	●	
?	?					■	■			●	●	●	
●	●	●	C		●		■			●	●	●	
●							■			●	●	●	
●	●	●	C		7		■			●	●		
●	●	●	C		8		■			●			
●	●				8	●	■	E		●	●	●	●
●	●	●	C		8		■	E		●	●		
●	●		C				■	E		●		●	
●	●	●	C		7	●	■			●			
●	●	●					■			●			
●	●	●	C,S	●	8	●	●	E		●	●	●	
■	■						■						
●	●	●			7		■					●	
●	●	●			8	●	■		●			●	
●	●						■						
●	■				●	5	■	●					
●	●	●			●		■	●					
■	■						■					●	
■	■						■					●	
●	●				●		■	●			●	●	
●	●	●	C	●	7		■	●		●	●	●	
■	■						■			●	●	●	

1:250,000 or smaller, from a set of Landsat scenes of the country. Besides, photomaps at a scale of 1:250,000 or smaller of areas otherwise inaccessible or poorly mapped can yield a measure of a country's agricultural land, forestlands, and rangelands; can enable a country to identify its surface water and resources; and can add important morphological features to its geological information. The country can locate such resources, simply discriminated by the sensors and corroborated by ground verification. "Generally, the data for these simple, easily acquired benefits can be extracted from one-time coverage scenes of an area or

Table 3-4. Remote Sensing Activities in Latin America

	Source of Information	Remote Sensing Infrastructure	Agency/Institutional Involvement — Interest or Involvement Identified					Identifiable Remote Sensing Programs/Projects — Basic	Type of Basic Data					
			Governmental Agencies	Military	Individual Scientists	Commercial/Private Sector	Academic Sector	Identified?	Cartographic	Aerial Photographic	Landsat	Other Spaceborne	Other Sensors	Automatic Data Processing
Argentina		3	●	●	●		●	●	●	●	●	●		5
Bolivia	●	1	●	●	●	■	●		●	●	●	●	●	
Brazil	●	1	●		●	●	●	■	●	●	●	●	●	●
Chile		1	●	●	●	●	●		●	●	●	●	●	
Columbia		3	●	●		●	●	●	●	●	●		●	6
Costa Rica		4	●		●			●	●					
Ecuador			●	●		●		●	●	●	●			
El Salvador		2	●					●	●	●	●	●		
Guatemala		2	●	●					●	●	●			
Honduras		3	●						●	●				
Jamaica		3	●			●			●	●				
Mexico		2	●	●	●	●	●	●	●	●	●	●		●
Nicaragua		4	●		ʼ				●	●		●		
Panama		2	●	●		●	●		●	●	●	●	●	6
Paraguay		4	●	●				●		●	●	●		
Peru	●	1	●	●	●	■	●	◉	●	●	●	●	●	●
Venezuela		3	●		●	●		●	●	●	●		●	

● Positive
■ Negative
◉ Proposed
1 National remote sensing committee or coordinating body established
2 No coordinating body established but lead-agency can be defined

3 Various official sources
4 No infrastructure
5 Internal capability in existence or developing
6 External capabilities
7 Visual only
8 Machine assisted

from a few scenes taken at different seasons" (National Academy of Sciences, 1977).

We have already discussed the value of a low-cost photographic method for processing satellite imagery. Attention should also be given to surveying and monitoring by digital analysis, as covered in the 1977 National Academy of Sciences report. Members of the National Research Council's Ad Hoc Committee on Remote Sensing for Development note that Landsat multispectral data in digital form open new possibilities through computer analysis and manipulation

Facilities and Personnel Available							Education and Training			Activities of External Organizations Within the Country			
Basic Capability		Data Collection			Data Interpretation								
Photographic Reproduction	Cartographic Capability	Aircraft Capability	Sensor Capability	Data Collection Military Only	Manual Interpretation	Automatic Data Processing	Academic Programs Offered?	Users Agencies Conduct Training Programs	External Seminar/Workshop Participation	International Agencies	Bilateral Assistance	Commercial/Private Sector Activities	Notes
•	•	•	C,S	■	8	•		I,E	•		•		
•	•	•	C		•	•	•	•	•	•	•	•	
•	•	•	•		8	•	•	I	•	•	•	•	
•	•	•	C		7		•	I	•		•	•	
•	•	•	C,S	■	8		•	E	•	•	•	•	
•	•								•	•	•	•	
•	•	•	C		8		■		•	•	•		
•	•	•	C	■	8		■		•	•	•		
•	•	•	C	■	8	•	■		•		•		
•	•				7		■	■	•		•		
•	•			■	7		■	■	•		•		
•	•	•	C,S	■	8	•		I	•		•	•	
•					7		■		•	•			
•	•			■			•		•	•	•	•	
•	•			•	7		■	■	•		•	•	
•	•	•	•		•	•	■	■	•		•	■	
•	•	•	C		8				•		•	•	

C Camera
S Scanner
O Other (radiometer, magnetometer, radar, etc)
I Independent (internal staff and expertise)
E Externally

Table 3-5. Remote Sensing Activities in Africa

	Source of Information	Remote Sensing Infrastructure	Agency/Institutional Involvement					Identifiable Remote Sensing Programs/Projects						
			Governmental Agencies	Military	Individual Scientists	Commercial/Private Sector	Academic Sector	Basic Identified?	Cartographic	Aerial Photographic	Landsat	Other Spaceborne	Other Sensors	Automatic Data Processing
Algeria	●	3	●	●	●				●	●	●		●	■
Angola	●	4				●			●	●	●			
Botswana		3	●		●				●	●	●		●	6
Burundi	●	4	●					■	■	●				■
Cameroon	●	3	●		●		●		●	●		●		■
Central African Republic	●	2	●		●	●	●		●	●	●	●	■	■
Chad	●	2	●		●				■	●	◉	◉	●	■
Congo Rep.	●	4	●					■		●	■			
Dahomey	●	4	●		●				●	●	◉	◉		
Egypt		2	●		●		●		●	●	●		●	●
Ethiopia	●	4	●		●		●	■	●	●	◉			
Gabon	●	2	●			●			●	●	●	●	●	■
Gambia								■	■	●	■			
Ghana	●	3	●		●		●		●	●	■	●		■
Guinea	●	2	●		●				●	●	◉			
Ivory Coast	●	4	●		●		●		●	●	●			
Kenya	●	3	●		●	●	●	●	●	●	●			
Lesotho		4	●		●		●		●	●	●			6
Liberia		4						■	●	●	■			
Libya	●	2	●		●				●	●	◉	◉	●	
Malagasy	●	3	●		●			■	●	●	■			
Malawi								■	●	●	■			■
Mali	●	2	●		●				●	●	●	◉	●	■
Morocco	●	3	●		●	●			●	●	◉	◉		
Mauritania	●	3	●		●				■	●	◉	◉	●	
Mauritius								■	●	●	■			
Mozambique								■	●	●	■			
Namibia								■	●	●				
Niger	●	3	●		●				■	●	●	◉		■
Nigeria	●	3	●		●				●	●	◉		●	
Rhodesia										●	?			
Ruanda		4						■	■	●				■
Senegal	●	4	●		●			■	■	●	◉	◉		
Sierra Leone		4	●		●			■	■	●	■			
Somalia	●								●	●	◉		●	●
Sudan	●	3	●		●				●	●	◉			6
South Africa	●	3	●	●	●		●		●	●	●		●	6
Swaziland											◉			
Tanzania	●	2	●		●				●	●	◉	◉		6

| Facilities and Personnel Available | | | | | | | Education and Training | | | Activities of External Organizations Within the Country | | | |
| Basic Capability | | Data Collection | | | Data Interpretation | | | | | | | | |
Photographic Reproduction	Cartographic Capability	Aircraft Capability	Sensor Capability	Data Collection Military Only	Manual Interpretation	Automatic Data Processing	Academic Programs Offered?	Users Agencies Conduct Training Programs	External Seminar/Workshop Participation	International Agencies	Bilateral Assistance	Commercial/Private Sector Activities	Notes
●	●	●		●	7		■	E	●		●	●	
							■		■			●	
●		■			7		■		●		●		
●	■	■	■	■	7	■	●	E	■	●	●	●	
●	■	■		■			■	E	●	●	●	●	
●	■	■			(7)		■		●	●		●	
●	●	●	C,S		8	●	◉	I	●	●	●		
							■		●			●	
●	■	■	■	■	(7)	■	■	E	■			●	
●	●				7		■	E	■	●	●	●	
							■		●			●	
●	■	■			(7)		■		●	●	●	●	
●	●	●	C		7		■	E	●	●	●	●	
					(7)		■		●		●		
●					(7)		■	E	●		■	●	
●	●	■		■			■		●		●	●	
							■		●			●	
●	■	■		■			■	E	■	●	●	●	
●	●				(7)		■		■		●	●	
	■	■		■			■	E	●	●		●	
●							■		●				
●	●						■		■				
							■		■				
	■	■		■			■	E	●	●	●		
●	●		C		(7)		?	E	●	●			
							?		■				
●		■		■			■	E	●	●		●	
							■		●				
					8	●	■		●			●	
●	●				(7)		■	E	●	●	●	●	
●	●	●	C,O		8	●	?	I	●	■	■	●	
	●				(7)		■		●				
●	●	●		●	●	●	■	E	●	●	●	●	

of data. Landsat scenes in digital form can be used to generate special-purpose images for visual analysis, or can be used for delineation of features of interest by computer-assisted recognition of spectral patterns, or "for a broad range of statistical analyses including the generation of a quantitative data base for part or all of a country's land area. Use of digital data and computer techniques permits rapid and variable-scale (i.e., magnification) analysis of spectral patterns, quantitative comparison of spectral responses in data acquired at different times, and rapid analysis of spectral differences between targets that are too small to be detected by the human eye. These capabilities are useful in monitoring and quantifying both natural and man-caused changes in the environment, and when the changes of interest are detectable from analysis of spectral signatures, computer techniques are often faster and cheaper than manual extraction of the same information."

Data such as soils maps, topography maps, population density maps, and geology surveys are being digitized, adjusted to an appropriate scale, and added to the data bank for overlay and correlation studies with Landsat data. The over-

Table 3-5. Remote Sensing Activities in Africa (continued)

	Source of Information	Agency/Institutional Involvement — Interest or Involvement Identified						Identifiable Remote Sensing Programs/Projects — Basic		Type of Basic Data				
		Remote Sensing Infrastructure	Governmental Agencies	Military	Individual Scientists	Commercial/Private Sector	Academic Sector	Identified?	Cartographic	Aerial Photographic	Landsat	Other Spaceborne	Other Sensors	Automatic Data Processing
Togo								■	●	●	■			
Tunisia	●	3	●		●				●	●	◉			
Uganda			●		●			■	●	●	■			
Upper Volta	●	2	●		●			■		◉	◉	●	■	
Zaire	●	1	●	●	●	●			●	●	5	6	●	6
Zambia	●							■	●	◉	■			

● Positive	3 Various official sources
■ Negative	4 No infrastructure
◉ Proposed	5 Internal capability in existence or developing
1 National remote sensing committee or coordinating body established	6 External capabilities
2 No coordinating body established but lead-agency can be defined	7 Visual only
	8 Machine assisted

Reproduced from *Resource Sensing from Space,* National Academy Press, Washington, D.C., 1977.

lays are then used to produce soil productivity maps, land-use capability maps, and a range of correlation studies.

The problem of Landsat data's being used in digital form in the less-advanced developing countries is considered in the context of possible contracting with foreign firms for both specially enhanced imagery and computer-assisted analysis and interpretation. Further, to conduct its operations when unable to do so itself, each such nation would find it useful to have some specialists familiar with digital analysis. Such individuals are able to commission needed services from abroad and to work effectively with overseas consultants.

The economics of remote sensing information systems has served as the theme of conferences at San Jose State University and has been the theme of articles in *Photogrammetric Engineering and Remote Sensing* (Watkins, 1978). The important issue of the economics of remote sensing lies largely outside the scope of the book, and readers are encouraged to consult the proceedings of the conferences at San Jose State, as well as the above-mentioned periodical, for information on remote sensing economics.

Facilities and Personnel Available			Education and Training	Activities of External Organizations Within the Country
Basic Capability	Data Collection	Data Interpretation		

Photographic Reproduction	Cartographic Capability	Aircraft Capability	Sensor Capability	Data Collection Military Only	Manual Interpretation	Automatic Data Processing	Academic Programs Offered?	Users Agencies Conduct Training Programs	External Seminar/Workshop Participation	International Agencies	Bilateral Assistance	Commercial/Private Sector Activities	Notes
●	●				(7)		■		■	●	●	●	
		■	■	■			■	E		●	●	●	
●	●	●	C	●	8	●	■	E		●	●	●	
●	●						■		■			●	

C Camera
S Scanner
O Other (radiometer, magnetometer, radar, etc)
I Independent (internal staff and expertise)
E Externally

The practical uses of remote sensing in land-use evaluations, planning, and design are being increasingly reported by developing nations. The Food and Agriculture Organization (FAO) of the United Nations reported in 1974 that it had begun using aerial photography 25 years earlier in its first field projects. At the outset, ordinary black and white (panchromatic) photographs were used; but in the 1960s, with the introduction of reliable color photography, and then later of infrared color photography, new applications in natural resource surveys occurred. Geometrically corrected aerial photographs, known as *orthophotographs,* and assemblages of these, known as *orthophotomaps,* are now being employed by the FAO in place of maps. Photography is used extensively for surveys of landform, geology, land use, vegetation, forest types, forest strata, and soils.

Aerial photography has been or was, at the time of the FAO publication, being used in numerous FAO-executed projects—for forest inventories in Argentina, Bhutan, Cameroon, Gabon, Malaysia, Peru, and Senegal, among others; an agroindustrial development project in Bolivia; an irrigation project in Dahomy; a land capability survey in Indonesia; river course mapping in the Republic of Korea; a land-use study in Morocco; and soil mapping in Ethiopia, Nepal, the Sudan, Syria, and Thailand. In addition to aerial photography, other remote sensing tools were looked at as well.

Radar was experimented with in a United Nations Development Program (UNDP)-FAO project in Surinam, considering the use of radar altimetry for determining tree heights. (Radar employs scanners instead of cameras.) Also, thermal scanning, which depends on emitted heat rather than on reflected light or near-infrared reflected radiation, was "beginning to prove useful to FAO for the study of cold and warm ocean currents which may influence fish concentrations, for detecting groundwater near the surface of the earth, and for providing information on surface and near-surface features which can influence agricultural practice" (Food and Agriculture Organization, 1974). In North America, thermal scanning was being used as an aid to forest fire control. Thermal imagery was employed by the FAO to locate submarine freshwater springs along the Lebanese coast. Side-looking radar, which is able to penetrate clouds, was proving essential for obtaining aerial coverage of high-rainfall tropical areas with long periods of cloud cover—for example, the west coast of South America.

To return briefly to the question of costs of remote sensing in developing countries, we find that in a resource inventory assessment in Costa Rica focusing on monitoring the conversion of forest to rangeland (Watkins, 1978), the program utilized low- and medium-level aerial photographs (natural color and color-infrared) to obtain data on forest cover and Level I land-use classifications (Anderson et al., 1976). (This classification is alluded to later in this chapter.) Conclusions reached were that remote sensing can meet most but not all resource assessment needs; project design is of paramount importance for utilizing remote sensing as a data source; present satellite systems are not adequate for preparation of a resource data base, whereas aerial photography will meet the majority of requirements for a data base; satellite systems are able to provide useful information for monitoring a resource assessment program.

Whyte (1976), in his important contribution to land appraisal, notes the application of aerial photography and space technology to resource sensing, with special focus on southern Asia and the southwestern Pacific. One example is the application of the technique to southern Asia, as discussed by Verstappen, where "in addition to the usual survey of geomorphology, geology, soils and water resources, reference is made to calculations of the relative period in areas under shifting cultivation and to the special features of Central Rajasthan; widespread line concretions at depths of 30–120 cm present addition of rainfall to groundwater, necessitating the construction of tanks for storage of surface water; their siting has a profound influence on the sociology of the region."

Further illustration is provided (based on Verstappen, 1966) from the areas of Java and New Guinea, where aerial photographs have been especially useful in the study of deltas. They reveal, if combined with essential field checking, varied features of terrain that cannot, or can only with great difficulty, be obtained by other means. These aspects are noted: (1) delineation of individual parts and elements of the delta; history of the delta; interpretation of dissected older parts; mapping of vegetation, soil, hydrology, land use, and mangrove coasts; (2) dynamic interpretations of deltas and delta coasts; changes in river courses; variations in frequency or force of onshore winds; coastal movements; and photogrammetry, mapping of shorelines, underwater features, determination of depth, and the like.

It is desirable to examine the question of the internationalization of remote sensing, and to note the active role of the United States Agency for International Development (USAID) in the transfer of remote sensing technology to the developing countries. Paul (1978) notes that in Peru locating was made, by Landsat digital analysis, of stands of aguaje palm, which is regarded as a valuable resource for cooking oil. In Pakistan, rates of river sediment loads and the accretion of coastal swamps that accompany these are being estimated by Landsat image analysis in order to locate a new port in an area of minimum silting. There is cognizance of the controversial aspects of satellite earth surveys. Among them is the utilization-to-acquisition ratio, which is used in some quarters to argue that "there are already volumes of environmental and Earth resources data and, even if they are limited in accuracy and coverage quality, the problem is that no one is analyzing how to use these existing data to deal effectively with the Earth's agricultural, land-use and environmental problems."

This statement can be supported by many involved in remote sensing. Paul (1978) cites the example of the Sahel program: The United States National Aeronautics and Space Administration had acquired imagery over Mauritania and Senegal during December 7–12, 1977. The Sahel program staff were concerned with determining the extent of desertification by using this one Landsat imagery sequence. The focus of discussion was a new data base and the monitoring of changing deserts in future months and years. It was pointed out to the group that dramatic changes had begun in 1972, with an unexpected drought and the loss of hundreds of thousands of lives and almost three quarters of the livestock in Upper Volta, Mauritania, Mali, and Chad. The important 1972 and 1976 Landsat imagery of the same areas that were imaged in 1977 was acquired, and the multitemporal analysis necessary to see the real losses in agricultural land in the early 1970s was done. The 1972 and 1976 Landsat imagery exemplifies data acquired but almost not utilized.

In the case of assistance programs for developing countries, USAID faced the problem of 30–40 countries requesting help from the agency. This high number dictated the need for economy of scale, resulting in the concept of regional remote sensing training and user assistance centers. Such centers attack regional problems associated with similar geomorphological or ecological conditions, as opposed to concerns in any one country. Where related institutions already existed, as in the case of the Nairobi, Kenya, Center for Services in Surveying and Mapping, the regional remote sensing center was made an operational arm of the existing facility. Elements of a regional remote sensing center are shown in Figure 3-8. Another such regional remote sensing center was created at Ouagadougou, Upper Volta. Others have been proposed elsewhere.

In the foregoing regional remote sensing center context, problems that have

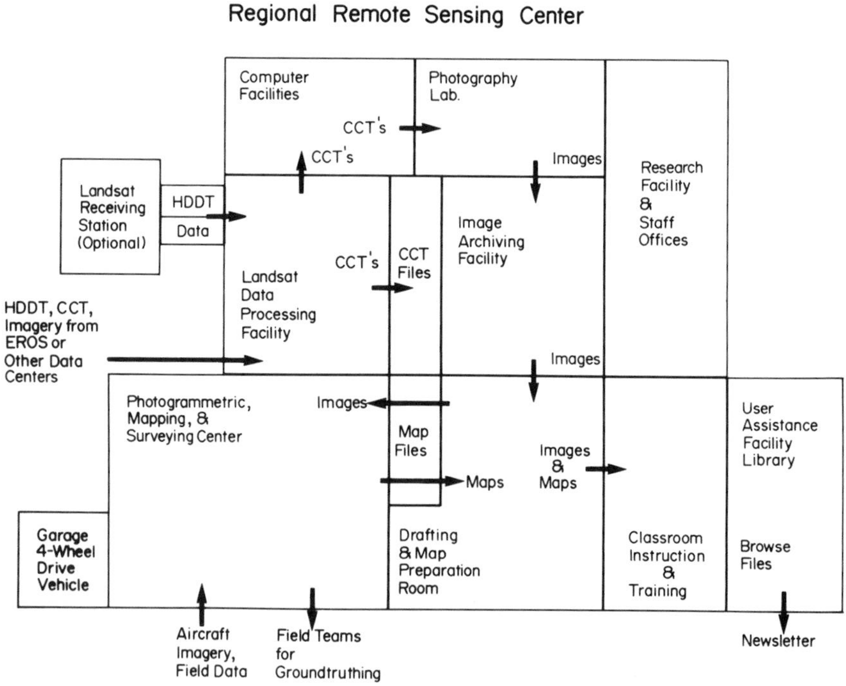

Regional Remote Sensing Center

Figure 3-8. Elements of a regional remote sensing center. Produced for Agency for International Development by Technology Application Center, University of New Mexico.

already been identified as of regional concern are agricultural production, desertification, deforestation, and hazard assessment. For the future, identified regional resource needs include deforestation and desertification survey techniques, coastal zone food resources, mineral resource surveys for fertilizer production, geothermal exploration techniques, appropriate and economical digital image processing systems and resource information systems, improved cartographic techniques, and crop production monitoring and estimation. The USAID has described its long-term goal as to "continue to integrate aerial and satellite acquired data to satisfy basic human needs of food and fiber production, water management, improved shelter, and energy use; and to further the modernization of developing economies" (U.S. Agency for International Development, 1980).

In Mali (Paul, 1978) USAID was inventorying rangelands and would be looking with Landsat-C "for new lands free of the centuries-old onchocerciasis (river blindness) to relocate tens of thousands of residents presently overcrowding cities in Benin, Togo and Upper Volta." In Morocco, the Agency, using Landsat, produced photomaps, geology maps, and a sample inventory of coastal erosion areas and freshwater upwelling in the coastal zone. In Kenya, Landsat imagery of the Lake Nakuru area was visually intepreted to reveal 67% of the census enumerative boundaries existing on Kenya administrative maps, boundaries indicative of major land-cover changes. The USAID and the United States Bureau of the Census succeeded in convincing the statistics bureau in Kenya to redefine some of the remaining 33% of these boundaries to conform to population densities recognizable by cultural land uses. These are agriculture and settlements interpreted from Landsat imagery.

In Bolivia, new land-colonization programs are reported to have resulted in the migration of tens of thousands of inhabitants of the altiplano to the eastern lowland jungles. The eastern rain forests are being decimated as a consequence, and slash-and-burn agricultural practices and new villages are reducing the forests at a more rapid pace than government planners had realized. In this case, Landsat computer-compatible tapes were digitally classified, and land-cover theme maps were developed to show the extent of deforestation. Rough estimates of numbers and densities of people in lowland areas have been provided from agricultural and village patterns recognized on the maps.

Results of a research project designed to develop a viable method for producing small-scale rural land-use maps in semiarid developing countries by using Landsat imagery obtained from orbital multispectral scanners (MSS) are reported by Van Genderen et al. (1978). Of interest is the fact that until the 1960s, detailed rural land-use surveys were carried out mainly in developed nations. Land-use maps with a wide variety of scales and classifications were produced, primarily to ascertain the spatial distribution of land use at a particular time. Useful analytical tools for general or reconnaissance evaluations by planners came about through the resulting land-use maps. Time-consuming and frequently expensive methods were employed; collection of information was from vertical black and white panchromatic aerial photography, as well as from field reports and a variety of statistical agencies. The planning (of economic development policies) of new independent countries since the second World War has required the use of medium- to small-scale land-use maps for broad overviews of regions; these provide the bases for more detailed, more diverse investigations at larger scales.

Van Genderen, Vass, and Lock, citing Landgrebe (1972), stress the problems that still exist in the utilization of information from remote sensing data, and remind us that the identification of image characteristics and lack of clarity caused by the quality and resolution factors of the data have presented problems in the interpretation of land use at medium to small scales. Their methods for arriving at guidelines for using Landsat data for developing countries' rural land-use surveys are divided into preoperational and operational stages (Figure 3-9). Both require careful consideration of possible imagery, base maps, and other reference material. With objectives defined, standard preprocessed Landsat MSS imagery can then be selected and purchased. The prospective user of imagery is cautioned concerning the wide choice available (four different scales, four different black and white positive or negative spectral bands, and transparency or paper print format). To avoid unwarranted expense, imagery should be selected so that seasonal variations in vegetation cover can be utilized in the accurate delineation and identification of land-use boundaries and categories. Stress is placed on recommended standards: Preprocessed imagery of each Landsat MSS frame for rural land-use mapping consists of one 1:1,000,000 false color composite transparency, if this exists, or 1:1,000,000 black and white positive transparencies of bands 4, 5, and 7 for producing color composites by the diazo process. In addition, one 1:250,000 color composite print or, if unavailable, one 1:250,000 paper print of band 5, is required. Figure 3-9 provides a good overview of the two stages and the general sequence to be followed; details should be sought in the original well-illustrated *ITC Journal* article. It is concluded that

> it is feasible to design a methodology that can provide suitable guidelines for the operational production of small scale rural land use maps of semiarid developing regions from Landsat MSS imagery utilizing inexpensive and unsophisticated techniques. The suggested methodology should provide immediate practical benefits to map-makers attempting to produce and use maps in countries with limited budgets and equipment. As the Landsat MSS imagery system permits regular synoptic coverage of the Earth's surface, it provides an ideal method for establishing a satisfactory data base and further monitoring of land use changes over large areas. (Van Genderen et al., 1978)

Consideration had been given to many preprocessing and interpretation techniques, and these were rejected as being inappropriate because of the high cost of imagery or equipment, or on the basis of being inadequate for operational use. Also, the imagery and interpretation techniques consisting of color composites and monocular magnification proved simplest, fastest, and most versatile. The criteria and hierarchical structure of the United States Geological Survey's classification system for use with remote sensor data, covered in some detail later in this chapter, were found to be (Van Genderen et al., 1978) "acceptable as a general basis for researchers and organizations wishing to develop systems for their own regions."

As should be evident from the examples in this section, remote sensing in its several forms is being actively investigated and employed for developing country uses. Appropriateness of particular approaches and technologies for needs of

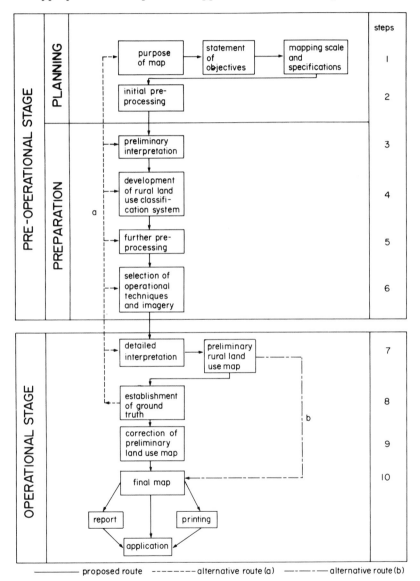

Figure 3-9. Diagrammetric representation of a proposed methodology for the production of small-scale rural land use maps from Landsat MSS imagery. *ITC Journal,* 1978-1, p. 40.

specific regions continues to be given attention by researchers. There is increasing recognition of the possibility of using remote sensing information seeking for physical planning and natural resource management purposes. It is encouraging to note vigorous consideration of the role, degree, and cost of remote sensing for needs in Africa, Asia, and Latin America. Further refinement of methods and techniques appropriate to specific developing country needs should result from this research. Holistic landscape analysis should ultimately benefit from such practically oriented focuses for such locations.

Remote Sensing and the Monitoring of Critical Indicators in Areas Subject to Desertification

Desertification, or degradation in productivity in dry lands is a critical world problem. Estimates of populations in areas recently undergoing severe desertification are listed according to livelihood in Table 3-6; a total of 78 million people is seen as threatened. Because of high income or other advantages, about a third may be regarded as being in a position to avoid the worst consequences of desertification, leaving about 50 million people who are "immediately menaced through the destruction of their livelihoods" (United Nations, 1978a).

The phenomenon of desertification is described in the Draft Plan of Action for the 1977 United Nations Conference on Desertification (United Nations, 1976)

> Deserts are areas of sparse or absent vegetation and low biological productivity primarily due to deficient rainfall. Desertification is seen as the extension or intensification of such conditions. In preparing for the

Table 3-6. Estimates of Populations According to Livelihood, in Areas Recently Undergoing Severe Desertification (In Thousands)

Region	Total population	Urban based	Cropping based	Animal based	Area (km^2)
Mediterranean Basin	9,820	2,995 31%	5,900 60%	925 9%	1,320,000
Sub-Saharan Africa	16,165	3,072 19%	6,014 37%	7,079 44%	6,850,000
Asia and the Pacific	28,482	7,740 27%	14,311 54%	6,431 19%	4,361,000
Americas	24,079	7,683 32%	13,417 56%	2,979 12%	17,545,000
Total	78,546	21,490 27%	39,642 51%	17,414 22%	30,076,000

Reproduced from *Desertification: Its Causes and Consequences.* Pergamon Press, New York. Compiled and edited by Secretariat of the United Nations Conference on Desertification. Nairobi, Kenya, August 29–September 9, 1977.

conference, attention was focused on tropical, subtropical, and temperate drylands. The Conference Plan of Action covers areas where desertification is occurring now and others vulnerable to future desertification, including arid, semi-arid and sub-humid areas . . . . Recognizing that the physical and biological effects of desertification are important to the world community mainly because of their impact on human beings, the Plan of Action focuses on the problems of people affected by desertification.

Human–environment interactions result in physical processes that lower productivity. The following are identified (Berry and Ford, 1977):

(a) general land degradation, including soil erosion;
(b) encroachment of brush/scrub and other unproductive forms of vegetation;
(c) encroachment of sand dunes on otherwise productive land;
(d) loss of soil nutrients;
(e) loss of ground moisture;
(f) decrease in soil biota;
(g) salinization of irrigated lands; and
(h) waterlogging.

Important recommendations have been advanced by Berry and Ford (1977) for a system to monitor critical indicators in areas prone to desertification. Differences exist about the underlying assumptions. Some suggest that including such processes as waterlogging and salinization overly generalizes the terms of reference for desertification, whereas others believe that emphasis on soil and human factors underplays the role of physical phenomena, such as climate, soil type, or natural vegetation. There are those who stress that emphasis on local land management fails to take account of externally introduced technology, commercial crops, or imposed patterns of land access. But Berry and Ford (at the time of the report each a codirector of the Program for International Development at Clark University, Worcester, Massachusetts) accept the United Nations definition and endorse the emphasis on the impact on people. They assume that desertification is a result of interaction between humans and the physical environment and increases when management of the environment is not adjusted so as to maintain the long-term balance between existing land and water systems. The causes of imbalance are complex and relate to both climate and ecosystem fluctuations. They also relate to current processes of social and demographic change.

Desertification involves both natural and human elements, leads to loss of land productivity and human well-being, and as a process deserving global attention, suggests the importance of monitoring. Monitoring desertification involves monitoring for scientific understanding, monitoring for awareness ("people undergoing desertification may fail to perceive its effects as they adapt to the changing landscape or it may be obscured by short-term climatic fluctuations. The willingness to take action to reverse or prevent desertification depends on such awareness at all levels of decision making—local, national and global. A monitoring system should be educative and directly enhance awareness of the problem at all levels of involvement") (Berry and Ford, 1977), and monitoring for action ("it is necessary to have early warning for preventative action and assessment for selection for remedial action. Remedial actions are seldom dramatic and they require careful observation against baseline conditions to evaluate their progress or failure. A monitoring system should assist in selecting priority areas for action and evaluating the effort"). Monitoring of desertification is a long-term process (10–15 years).

Identification and inclusion of variables related to human activities and conditions is regarded as critical in the monitoring system. Climatic fluctuations (not the primary determining factor in desertification, but serving to dramatize or obscure desertification) should be included. Local, national, regional, and global functioning of the monitoring system is considered essential. Local and on-ground monitoring is needed; satellite imagery as a data collection and analysis tool is supplementary to them.

At the global level is envisaged a small international coordinating body, to be housed within the Global Environmental Monitoring System (GEMS), which would monitor five global indicators; albedo, dust storms, rainfall, soil erosion and sedimentation, and salinization. At the *regional level,* several inter-nation affiliations are seen in the picture, varying in responsibilities, according to regional needs, building on existing regional institutions linking the global activities to national/local work and which would monitor the regional indicators of productivity, standing biomass, climate, nutrition and salinization.

At the *national/local level,* activities would be coordinated through Man and the Biosphere national committees which would devise a representative sample of village and nomadic communities to monitor productivity, human well-being and human perception.

Berry and Ford's monitoring of desertification draws upon data already being collected and analyzed in other monitoring systems. The essence of the monitoring system is the linking of a variety of techniques, including on-ground monitoring by local people, national and regional professionally-organized activities, and the use of satellites and other existing global data gathering systems for a world-wide overview.

Four prototype regional models, representing very different physical and social settings in regions prone to desertification, were introduced. Local institutions involved were not consulted in constructing the models. Relationships mentioned provide examples of how a select number of indicators might be used in an integrated monitoring system.

1. The West African Sahel constitutes an example of general ecological diversity with a long history of interstate cooperation.
2. Semiarid East Africa exemplifies ecological diversity with a long history of interstate cooperation, but where a regional monitoring center may take a more coordinated role.
3. Southwest Asia typifies ecological diversity and large nation states with very different sets of resources.

4. Semiarid United States (possibly to include Mexico's dry lands as well) manifests some ecological diversity with well-developed institutions that have not joined together to monitor, analyze, or recommend action on desertification.

The activities of the United States Agency for International Development, mentioned earlier in this chapter, include a focus on the use of remote sensing for monitoring the rate and direction of desert encroachment. As the USAID notes (1980), since Landsat has been in existence, an extensive historical record of land-use changes in the Sahel region, which runs the width of Africa and crosses the traditional migration routes of many cultures (as well as the borders of several countries), has been compiled. In Figure 3-10, a sequence of three photos shows an area of Mauritania and Senegal during the period from 1972 to 1977. Using time-sequence photography, the retreat and advance of interdunal vegetation has been monitored and recorded. This phenomenon represents a

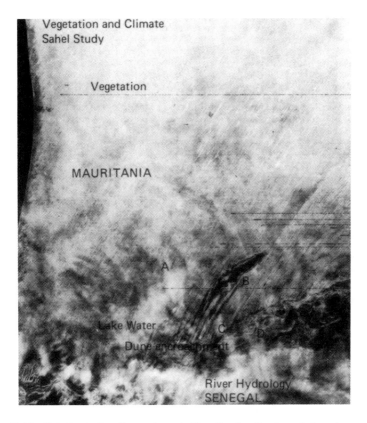

Figure 3-10. Sahel study, Senegal and Mauritania. Produced for Agency for International Development by the Remote Sensing Institute, South Dakota State University.

Figure 3-10 (cont.)

critical consideration relative to grazing for both wild and domestic livestock. Lake and river levels and cropland conditions have also been monitored, and potential groundwater locations designated for exploratory drilling. The stippled area in the Landsat scene of Mauritania is considered a promising source of groundwater.

In late 1982, reports were published about the interpretation of radar revealing a whole new body of geological knowledge about what lies under the Sahara (McCauley et al., 1982; Wilford, 1982).

> The shuttle imaging radar (SIR-A) carried on the space shuttle Columbia in November 1981 penetrated the extremely dry Selima Sand Sheet, dunes, and drift sand of the eastern Sahara, revealing previously unknown buried valleys, geologic structures, and possible Stone Age occupation sites. Radar reponses from bedrock and gravel surfaces be-

Figure 3-10 (cont.)

neath windblown sand several centimeters to possibly meters thick delineate sand- and alluvium-filled valleys, some nearly as wide as the Nile Valley and perhaps as old as middle Tertiary . . . . Stone age artifacts associated with soils in the alluvium suggest that areas near the wadis may have been sites of early human occupation. (McCauley et al., 1982, p. 1004)

John McCauley of the United States Geological Survey and his fellow researchers have concluded that "the potential for mapping ancient drainage patterns—and by inference, potential sources of near-surface ground water—is sufficient to arouse excitement among earth scientists, who now have a new means of exploring the deserts of the earth."

In a different vein, some slightly earlier work (Otterman et al., 1978) studies, through ground observations and numerical analysis of Landsat imagery (from computer-compatible digital tapes), the effects on the earth's surface of forming a 5.5 × 5.5 km exclosure in the arid region of Sinai. Removal of anthropogenic pressures over a period of about two years resulted in marked increase in vegetation, in both plant cover and number of species. As regards the management of semiarid ecosystems in Israel, Noi-Meir and Seligman (1979) point out, with the assistance of aerial photographs from 1944 and 1970, the sharp line of contrast between the Sinai and Negev in the 1970 photograph. No such line appears in the 1944 picture. During this period, on the Negev side of the border, there was little or no grazing or shrub removal. A dense shrub vegetation (mainly *Artemisia monosperma*) and litter of shrubs and annuals accumulated to result in a darker color in the photograph on that side of the border. This came about as a consequence of the border being effectively closed to the Sinai Bedouin and their livestock. Another coverage by Otterman et al. (1975) reported on a comparative study of vegetation, albedo, and temperatures in western Negev and Sinai ecosystems. Satellite imagery is referred to by the researchers.

Regional-Scale Land Use and Natural Resources

As we have noted, in the attempt to create regional-scale inventories that are both accurate and timely, remote sensing technology is being increasingly employed. The Land Use and Natural Resource (LUNR) Inventory of New York State of the 1960s is an example of the use of this technological capability (Hardy, 1970). The LUNR Inventory produced an information system designed to provide resource managers with data about the physical resources of the state. The primary element of the LUNR Inventory is a two-part overlay depicting 130 categories of land use. They can be aggregated to 11 general categories: agriculture, forestlands, commercial and industrial, extractive industry, nonproductive, outdoor recreation, public and semipublic land uses, residential, transportation, water resources, and wetlands.

The LUNR information was obtained from interpretation of aerial photography of the whole state dated 1967 and 1968, and the minimum area mapped is about 1 acre. The overlays are designed to be used with a topographic map as a base, at a scale of 1:24,000. The LUNR Inventory results were verified in field observations.

Both private and governmental users have access to the LUNR Inventory from a centralized data bank. The system is housed at the Resource Information Laboratory of Cornell University, directed by Dr. Ernest E. Hardy, at Ithaca, New York. Computer-generated data are also available, and all of the land-use categories available in overlay form can also be displayed in a computer-tabular or computer-graphic format. In the early 1970s, the inventory of six counties in the Catskill region of New York State was updated from more newly acquired photography.

As noted by Johannsen and Sanders (1982), intrastate use of natural resource data was studied by the Council of State Governments. Several serious problems were revealed, including:

significant duplication of data gathering activities by state agencies;

poor documentation of collection techniques and data definitions, with resulting misuse of data;

difficulty in identifying data sources, particularly for older data;

difficulty in transferring data between agencies because of different standards and formats;

failure of some state agencies to make their data accessible to other agencies;

decision making delayed by a lack of relevant data.

At the same time, there are demands for more and better data than ever before. Johannsen and Saunders (1982, p. 619) further mention that

Natural resource decision-makers in 32 states, beginning with New York in 1966, have seriously explored or instituted comprehensive, integrated approaches to resolving their data needs. Although the titles vary and short-term objectives differ, the concept is the same—a natural resource information system (NRIS).

An NRIS typically incorporates a centralized data bank or a linked network system that is accessible by all state resource agencies. Most systems incorporate federally generated data as well as data supplied by state agencies.

A listing of state-level natural resource information systems in the United States as of June 1980 (Table 3-7) reveals a variety of statuses, from those that are operational to others partially operational to yet others still in the research and design stages, while at least one ceased operations after gubernatorial veto of funding.

The Canada Land Inventory and the Canada Land Data System

In Canada, significant strides have been made over the last two decades in identifying land capabilities for agriculture, forestry, recreation, and wildlife. These have been assessed through the Canada Land Inventory (CLI). Included in this program are 2.6 million km^2 in the Canadian Atlantic provinces, as well as considerable areas of Quebec, Ontario, Manitoba, Saskatchewan, Alberta, and British Columbia. Development of the CLI involved federal and provincial governments, along with universities and the private sector.

The CLI was the result of increasing recognition in Canada that the land could not support all the demands being made upon it, and that land resources are not inexhaustible. The latter perception had existed since early in Canada's

Table 3-7. State-level Natural Resource Information Systems in the United States—as of June 1980[a]

State and System Name	Address	Year Started[b]	Status[c]
Alabama Resource Information System (ARIS)	Office of State Planning and Federal Programs 3734 Atlantic Highway Montgomery, Alabama 36130	1974	Partially operational. Data base is small; development by Auburn University continues.
Alaska	Department of Natural Resources 323 E. 4th Avenue Anchorage, Alaska 99501	1979	In start-up phase. Will ultimately store, sort, analyze, overlay, and compare all types of map data.
Arizona Resource Information System (ARIS)	State Land Department Information Services Division 1624 West Adams, Room 300 Phoenix, Arizona 85007	1972	Centralized source of maps, air photos, and satellite imagery in Arizona. Spatial data not yet computerized.
Arkansas Resource Management Information System (ARMIS)	Department of Economic Development #1 Capitol Mall Little Rock, Arkansas 72201	1979	Contract let to develop a system. Not yet functioning.
California Environmental Data Center	Office of Planning and Research Governor's Office 1440 Tenth Street Sacramento, California 95814	1978	Central system for all environmental data. Personnel assist and coordinate smaller systems developed by local governments.
Colorado	Department of Local Affairs 520 State Centennial Building 1313 Sherman Street Denver, Colorado 80203		Two specialized data systems are functioning. State is exploring more comprehensive system.

Connecticut Natural Resources Center	Department of Environmental Protection 165 Capital Avenue Hartford, Connecticut 06115	1973	Data are collected, collated, mapped, published, and distributed. Technical assistance is provided users. Data are not computer-stored.
Florida General Environmental Data	Department of Environmental Regulation, Office of Program and Data Analysis 2600 Blair Stone Road Tallahassee, Florida 32301	1978	Numerous subsystems relating to water and air quality data and permits. Goal is to merge all systems into GEDS by 1981.
Georgia Resource Assessment Program	Department of Natural Resources 270 Washington Street, SW Room 700 Atlanta, Georgia 30334	1974	Partial data base. Still conducting demonstration projects with Georgia Institute of Technology and federal agencies. Emphasis has been on data from remote sensing.
Iowa Water Resources Data System (IWARDS)	Iowa Geological Survey 123 N. Capitol Street Iowa City, Iowa 52242	1974	Emphasis on water, geologic, and climatic data. Recently acquired equipment for classification, display, and analysis of Landsat imagery.
Kentucky Resource Information System (KRIS)	Office of Policy and Program Analysis Capitol Plaza Tower Frankfort, Kentucky		In concept and design stage. Hope to develop a linked-network system with initial emphasis on surface mining and reclamation problems.
Louisiana Areal Resource Information System (LARIS)	State Planning Office 4528 Bennington Avenue Baton Rouge, Louisiana 70808	1979	Data base consists of land use, political boundaries, and hydrologic boundaries. Soils data being added. No in-house graphics ability.

Table 3-7. (continued)

State and System Name	Address	Year Started[b]	Status[c]
Maine	Maine Forest Service 255 Nutting Hall Orono, Maine 04469		A system is being developed to store, manipulate, and display natural-resource information.
Maryland Automated Geographic Information System (MAGI)	Department of State Planning 301 W. Preston Street Baltimore, Maryland 21201	1974	24 computerized files of physical, cultural, and areal data. Considerable software to manipulate and display data. Used for state and regional planning.
Minnesota Land Management Information System (MLMIS)	Land Management Information Center, 15 Capitol Square 550 Cedar Street St. Paul, Minnesota 55101	1967	Depository for 13 files of geographically based information. Information center provides data and technical assistance to all state agencies.
Mississippi Automated Resource Information System (MARIS)	Mississippi Research and Development Center P.O. Box 2470 Jackson, Mississippi 39205	1978	Data base is still being assembled. No projects completed as of August 1979.
Missouri Geographic Resource Center	University of Missouri 240 Electrical Engineering Bldg. Columbia, Missouri 65211	1980	A joint project of the University of Missouri-Columbia and several state and federal agencies.

System	Address	Year	Description
Montana Geo-Data Information System	Department of Community Affairs Research and Information Systems Capitol Station Helena, Montana 59601	1972	System contains considerable computer-stored files of resource, social, and economic data. Numerous demonstration projects completed.
Nebraska Natural Resources Information System (Natural Resources Data Bank)	Natural Resources Commission 301 Centennial Mall South P.O. Box 94876 Lincoln, Nebraska 68509	1970	Primarily concerned with storing and processing data relating to soil and water resources of state.
New Jersey (unnamed)	Division of State and Regional Planning, Bureau of Planning and Automated Systems 88 E. State Street Trenton, New Jersey 08625		Municipal boundaries, watershed, and land use data digitized to assist with statewide 208 water planning.
New York Land Use and Natural Resource Information System	Resource Information Laboratory Box 22 Roberts Hall Ithaca, New York 14853	1966	This system, operating out of the Resource Information Laboratory of Cornell University, provides resource maps and other products to users within the state.
North Carolina Land Resources Information Service (LRIS)	Department of Natural Resources and Community Development Box 27687 Raleigh, North Carolina 27611	1977	System is adding data on project-by-project basis because it is obligated to recover costs from user fees. Very versatile system.
North Dakota Regional Environmental Assessment Program (REAP)	Legislative Council State Capitol Bismarck, North Dakota 58505	1975	Program ceased operations in 1979, after gubernatorial veto of funding.

Table 3-7. (continued)

State and System Name	Address	Year Started[b]	Status[c]
Ohio Capability Analysis Project (OCAP)	Department of Natural Resources Division of Water Fountain Square, Building E Columbus, Ohio 43224	1973	System designed to provide data and technical assistance to local governments. Data generated and stored on a county-by-county basis.
Oklahoma Graphics Data System	Oklahoma Foundation for Research and Development Utilization P.O. Box 1328 Edmond, Oklahoma 73034		Cooperative project of several independent agencies. Focuses on forest cover data and biomass studies.
Pennsylvania Land Use Data Analysis (LUDA)	Department of Environmental Resources, Bureau of Environmental Planning 101 S. 2nd Street Harrisburg, Pennsylvania 17120		A system based on polygons (U.S.G.S. system), with capability of providing maps and data by various land classifications.
South Carolina (unnamed)	University of South Carolina Computer Services Division Columbia, South Carolina 29208		System still in research and design stage. Several demonstration projects completed.
South Dakota Resource Information System	State Planning Bureau Planning Information Section 415 S. Chapelle Pierre, South Dakota 57501	1974	Land cover, soils, and geologic data used to produce land capability maps for local and regional planning.

System	Agency	Date	Description
Tennessee Areal Design and Planning Tool (ADAPT)	Tennessee Department of Public Health, Division of Water Quality Control 621 Capitol Hill Building Nashville, Tennessee		Division of Water Quality in the Department of Public Health has examined this approach, which uses triangular grid cells.
Texas Natural Resources Information System (TNRIS)	Department of Water Resources P.O. Box 13087 Austin, Texas 78711	1972	Most extensively developed system. Over 200 computerized files of data. Remote terminals across state. Linked-network system.
Virginia Resources Information System (VARIS)	Office of Commerce and Resources Richmond, Virginia 23219		Feasibility study will be presented to General Assembly in early 1980 by Executive Branch Task Force.
Washington Gridded Resource Inventory Data System (GRIDS)	Department of Natural Resources		The Department of Natural Resources is associated with this project, but the current status is unknown.

[a] This information was gathered through personal conversations with NRIS personnel, reports distributed by the states, and publications from other sources. The author would appreciate being notified of inaccuracies.

[b] The exact year an NRIS came into existence is sometimes difficult to ascertain. Many started as small experimental projects affiliated with universities or federal agencies. Only a minority can be given a precise date, such as legislative action authorizing the creation of a system.

[c] The current operating status of the systems is extremely diverse, varying from small experimental programs in some states to large, well-established, and highly visible programs providing services to state agencies, industry, state residents, and out-of-state users.

· Reproduced from *Remote Sensing for Resource Management*. Edited by Chris J. Johannsen and James L. Sanders. Soil Conservation Society of America, Ankeny, Iowa.

history. A Senate Committee on Land Use proposed the inventory in 1958, and an endorsement was received by a 1961 Conference on Resources for Tomorrow. It was decided to develop a nationwide land data base for multidisciplinary land-use planning. The federal government began the CLI program in 1963 under the Agricultural and Rural Development Act (ARDA).

The Land Capability Rating System used seven classes to rate land capability for each sector involved, agriculture, forestry, recreation, and wildlife. Lands in Class 1 have the highest and those in Class 7 the lowest capability to support land-use activities of each sector. There are also subclasses to identify specific limiting factors for each class.

Aerial photography and existing information were used to map land areas with uniform characteristics. Field checking was done, and all information known or inferred about an area was interpreted, and then the results were used for the capability ratings. The classification is independent of location, accessibility, ownership, distance from cities or roads, and present use of the land. However, an assumption is made of good management.

More than 1000 CLI land capability maps have been published, at scales of 1:250,000 (in the case of British Columbia agriculture and forestry at 1: 125,000) and 1:1,000,000. Fifteen thousand field maps at a 1:50,000 scale were employed to prepare the CLI maps. Generalized information was recorded in the Canada Geographic Information System (CGIS). This computer system was created to store, analyze, and manipulate map data and related information for regional land-use planning and management. It can combine all maps to provide a nationwide data base for Canada, compare maps of different sectors, retrieve data for specific maps, calculate areas of specific map units, and compare and overlay other data with the CLI.

Specialized CLI maps include generalized land use, sportfish, critical capability areas, and land capability analysis. In the latter case, analysis was done in selected areas in British Columbia and Quebec. The land capability analysis maps are published at a scale of 1:250,000 and identify the most appropriate use in terms of physical capability.

The CLI data have been useful in national or provincial programs of land-use planning; rehabilitation of abandoned farmland; identification of critical areas of wildlife habitat management; appraisal of land value, especially farmland; parks location and evaluation; environmental impact assessment; and identification of flood-prone lands. It is recognized that the CLI data may be used for an initial overview, and are not designed for use in detailed or site planning. The CLI data identify broad areas of either high potential or severe limitations.

It should be noted that the Canada Land Data System (CLDS) is a computer system that handles land resource information. The CLDS can manipulate and produce land data in various forms to facilitate land-use planning and management at national, provincial, regional, and municipal levels. Its main component is the Canada Geographic Information System (CGIS). Types of data available include:

more than 3000 digitized maps. These maps include those of the CLI, which
    detail land capability for forestry, agriculture, recreation, and wildlife—
    waterfowl and ungulates—for one third of Canada;
unpublished CLI land-use maps;
updated land-use data for specific areas;
1971 and 1976 census and administrative boundaries;
watershed boundaries;
pollutant transfer maps;
information on federal land holdings;
ecological land data for various areas.

An index of information available can be obtained from the Lands Directorate,
which administers the system. Hence, the country of Canada is classified as to
landscape capability, and all of it is on computer, a major achievement for such
a vast area and an extremely valuable tool for landscape planners.

Beyond the CLI and CLDS, further Canadian contributions and achievements
in the late 1970s warrant mention, particularly the important work of George
Angus Hills. An article by Peter Jacobs (1979b) pays tribute to Hills's pioneering
efforts in the classification of landscape for its ecological potential. A valuable
selected list of Hills's more important writings from 1937 to 1979 is provided at
the end of the article; the list includes the noted 1960 Glackmeyer Report of
Multiple Land-Use Planning, for the Ontario Department of Lands and Forests,
Toronto, as well as the 1961 Research Report on the Ecological Basis of Land-
Use Planning, done for the same agency.

Additional articles pertaining to regional planning and landscape analysis
approaches used in Canada are described in the special issue of *Landscape
Planning* (6(2), 1979). Included, among others, are consideration of landscape
planning in Canada (Jacobs, 1979a) and a synthesis of mountains, people, and
institutions (Chambers, 1979) in which the author concludes there is a close
linkage of "environmental" problems to a serious dichotomy between the structure of natural systems and the structure of our society). Dorney and Hoffman
(1979) cover the development of landscape planning concepts and management
strategies for an urbanizing agricultural region. They examine historical and cultural land-use trends in looking at the difficulties of developing landscape
management strategies for the urbanizing portion of southern Ontario.

Natural Resource Change Analysis

As with other natural resource inventories geared to particular dates of aerial
photography, the New York State LUNR Inventory is used (among other purposes) for natural resource change analysis. Such analysis is carried out to monitor the land-use change that has taken place in one area over a period of time.
Photographs for two time periods are compared as regards land use on each.
This comparison reveals where and what kinds of changes have occurred, and

their extent. If the earlier inventory materials can be used for the baseline year, one-half of the analysis is already done. In the case of the LUNR Inventory, overlays provide the graphic display and the LUNR computer printout provides the statistical display of the categories. Then, by using the LUNR overlay in conjunction with the newer air photographs, the old data can be updated.

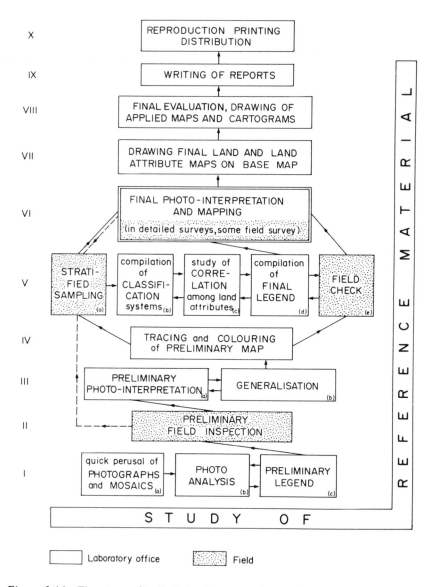

Figure 3-11. The steps of holistic land survey using aerial photographic interpretation. Zonneveld, I. S. 1979. Use of aerial photographs in geography and geomorphology. ITC Textbook in Photo-interpretation, Volume VII, p. 81.

Aerial photographs play a large part in any natural resource inventory. For that reason, they should be chosen carefully, when a choice is possible. The scale of the photos should provide enough detail and be compatible with the base maps. The quality should be assessed carefully. The suitability of the film type should be determined. Other factors to be considered are the size of the area to be inventoried and the number of categories to be interpreted. Naturally, the larger the area and the more categories to be interpreted, the more time the project will take. The season of photography is also important. Engineers and tax assessors like spring coverage. For land use, late summer and early fall coverage is desired.

In the above example, final products include the graphic display for each time period as well as the statistical summaries, and the latter may be organized by township, county, topographic map, or whatever unit is needed. Producing a statistical summary increases the time needed by approximately two thirds. As an example, for a town of approximately 50 square miles, production of mapped materials with 55 LUNR Inventory categories takes approximately three man-days, while the statistical summary takes approximately five man-days.

As already noted, the speed and sophistication with which landscape inventories are carried out have materially increased as a result of satellite and other remotely sensed data. The methods of analysis may be visual interpretation or advanced computer analysis (labor intensive to machine intensive), and many incorporate recently developed low-cost techniques for processing satellite imagery (see also Figure 3-11).

Remote Sensing and a Classification System for Land-Use Planning

In their United States Geological Survey Professional Paper, Anderson et al. (1976) address the framework of a national land-use and land-cover classification system for use with remote sensor data. Their system is designed to meet the needs of United States federal and state agencies for an up-to-date overview of land use and land cover throughout the country "on a basis that is uniform in categorization at the more generalized first and second levels, and furthermore, will be receptive to data from satellite and aircraft remote sensors." The proposed system uses the features of existing widely used classification systems that are amendable to data derived from remote sensing sources. It is left open ended to allow federal, regional, state, and local agencies flexibility to develop more detailed land-use classification at the third and fourth levels. This is done to enable those agencies to meet their particular needs and at the same time to remain compatible with each other and the national system.

For regional and city land-use planners, it is important to have information about the distribution and area of agricultural, recreation, and urban lands, the changes that are occurring, and the proportions of such changes. Such information is, of course, important to governmental officials at various levels, as well as to legislators involved in the determination of land-use policy.

In the past, highly time-consuming field surveys had to be relied on to obtain

Table 3-8. Aggregation of Land Use/Land Cover Types

Level I	Level II	Level III
1. Urban or built-up	11. Residential	111. Single-family units
		112. Multi-family units
		113. Group quarters
		114. Residential hotels
		115. Mobile home parks
		116. Transient lodgings
		117. Other

Reproduced from U.S. Geological Survey Professional Paper 964.

needed information. Now, remote sensing offers a faster and more efficient method.

The outline of Levels I, II, and III of this land-use classification system is reproduced in Table 3-8. Level I could be regarded as uniquely identifiable with 90% accuracy using the Landsat imagery. This level is designed for use with very small-scale imagery. Level II is designed for use with small-scale aerial photographs. Level I information, while gathered by Landsat satellites, could also be interpreted from conventional medium-scale photography or compiled from a ground survey. On the other hand, Landsat data may be used for accurate interpretation of some Level II categories. The authors provide an example of subcategorization of residential land as keyed to the standard land-use code.

In this system, users desiring Level III information would need to acquire much information to supplement that obtained from medium-scale aerial photography. At Level IV, in addition to information obtained from large-scale aerial photography, great amounts of supplemental information would be required.

A land-use and land-cover classification system that can effectively employ orbital and high-altitude remote sensor data should meet the following criteria (Anderson, 1971).

1. The minimum level of interpretation accuracy in the identification of land-use and land-cover categories from remote sensor data should be at least 85%.
2. The accuracy of interpretation for the several categories should be about equal.
3. Repeatable or repetitive results should be obtainable from one interpreter to another and from one time of sensing to another.
4. The classification system should be applicable over extensive areas.
5. The categorization should permit vegetation and other types of land cover to be used as surrogates for activity.
6. The classification system should be suitable for use with remote sensor data obtained at different times of the year.
7. Effective use of subcategories that can be obtained from ground surveys or

from the use of larger-scale or enhanced remote sensor data should be possible.

8. Aggregation of categories must be possible.
9. Comparison with future land-use data should be possible.
10. Multiple uses of land should be recognized when possible.

Remote Sensing and Vegetational Analysis

Investigations involving vegetation and biomes have made use of remote sensing in a number of ways, including assessments of damage from insect infestations and fires; spread of desertification; agricultural crop harvest prediction; wetlands mapping; natural habitat analysis; quantification of urban tree stress. Although such investigations are specific to the parameters noted, they can serve as important parts of analysis that is holistically oriented, ultimately considering the total impression of the landscape (see Tables 3-9, 3-10). In considering survey methods for range management in the third world with special emphasis on remote sensing, Zonneveld (1978) noted that the possibility of extrapolation and intrapolation from terrain to vegetation data and vice versa are the key to the efficiency of the landscape-guided method. Another advantage is that even if vegetation is the most wanted information, data on soil, terrain relief, hydrology, and the like can be gathered with little extra effort. At the same time, the vegetation data are correlated mainly with environmental factors. The landscape method focuses primarily on the statistical aspects of the land, because landform relief and the not too dynamic parts of the vegetation are used in the photointerpretation and classification. Soon, however, it may be possible to combine the advantages of the landscape method with other methods that can directly indicate the change of vegetation in an efficient way: sequential satellite recording. Such systems are not yet operational for third world countries. Only "haphazard orientations have been made that could not be very effective, because of administrative and technical constraints" (Zonneveld, 1978). Landscape configuration, having once been assessed by a landscape-directed range-ecological survey, is considered acceptable as a permanent base. Further, sequential data from remote sensing, combined with field sampling on some permanent plots and on more temporally recorded areas (selected, preferably stratified on the base of remote sensing images) would make it possible to make seasonal and other changes visible by extrapolation.

Detection of Vegetational Disorders

The possibility of detecting vegetational disorders, both pathogenic and non-pathogenic, via remote sensing has spurred the interest of several researchers. The goal is effective previsual detection of vegetational maladies, including stress-induced difficulties caused by water and nutrients. The ability to monitor large-scale impacts of land uses and of air pollution on vegetational systems is an implicit objective in some of this activity.

To be more specific, stress in trees and other plants is defined as any disturbance of the normal growth cycle brought about by any living entity or environmental factor that interferes with the manufacture, translocation, or utilization

Table 3-9. Environmental Parameters of Habitats to be Evaluated

Evaluation by ERTS Sensors
    Grazing conditions
    Snow
    Ocean current
    Cover type
    Floe-ice conditions
    Vegetation type
    Vegetation changes
    Human activity
    Geomorphology
    Fish schools (possibly shallow-schooling nekton)
    Ocean current direction
    Effect of animals on forage
    Characteristics of staging and rest areas
    Water turbidity (qualitatively)
    Flooding
    Drainage of swamps

Evaluation Best Accomplished by Contact Sensors
    Air temperature (near-surface and aloft)
    Humidity
    Water temperature
    Salinity
    Water pressure
    Light intensity (in water)
    Altitude
    Noise level
    Magnetic field
    Light intensity (in air)

Evaluation Needed by Field Party or Field Station
    Wind speed (at surface)
    Snow depth
    Wind azimuth (at surface)
    Current velocity
    DDT level (of water)

From Alexander et al. (1974). Courtesy of Dames and Moore.

Table 3-10. Hierarchy of Landscape Units in Relation to Map Scale

Explanation	Suggested Unit Names	Approx. Map-Scale Range
I. Large size landscape units (approx. 10,000 ha or higher)	Bioclimatic region or zone, Biogeo-climatic zone, Life zone, Vegetation zone, Ecoregion, Biome, Ecological zone	National (small map scale) 1:1 million and smaller in map scale
II. Intermediate size landscape units (approx. 500 to 10,000 ha)	Land system, Land form, Forest cover type, Plant formation, Forest land type, Forest ecosystem type.	Regional-overview (medium map scale) 1:1 million to 1:100,000 (or occasion-ally up to 1:50,000)
III. Small-sized landscape units (approx. 1 to 500 ha)	Forest site type, Forest habitat type, Land type	Subregional or Local-detailed (large map scale) 1:100,000 (or 1:50,000 up to 1:1,000)

From Assessing Tropical Forest Lands: Their Suitability for Sustainable Purposes, edited by Richard A. Carpenter. Tycooly International Publishing Ltd, Dublin, 1981.

of food, mineral nutrients, and water in such a way that the affected tree changes in appearance (Agrios, 1969). In urban situations many factors can cause tree stress, including soil compaction, salts used for deicing, mechanical injury, air pollution, insect pests, plant disease agents, temperature extremes, and drought.

Quantification of Urban Tree Stress

Remote sensing is being used in urban environments for a wide range of purposes in vegetation surveys and vegetation management, often in close relation to land-use planning and design, and the detection of urban tree stress has recently been considered by researchers at the College of Environmental Science and Forestry at Syracuse, New York (Lillesand et al., 1978). Microdensitometric analysis of aerial photography (measurement of the density of small film areas with an instrument called a microdensitometer) was involved and included four major phases, which illustrate the sequence followed in such work: data acquisition, data reduction, stress prediction model development, and stress prediction model testing (Figure 3-12).

In the first of these stages, color and color infrared aerial photographs were taken over representative test areas at the same time that field crews collected ground data. Ground data, which consisted of various indicators of stem, crown, foliage, and site conditions of individual trees, were collected for some 1400 trees appearing in the photographs.

Next, based on the ground data, various "stress indexes" were formulated. These numerically express the stress conditon of each tree sampled in the study. The indexes stemmed from analysis of the ground data set, irrespective of the photographic imagery. Three types of stress indexes resulted: "trunk and limb" index, "foliage" index, and "crown and branch" index. In addition, optical density of the image of each tree crown observed in the field was measured from the photography by using a spot microdensitometer.

Following this, the ground and aerial data were related to develop a series of stress prediction models which could be used to predict the stress indexes for trees appearing in the photography that were not analyzed on the ground.

Finally, the stress prediction models were tested by applying them to a random sample of trees within the test area. They were also tested by repeating the 1975 ground and aerial data acquisition effort during the 1976 growing season. To determine the applicability of the models in other urban environments, they were also tested on a preliminary basis in Rochester, New York, during 1976.

General conclusions derived from the study (which refer specifically to the limited data and conditions of their experiment) are as follows.

1. Broadband densitometric measurements extracted from conventionally acquired photography and analyzed in accordance with multivariate statistical

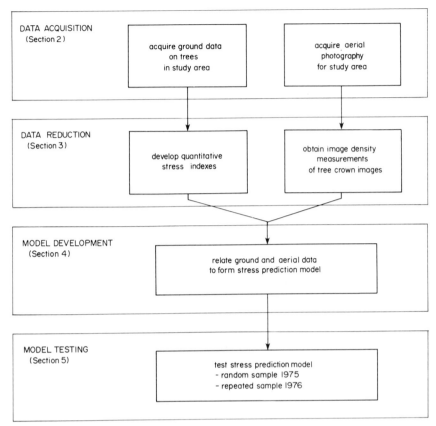

Figure 3-12. Major elements of tree stress study. Reprinted with permission of authors.

methodologies can be used to detect and quantify urban tree stress. Color infrared is superior to normal color film for this purpose.
2. Under many conditions, aerial techniques appear to be more reliable and efficient than ground survey techniques. A combined correlation of stress symptoms and photographic data shows the photographic data to be as good a predictor of tree stress as ground observations. The photographic technique provides an objective and efficient prediction of tree stress. Ground survey procedures are subjective and tedious.
3. The timing of aerial photography with respect to rainfall and the seasonal development of tree stress symptoms has substantial bearing on the ability of aerial data to predict stress levels. Photographic data acquired after two weeks of dry weather correlated most strongly with stress conditions. This enabled previsual detection of stress. Shorter periods of dry weather led to reduced correlations between photographic and ground data, regardless of the date of photography.

To be sure, the conclusions represent a working hypothesis, and the researchers caution against drawing inferences for conditions differing from those of their study. Nevertheless, the approach used deserves further exploration and application. The successful use of aerial photographic methods for detecting and quantifying urban tree stress can materially assist planners, designers, foresters, and managers of landscape resources.

Aerial Photographic Detection of a Specific Plant Disease

The ability to detect particular tree diseases, rather than more generalized stress symptoms, has been given attention. One group of researchers investigating detection of Dutch elm disease has expressed reservations about the efficacy of the tool for this purpose. Landscape-ecological surveys that include specific disease (or malady) identification as one of the objectives in a holistic approach will want to consider these limitations of remote sensing capabilities.

Hammerschlag and Sopstyle (1975), then respectively the chief of the Ecological Services Laboratory, United States National Capital Parks (NCP), and a photographic technologist at the Chesapeake Bay Ecological Program office, report on the use of several forms of aerial photography in pursuit of a technique that could provide early detection of Dutch elm disease. They cautioned, in 1975, that although "remote sensing, particularly photographic, has long been touted as an advantageous tool for detection of plant disease or stress, . . . despite a long history of almost 50 years of intrigue, investigations, and technological development, the degree of success stories remains small and examples of practical utilization even less numerous." The National Parks Service in Washington, D.C. (National Capital Parks) is the custodian for the major visitor areas of the nation's capital. "Since elm trees have been the shade trees of choice over the years, the National Capital Parks service finds itself operating a very intense, multifaceted Dutch elm disease control program. The NCP, in conjunction with NASA, Wallops Flight Center, initiated its own investigations of remote sensing systems for early detection of plant stress which could be used in particular to augment NCP's Dutch elm sanitation program. These investigations were limited to photographic techniques at this time because of the need to resolve differences within or between tree crowns. An early plant stress detection system could prove useful in solving myriad other problems for the National Park Service" (Hammerschlag and Sopstyle, 1975).

They reveal that the two most promising techniques tested were multispectral photography, with object enhancement and biband ratioing, coupled with scanning microdensitometry. [Readers will find technical coverage of these techniques in Lillesand and Kiefer (1979).] Hammerschlag and Sopstyle note that for practical purposes, the multispectral system has the advantage of providing a readily interpretable image in a relatively short time. In addition, they emphasize that color infrared film would be optimal for a "long-term detection of loss of plant vigor, which results in a physical change in a plant canopy, but should find minimal practicality for early detection of specific sources of plant stress such as Dutch elm disease." The specific intent and focus of vegetational analysis as part of larger landscape-ecological analysis will determine the selection of appropriate techniques. Persons involved should be guided in part by the observations made in Washington, D.C.

Documentation of the Impact of Breakwaters and Other Bodies on the Structure of the Coast

Detached breakwaters, groins, and other artificial bodies are known to have a decided impact on the structure of coastal areas. In various parts of the world, recreation planners (and others concerned with the dynamic character of coastal locations where such artificial elements are employed) often need visual documentation of changes that occur. Such documentation results from analysis of change that has taken place in a particular area over a period of time (as was discussed earlier in this chapter with regard to land-use change). Aerial photographs from two or more time periods make comparison possible, and thus become an important tool in detecting alterations.

The Mediterranean coastal area of Israel has been monitored for such changes (Nir, 1976). Aerial photographs were extensively used in the analysis of the impact of the artificial bodies. The study effectively illustrates the value of aerial photographs in recording the influence over various periods of time in several coastal places from north to south in Israel. The photographs were used in order to detect erosion, shifting of sands, destruction of kurkar cliffs ("kurkar" is a local name for carbonate-cemented quartzitic sandstone) found in close proximity to the sea in the coastal plain, and other phenomena; and subsequently maps were developed portraying the changes along the coast.

Such artificial bodies as detached breakwaters and groins at different distances from the coast, were constructed on the Mediterranean coast of Israel beginning in 1960, and 15 breakwaters were in place from Nahariya to Bat Yam (see map, Figure 3-13). The first declared objective of emplacing such breakwaters was to create marine structures to widen the beach and protect swimmers against the waves during the summer season. During the months in which swimming occurs, the average wave height is only between 80 and 150 cm.

The changes that occur in a location, and especially those near the breakwaters themselves, were studied at different sites, assisted by sporadic mapping and especially by aerial photography from previous periods. Photographs taken before, during, and after erection of the structures provided a comparative basis for a study of change. This study revealed that during and following construction much sand accumulated between the breakwater and the shore in bodies known as tombolos. As a result, the coastal landscapes were harmed in a most discernible way. A removal of sand occurred and the width was narrowed. Within a few years' time after this, improvement occurred, and the nearby coastal areas returned to more or less their original condition (i.e., their condition before construction took place). The filling-in of this sand is done at the expense of more distant coastal areas. In one way or another, the temporary deficiency is divided somewhere along the greater length of the coastline. Thus an "averaging out of the quantity" takes place on the general coastline.

During the intermediate period, between the beginning of construction of the artificial structure until the time when partial or full restoration of the coast of

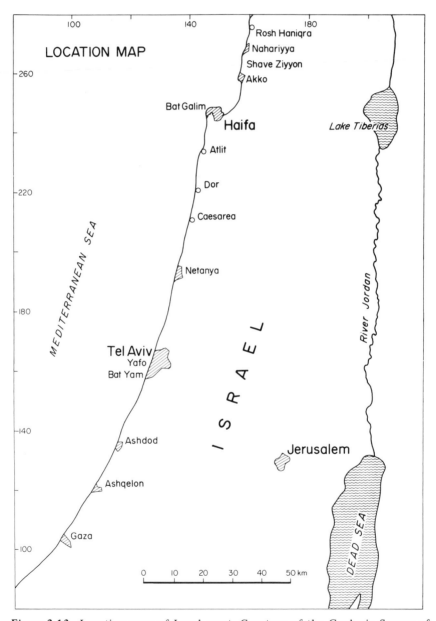

Figure 3-13. Location map of Israel coast. Courtesy of the Geologic Survey of Israel.

the area occurs, all the coasts suffer. (The coast that suffers destruction is the one found "downstream" from the main sand transport—beyond the structure.) The destruction and removal in such areas are very accelerated. Indirectly, the kurkar cliffs, since they are composed of soft rock, also suffer, and are damaged at an even more accelerated rate.

The sandy areas that accumulated in the tombolo created in the "shadow" of the different breakwater bodies or in the anchorage areas built for the same purpose reached 190,000 m^2. In the same areas, sand accumulation reached 270,000 m^3, much greater than that of the annual transport of sands along the coast of Israel from the Nile Delta via the north Sinai beaches and shallow shelf. This accumulation occurred in seven separate areas from Bat Yam to Achziv. From the standpoint of focal points for seaside swimming, this represents a recognizable and possibly positive increase. However, the researchers note that since the tombolo reaches 50% of its final area in the first year this quick accumulation severely impacts nearby coastal areas.

In the second year after construction takes place, a decided slowing down occurs in the process of tombolo growth, and three or four years are required before a maturation stage is reached. This condition bespeaks very small and insignificant changes in the area, except for the movement of the coastline as a result of winter destruction and summer deposition.

It is evident that aerial photography (see Figure 3-13-3-18) was utilized in a major way in this study. After their analysis the authors concluded that despite improvement at this or that coastal spot, other more flexible methods (from among those now existing) should be sought to provide tranquil waters for swimmers. More detailed study of the areas' existing structure is proposed in order to obtain a clearer and more quantitative picture of the impact.

Utilizing Aerial Views in Holistic Preservation Planning, Decision Making, and Public Education

The New York State Council on the Arts has been actively involved for over a decade in a program of architectural preservation. The Council was instrumental in producing a series of books concerned with architecture worth saving in several parts of the state. An exceptional publication (Jacobs, 1979) departs somewhat from the usual architectural emphasis. This document is, in effect, geared to the would-be environmental taxonomist, one showing how architectural data, whether in the form of photographs, drawings, or structural and historical information, might be treated within a taxonomic classification developed for surveying visible resources or a rural area. It is, in a real way, a catalog for the environment.

The late Stephen Jacobs, renowned architectural preservationist at Cornell University's College of Architecture, Art and Planning, Ithaca, New York, was long involved in a theoretical approach to the study of surroundings with an eye to creating a model for eventual group action. Numerous practical demonstrations of his approach and its effectiveness can be found, among them several

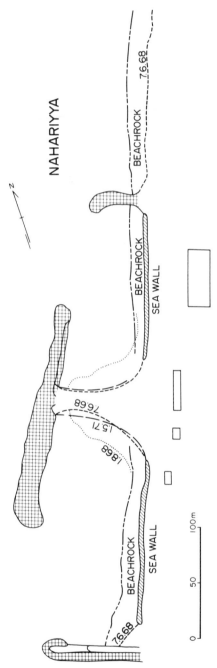

Figure 3-14. Map of tombolo area, Nahariya Beach. Courtesy of the Geological Survey of Israel.

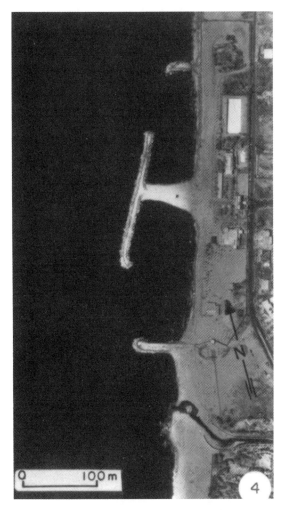

Figure 3-15. Breakwater of Nahariya. Photo (4) Aerial photograph of break-water and its surroundings: clearly seen are two groins that close the break-(water) from the north and from the south. Also seen is the inclination of the break to the coast. Photographed 1/5/71.

locations in downtown Ithaca, New York. The methodology used by Jacobs will be given detailed attention later on in this chapter. Of special interest here is the utilization of aerial views in holistic preservation planning, decision making, and public education.

The cataloging of rural Wayne County's environment emphasized the survey-ing of readily observable features of the physical environment using a system that is easily stored and evaluated. Although the great majority of the classified items are architectural, the balance could be adjusted in favor of other ob-servable features of the Wayne County landscape. The work involves a

Figure 3-15 (cont.). Breakwater of Nahariya. Photo (5) Aerial photograph of breakwater and its surroundings: clearly seen is the shallow section in the vicinity of the break (light in the picture), and several stony areas (dark in the picture). Photographed 11/5/71.

comprehensive, integrating approach, as will be noted in the coverage under methodologies. It is built upon recognition of the interrelationship of a physical setting and the record of human occupancy superimposed on it. The geological conformations, for example, which determine soil and drainage patterns and supply mineral resources, strongly relate to the character of particular human settlements. As Jacobs (1979) notes, "Rich soils make possible fine farm buildings, prosperous rural communities, and lively market towns. Poor soils produce relatively little of architectural or cultural interest. Quarries and mines exploit underground resources, while cut stone or cobblestone structures suggest the influence of local materials on construction and design." The basic classifications employed constitute a framework for the documentation and understanding of both activities and material accomplishments of the area, ranging from the period of the first settlers to that of the current inhabitants. Thus, the result is

Figure 3.15 (cont.). Breakwater of Nahariya. Photo (6) Aerial photo of break-water and its surroundings. Near the southern groin is seen an accumulation of sand, that apparently shows the direction of transport toward the north. This phenomenon is also seen in other aerial photographs of this area. Courtesy of the Geographical Survey of Israel.

a striking treatment of natural patterns, social development, and economic development.

Photographs and maps are extensively used, and aerial photographic views are essential to the effective cataloging. For example, aerial views illustrate the unique Chimney Bluffs area along Lake Ontario and the undulating shoreline, and accompany presentation of the underlying natural resources (Figures 3-19–3-21). The long, rounded drumlin hills characteristic of parts of Wayne County are shown in Figure 3-22. An air view of the Clyde River is used in the "eco-logical" section to suggest the character of the area before settlers cleared farms in the thickly wooded county (Figure 3-23). In dealing with social development, some settlements at various historical periods are represented by paintings from earlier periods and updated with recent aerial photography (planned community,

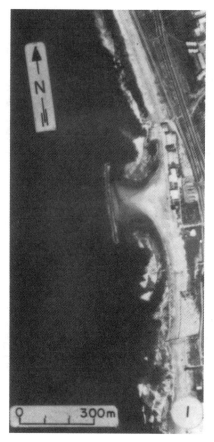

Figure 3-16. Breakwater of Carmel Beach. Photo (1) Aerial photograph show-ing the breakwater and its surroundings immediately after its establishment. The tombolo is still at the beginning of its development. Photographed 10/2/69.

Figure 3-24), or with maps and aerial photographs, as in the case of a highway junction village (Figures 3-25 and 3-26). Thus, the analysis and presentation of the landscape ecology of a rural area is enhanced in a major way by the use of high-quality aerial views. The impact on the planning and decision-making, as well as public education, is quite sharpened by the clear identification afforded by such photographs of Wayne County from the air.

User Needs in Physical Planning; in Possible Utilization of Remote Sensing

In determining the place of remote sensing in landscape-ecological surveys, canvassing of the physical planners, agencies, and decision makers responsible for regional efforts and interacting with them in order to ascertain their data requirements are essential. Identification of just how and where remote sensing in its several forms may practically serve these users can result from direct

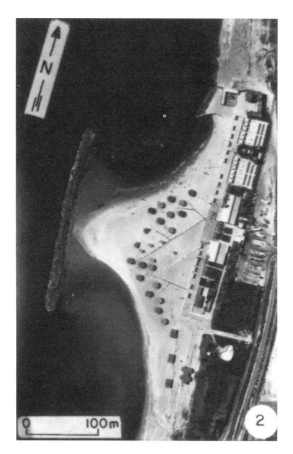

Figure 3-16 (cont.). Breakwater of Carmel Beach. Photo (2) Aerial photo of the breakwater and its surroundings. The tombolo is very developed, but is not at the peak of its growth. Photographed 11/5/71.

surveying of such individuals and offices. One of the authors led such an investigation in Israel to determine the need, potential, and feasibility for utilization of remote sensing in land-use planning and natural resources management (Lieberman, 1978). Particular attention was paid to how remote sensing might be used in a yet-to-be-developed comprehensive land-use and natural resource information system for physical planning needs in Israel. An attempt was made to relate the investigation to other studies then being conducted, at the Technion-Israel Institute of Technology's Center for Urban and Regional Studies and at the Israel Ministry of Agriculture Branch for Soil Conservation's Survey and Mapping Division, on inventories, information systems, and methodologies for use in land-use planning in the country.

The year-long investigation was conducted through personal conferences with researchers and academicians; government and private professionals engaged in land-use planning, "environmental planning," and resources management; officials in several government bureaus and agencies; and administrators and re-

Figure 3-16 (cont.). Breakwater of Carmel Beach. Photo (3) The tombolo at its peak of development and covering about 70% of the protected area protected in an original way by the breakwater. There is a certain asymmetry with the northern side somewhat larger than the southern, this by direct dependence on a small groin found on the boundary of the northern area on the one hand and in the direction of the movement of sands on the other hand. Photographed 8/25/73. Courtesy of the Geological Survey of Israel.

Figure 3-17. Carmel Beach. Photo (4) A small groin that was built on the bathing beach open from the south to the structure of the Carmel coast. This groin was established after the coast narrowed and most of its sand was carried both to the tombolo at the time of southwest storms, and also to the south. On the beach are seen particles of lime and chalk, that are characteristic of this beach and/or other beaches along the Carmel coastal plain. Courtesy of the Geological Survey of Israel.

searchers in the government primarily concerned with surveying and mapping land use and natural resources. Also met with were researchers involved in studies of the applications of remote sensing for civilian purposes in Israel and elsewhere. A colloquium on the potential for utilization of remote sensing resources in land-use planning was organized and implemented, to which individuals from the above disciplines and locations were invited. At the colloquium, participants were asked to react to the specific pertinence of remote sensing to perceived needs in their areas of professional focus. There was a questionnaire survey of colloquium participants in which questions were raised on the data requirements and information needs of these individuals and offices responsible for regional efforts in Israel. There was a review of publications and documents emanating from government agencies, private planning firms, research units, and institutions (to gain a sense of how remote sensing was being viewed and/or utilized in the context of the responsibilities of that entity and its roles and responsibilities).

The study of user needs involved conducting:

1. an inventory assessment of the current state of remote sensing data, its location availability, and application for civilian planning purposes, as well as of anticipated expansion and intensification of remote sensing efforts.
2. A survey of the current land-use and natural resource information needs of planners and decision makers and the forms in which the information is needed. The year-long investigation heavily focused on these two areas. In addition, consideration was given to the question of costs involved in utilizing remote sensing as part of a land-use and natural resources information system.

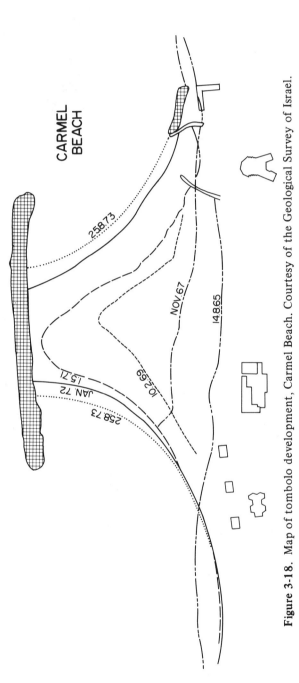

Figure 3-18. Map of tombolo development, Carmel Beach. Courtesy of the Geological Survey of Israel.

Figure 3-19. Chimney Bluffs. Rising 175 feet above the lake level. Chimney Bluffs has been set aside for use as the county's first state park. Glacial deposits of mixed clay, sand, gravel, and stones, the bluffs are attacked by storm waves that erode them into pinnacles and prows. Photo by H. Lyon.

The survey of information needs and forms utilized questions dealing with subject matter categories, scale needs, forms of data, area of concentration, type of data, and the nature of remote sensing involvements of the office concerned. The questions raised may serve as a model set for canvassing purposes in other countries and regions (with necessary modifications for local conditions, of course). They are presented here, along with selected responses.

Five questionnaire responses have been selected for this chapter, which reflect professional interests and needs of: (a) two geologists; (b) landscape architects; (c) a geographer; (d) a regional planner working for a government institution involved in settlement study patterns; and (e) the Mapping and Surveying Division of the Ministry of Agriculture. In cases where no reply was made, the letter has been left out.

Many other highly detailed responses would be required from a diverse group of users in order to develop a substantive portrayal of how the question of information needs is viewed. Nonetheless, the few selected responses provided here from the survey in Israel represent a range of disciplines and orientations and thus provide a useful index to professional desires.

1. Do you conduct regional studies (50–250 km)? If so, what is a typical area covered (approximately to the nearest square km)?

Figure 3-20. Chimney Bluffs and East Bay, Huron. The eastern section of the county's shoreline is characterized by high bluffs and deep bays with barrier sandbars. Chimney Bluffs, in the foreground, is the county's most spectacular natural feature. East Bay, one of a series of broad estuaries formed by streams flowing north into the lake, seen in the middle distance, is the terminus for Mudge Creek. Photo by H. Lyon.

Responses to Question 1:
(a) We are engaged in geotechnical mapping. 50 km² or less would be appropriate for urban geotechnical mapping.
(b) 50 km²–500 km²
(c) 1400 km²
(d) i. "Rurban" (rural urban): Macroregion from Netanya to Ashdod
    ii. Rural development: Yamit–Beersheba (1250 km²)
(e) 100–150 km², according to the size of the regional council or region of interest (see Figure 3-27).

2. What is the size of the smallest datum that you would use (i.e., for vegetation it may be approximately 1/4 acre and less)?
   Responses to Question 2:
   (a) 1 m
   (b) 2 dunams (1 dunam = 1/4 acre)
   (c) 4 dunams
   (d) one moshav farmstead—about 10 dunams [A moshav is a cooperative

Figure 3-21. Ontario Lake Shore—west, Sodus. The undulating shoreline of the western portion of Wayne County links a series of low headlands. Shallow coves and steep shores provide inadequate refuge for boats caught in lake storms. Rich deposits of silt form the lower-lying areas, which are largely devoted to apple-growing. Large fruit farms have been developed here, and have profited. Orchards in different stages of development are interspersed with wooded uplands and stream beds. When the leaves change color in the autumn, a quilt-like pattern of textures and colors emerges. Photo by Plant Pathology Department, Cornell University.

       farming village, with individually farmed plots, individual residences, and some joint marketing and machinery-purchase arrangements.]

  (e) according to survey scale

      1:5,000   5 dunams

      1:10,000   10 dunams

      1:50,000   50 dunams

3. In what forms do you want the initial and subsequent data (topography maps, computer tapes, acetate overlays, etc.)?

    Responses to Question 3:

  (a) generally topography maps will do

  (b) topography maps; air photographs—aerial overlays

  (c) "not clear"

  (d) topography maps; computer tapes, acetate overlays, field description data, vegetation transects, all kinds of field measuring stations data, soil laboratory analysis

4. What is your area of concentration (e.g., regional recreation planning, water quality analysis, etc.)?

Responses to Question 4:

(a) Geotechnical mapping; sedimentary processes

(b) Impact analysis, regional planning, recreation planning, urban open-space planning, visual analysis

(c) Converting agricultural land into built-up areas

Figure 3-22. Drumlins. Long, rounded hills known as drumlins dot the land-scape of Wayne County, emerging in clusters and groups to define shallow valleys or surround marshy pockets. Most of the bluffs on the Lake Ontario shoreline of the county are drumlins. Best-known are the groupings in Palmyra and Savannah, which have been extensively studied by geologists. As much as a hundred feet high and covering an area of two square miles, the larger drumlins provide an undulating horizon or a textured background for a more intimate scene in most parts of the county. The gentler south and east slopes are usually cultivated. The more exposed western slopes, the steep northern ones, and the crests are often left wooded. Where animals have cleared out the underbrush by grazing, the tree trunks are often silhouetted against the light as a narrow band following the slope and are defined by the mass of foliage or branches above. Except for an occasional "rocdrumlin" formed around a stone outcropping and longer than the others, drumlins contain a mixture of clay, pebbles, and boulders. They usually are built up of four concentric beds, of which the lowest is gray-blue in color. Drumlins were formed by the southward movement of glaciers and were left behind when the ice sheet melted. Photo by H. Lyon.

(d) Urban expansion—processes and control (rural–agricultural conservation); land settlement

(e) Soil evaluation and potential, crop suitability, agroclimatology, drainage, pasture vegetation, urban expansion

5. Do you use consultants and/or regional maps of other disciplines in your analysis? If so, what disciplines?

Responses to Question 5:

(a) topographical, hydrographical, soil (map form)

(b) soil maps, geological maps, hydrological maps, vegetation maps, consultants; ecologists, psychologists, and others

(c) cadastral maps

(d) Israeli Survey Department maps, Public Works Department (Ma'atz) plans, Ministry of Agriculture aerial photographs

(e) special consultants for each discipline

6. Do you use any of the following natural resource data? If not, please state the data that you use. (Mark where appropriate.)

topography slope

topography orientation

vegetation type

vegetation edges (ecotone)

Figure 3-23. Air view of section of Clyde River. Suggests the character of the county before settlers came to clear farms in the thick woods and drain many of the low lands. Photo by H. Lyon.

Figure 3-24. Southern view of Clyde. Clyde Village was "a place of much business" with about 130 dwellings when this sketch was made from the south bank of the river. It was published in 1846, along with views of Lyons and Palmyra, in Barber and Howe's "Historical Collections of the State of New York." In the foreground is the old covered bridge. It probably replaced the first span across the Clyde River, a bridge at Sodus Street, constructed in 1810. The tall steeple belonged to the Presbyterian Church. In the center is the Methodist Church, with the steeple of the Baptist Church on the right. Today, none of this remains.

water (if so, state type, e.g., fish ponds)
wildlife type
wildlife quality
unique resources
geology—surface
geology—subsurface
soils
other (specify . . .)
Responses to Question 6:

(a) topography slope
     topography orientation
     water [failed to state type]
     geology—surface
     geology—subsurface
     soils

(b) topography slope
     topography orientation
     vegetation type
     vegetation edges (ecotone)
     water: wells, springs, wadis, fish ponds sea (surface-)
     wildlife type
     wildlife quality
     wildlife habitat
     unique resources
     geology—surface
     soils
     other: groundwater, single plant specimen, scenic data

(c) topography orientation
     vegetation type
     (?) vegetation edges (ecotone) [marked with question by responder]
     geology—surface
     soils

(d) topography slope
     topography orientation
     vegetation type
     water
     unique resources
     soils
     other: archeological sites

By 1866 the wooden bridge was "broken down", and replaced by an iron structure. The early churches of the 1820's and 1830's, in style transitional from late
Georgian to classic revival, were replaced by masonry structures with stained
glass windows and medievalizing detail in the post-Civil War era. Below is
an aerial view of the Village of Clyde, and its main north-south artery (New York
State Route 414) today. Sketch courtesy of Wayne County Historical Society.
Photo by H. Lyon.

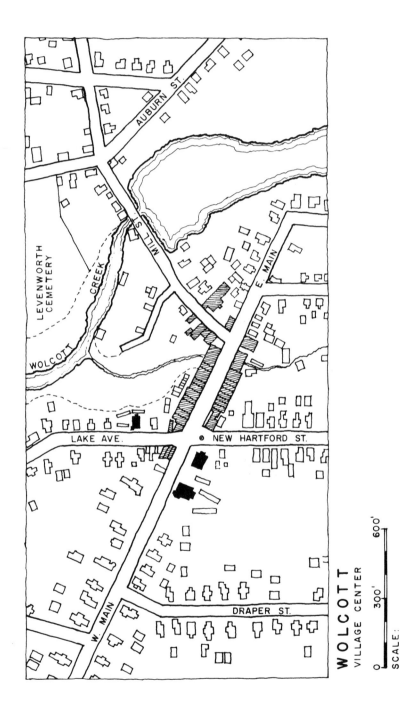

WOLCOTT
VILLAGE CENTER

SCALE: 0   300'   600'

(e) topography slope
topography orientation
vegetation type
water: lakes, ponds, wadis, rivers, drainage canals
geology—surface
geology—subsurface
soils
other: agrotopoclimatology

7. What existing cultural conditions are most important to your needs?
present ownership
distance from present development
present use
possible future use
existing legislation and financing
project demand
cost of land
present property taxation
other
Responses to Question 7:
(a) generally none
(b) present ownership
distance from present development
present use
possible future use [underscored by responder]
existing legislation and financing
project demand
cost of land
(c) distance from present development
present use
possible future use
existing legislation and financing
cost of land
present property taxation

Figure 3-25. Map of Wolcott Village Center. Wolcott Village is the trading center for a large farming area. Five miles south of Lake Ontario, the village is located at the junction of the towns of Wolcott, Huron, Butler and Rose, all of which were originally part of the town of Wolcott. The other incorporated village in the northeast portion of Wayne County is the smaller Red Creek, on its eastern border. A spiderweb of radial roads extends out from Wolcott Village in all directions, providing access for the residents of a picturesque but sparsely settled countryside. Today agricultural activities still predominate. In the northeast corner of the village large warehouses have been constructed along the railroad line. But the automobile has pre-empted the space behind the Main Street stores, and a service station is set between the nineteenth-century houses on Main Street west of Four Corners. There the village's pride, a colorful sculptural group, dominates the scene. Map courtesy of Wayne County Historical Society.

(d) present ownership
     distance from present development
     present use
     possible future use
     existing legislation and financing
     project demand

Figure 3-26. Aerial view of Wolcott Village. Wolcott Village is the principal center of the northeastern portion of the county. From the east, the Ridge Road (New York State Highway 104) enters the village on Mill Street, crossing the dam which backs up Wolcott Creek to form a long narrow millpond to the south. Just beyond a brick foundry building of 1877 (now the property of the Rochester Gas and Electric Co.) it angles to the northwest, skirting the pothole and gorge below the mills. Here is the block-long commercial heart of the village, on Main Street. The old business district terminates at the green fronting on New Hartford Street, known as Whiskey Hill Road (County road 133), south of the village. Houses cluster along this, one of the few radial roads laid out to run straight to the village limits. The cemetery is on the east of the wooded valley, occupying the bluff formed by the bend in the stream north of the mill. On the west, Main Street crosses the New York Central Railroad tracks and continues northwest as the Lummisville Road to Bonnicastle. It is out of the picture before the Ridge Road branches off, aiming due west for the Sodus Bay Bridge. A warehouse area extends along the tracks either side of the intersection, with the irregular tower of the Agway building to the north. Coming from North Rose the railroad moves northeast, in the direction of the site of Furnaceville, before swinging east toward Red Creek. 1976 Notation: A new Route 104 has been built, bypassing Wolcott Village. Photo by H. Lyon.

(e) present ownership
      present use
      possible future use
      other: urban expansion; rainwater storage
8. Generally, there are several elements considered important as guides for the
   spatial allocation of activities. The factors include type of activity, distance

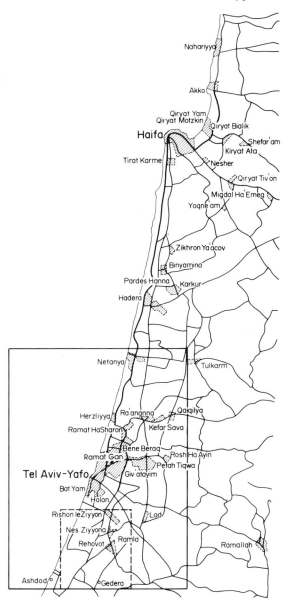

Figure 3-27. Map of macro-urban region from Netanya-Ashdod; micro-region
from Rishon Le Zion-Gedera, Israel.

from other activities and settlements, availability and diversity. Are there other factors that you consider important?

Responses to Question 8:

(a) generally of no interest
(b) those helping interpretation and prediction of mutual existing or expected influences of activities (impact hazards)
local climate
screening effects
(c) institutional barriers (governmental demand and supply)
institutional organization
types of settlements
accessibility
(d) none listed by responder

9. What land use or natural resource data not presently obtainable would you like to see available?

Responses to Question 9:

(a) cooperative information about proposed development plans
climatic maps
(b) environmental damages and hazards

10. What are your needs for remote sensing inputs?

Responses to Question 10:

(a) base for the interpretation of data, diseases within vegetation types as classified by agricultural and forest sciences
(b) interpretation of water quality, turbidity, shoreline movements
(c) repetitive coverage
detection of changes
(d) a 4- to 8-channel scanner set
a remote sensing control center

11. What have you (your office) done in remote sensing?

Reponses to Question 11:

(a) Mediterranean currents of Earth Resources Technical Satellite imagery [reply made by responder interested in sedimentary processes]
(b) aerial photographic interpretation only
(c) nothing directly. Planning and research data were obtained from other governmental bodies.
(d) intensive use of black and white aerial photography; intensifying the use of color infrared aerial photography

12. Over the next 10 years, in what ways do you see remote sensing assisting in your work?

Responses to Question 12:

(a) indication for site selection in recreation planning, for power plants, agricultural activities
(b) depends on Landsat images
(c) observing changes in land use over time (urban and agricultural)
integration of local and regional master plans
supporting data for opening of new lands (rural settlement planning)

(d) conventional use of low-altitude scanners and satellites
     information with high-resolution pattern recognition in aerial images

A Concluding Note on Remote Sensing

In our consideration of remote sensing in its several forms as a tool for holistic landscape evaluation we have shown it to have a special place in surveying land use and land cover, as well as specific focuses (such as vegetation stress, plant disorders) that constitute parts of such landscape evaluation.

As Allan (1980, p. 36) points out,

> the use of aerial imagery leads naturally to an ecological approach to the recording and analysis of environmental and socio-economic factors. Aerial images are crowded with a mass of indicators of natural and man-made phenomena; there is no escape from the complexity and no avoiding the clear relationships between geomorphology, soil, water, vegetation and land use. It is natural, therefore, that integrated studies and land evaluation should have grown out of the relationships first identified between environmental scientists and later environmental and social scientists, all of whom have made use of the same data source, the aerial image, when seeking information as a basis for establishing options in resource management.

As new techniques are developed and made operational, the potentials of this tool should be carefully weighed for particular needs, and valuable use made of it in holistic landscape evaluation.

The Untermain Sensitivity Model:
An Aid for Holistic Land-Use Planning

The sensitivity model of F. Vester and A. von Hesler (1980) has already been identified as one whose orientation is based on the laws and regularity of biological systems. In its focus on the biocybernetic system, it provides a basis for holistic land planning and represents a truly fresh approach. The model stressed each step in the construction of an instrument for planning, dealt with real problems of planning, and was developed with existing communities and actual data in the Untermain region of Germany, east of the City of Frankfurt.

The Untermain model stresses the interrelation between variables, not the variables themselves, and it is concluded that a few relevant data may be employed to understand highly complex systems. In the model, it is sufficient to have macroscopic information at hand in order to arrive at criteria for judging viability, system stability, and biocybernetic basic roles.

We have already noted that the model is an important tool for holistic land-use planning and has importance for landscape ecology as a total human ecosystems (THE) science. It is an excellent teaching instrument (didactic methodological instruction) for systematically confronting complex systems through programmed education. It is its potential application to regional land-

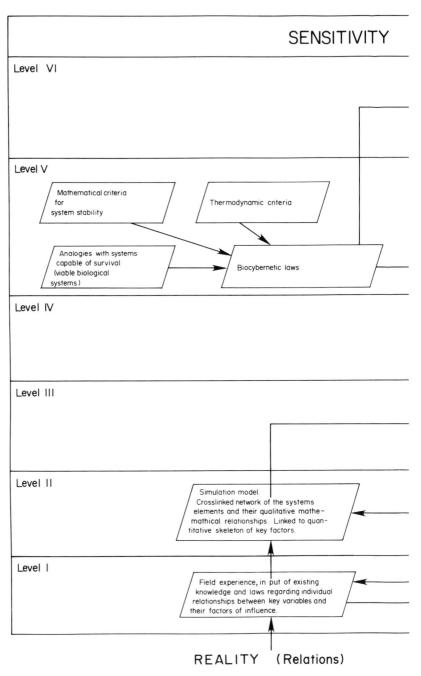

Figure 3-28. Diagram of the sensitivity model. From Sensitivitätsmodell,

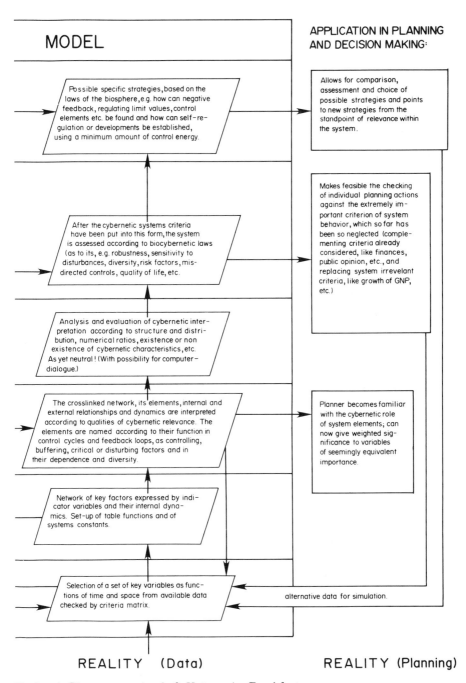

MODEL

APPLICATION IN PLANNING AND DECISION MAKING:

Possible specific strategies, based on the laws of the biosphere, e.g. how can negative feedback, regulating limit values, control elements etc. be found and how can self-regulation or developments be established, using a minimum amount of control energy.

Allows for comparison, assessment and choice of possible strategies and points to new strategies from the standpoint of relevance within the system.

After the cybernetic systems criteria have been put into this form, the system is assessed according to biocybernetic laws (as to its, e.g. robustness, sensitivity to disturbances, diversity, risk factors, misdirected controls, quality of life, etc.

Makes feasible the checking of individual planning actions against the extremely important criterion of system behavior, which so far has been so neglected (complementing criteria already considered, like finances, public opinion, etc., and replacing system irrevelant criteria, like growth of GNP, etc.)

Analysis and evaluation of cybernetic interpretation according to structure and distribution, numerical ratios, existence or non existence of cybernetic characteristics, etc. As yet neutral! (With possibility for computer-dialogue.)

The crosslinked network, its elements, internal and external relationships and dynamics are interpreted according to qualities of cybernetic relevance. The elements are named according to their function in control cycles and feedback loops, as controlling, buffering, critical or disturbing factors and in their dependence and diversity.

Planner becomes familiar with the cybernetic role of system elements; can now give weighted significance to variables of seemingly equivalent importance.

Network of key factors expressed by indicator variables and their internal dynamics. Set-up of table functions and of systems constants.

Selection of a set of key variables as functions of time and space from available data checked by criteria matrix.

alternative data for simulation.

REALITY (Data)

REALITY (Planning)

Regionale Planungsgemeinschaft Untermain, Frankfurt.

use planning that we will concern ourselves with here, looking at the simulation, interpretation, evaluation, and strategy stages.

The sensitivity model can be divided into six levels (Figure 3-28) and through that division a sense of the model's ultimate goal of optimal synthesis of viability of human ecological systems becomes clearer. It employs a system-relevant variable sequence, includes special interpretation models, and is of course based on feedback behavior.

Simulation Model

In the simulation model, the interactions of economy, communal life, population, human ecology, natural balance, infrastructure, and land use, previously identified, are at this point converted into an aggregated model. The variables are labeled with their roles and placed in a form that will allow the input of quantitative values. This also occurs with the interdependencies, previously shown as arrows, for which is required a more detailed mathematical description of their nature and dynamics. Quantification is done on the basis of data available, and these are then used. Examples include allowing for the real area ratios, limit values for the number of commuters, ecological regulations or tax burdens, and the threshold values for attractiveness resulting from availability of recreation facilities or accessibility. Quantitative probes thus link the model with reality.

Objectives in the building of the model are to create one that is satisfactorily aggregated and comprehensible, that will facilitate overall coverage and an overall cybernetic interpretation. This will occur by means of suitable programs and yet leave room for human action and reaction. Further condensing of the very detailed pattern of interactions again becomes necessary. Mathematical functions are allowed for.

Affected sectors can be represented by a single variable (indicator or block variable). This is especially the case for component sectors where a small bundle of factors can be represented by a small internal pattern of interactions. Later on, the aggregated model can be broken down into the original bundle of factors, making possible the study of specific internal relationships.

The criteria matrix is employed to check whether selection is adequate, including the probes, aside from the degree of aggregation. Specifically, does it meet the criteria of a cybernetically relevant set of variables? Even after aggregation, collected data still have their own specific effects.

The visual simulation model that formed the basis for the computer programs was developed from resultant variables and functions aggregated to produce a model like the original pattern of interactions. In the original pattern, types of function are shown against the curves in the interest of better comprehensibility.

A special computer program was developed, then slightly simplified for use with a commercially available programmable desk dialogue computer. The program is designed so that the interpretation and evaluation models are directly based on it. Thus, they can link and communicate with the simulation program.

Interpretation; Interpretation Models

Specific interpretation considers quantification of feedback loops, of dependency, of throughput, of diversity, and of the indices of influence. In the case of feedback cycles, "a knowledge of such cybernetic criteria allows recognition of what depends on the proper functioning of a feedback cycle, which elements have sufficient parallel elements and can therefore be 'switched off' for a time and which are unique and must therefore be treated very gently" (Vester and von Hesler, 1980). Dependencies cover relationships of various types, such as dependency on transportation or on energy. These should be negatively assessed. Throughput is regarded as a decisive step in evaluating the system, and is heavily influenced by the tendency toward growth. The example provided is of a system dependent on almost-exhausted resources. If the consumption of these is at a rate so low that complete exhaustion will first occur in 1000 years, then the system is not especially endangered by the shortage of the resources. Diversity is a measure of the number of differing types of elements in a system.

The purposes of the interpretation models are to describe the actual cybernetic status and to simulate a nominal status, so as to cybernetically interpret changes that result from a given measure. Interpretation models are available for the influence index, feedback, diversity, dependency, and throughput. The interpretation model *influence index* determines the extent to which factors of a simulation model function as active, passive, inert, and critical elements. Later, this affords the possibility of evaluating buffering or amplification of disturbances. The interpretation model *feedback* covers negative and positive feedback loops of the simulation model as regards their direction and strength, and determines their dominance. This results in indications of the system's ability to regulate itself. The interpretation model *diversity* covers diversities of the different sectors. The interpretation model *dependency* covers external and internal dependencies of the system. It compares their strengths. Later, in a biocybernetic evaluation stage, there occurs a processing of this information along with that of the interpretation models "feedback" and "diversity." The interpretation model *throughput* deals with limit values and functions of material, energy, and information throughput. In addition to evaluating overall system loadings, it evaluates the cybernetic maturity of the system.

The interpretation models make possible insight into system behavior in the light of several cybernetic criteria. They do not, however, allow definite evaluation as to the stability, flexibility, and viability of the system in question. Thus, the evidence afforded via the interpretation models has to be biocybernetically evaluated in order to check its validity from the viewpoint of overall system behavior. Such evaluation is executed against the backdrop of three sets of laws: the aforementioned eight cybernetic laws of viable systems; certain thermodynamic laws to which stability, fluctuation, and organization of open and complex systems are subject; and mathematical laws pertaining to structure. In the latter laws, it is, for example, possible to assess the optimal degree of cross-linkage of a system, relationships between its substructures, or the irreversibility of catastrophic events.

The developers of the Untermain sensitivity model (Vester and von Hesler, 1980) note that "one of the first findings from this feedback with laws governing viable systems is that there are always several 'optimal' possibilities within a viable pattern of interactions, whilst practically only compulsions exist below the limit of viability or when the tendency to stabilize has decreased." Evaluation of functional states is important for intact parts of a region; this generally reveals a range of alternatives among which chances of finding one or the other political, financial, or technical solution is far greater. "Restoring component systems to health once again produces a stability output for the system as a whole; this can be proven by a simple simulation run, but it is also to be observed in all living systems."

Evaluation Models

Systems behavior, which has to this point been interpreted cybernetically, will now undergo evaluation. Orientation in this case is done along the same lines, namely, to put the goal each system strives for (its survival) within the general ecological context.

Five evaluation models are employed to evaluate and assess a system regarding its stability, burdens, and flexibility, its risks and chances. The evaluation models are cybernetic scale, system burdens, irreversibility, self-regulation, and stability factor. The *cybernetic scale* model evaluates cybernetic maturity directly from variables and characteristics. It includes a flowchart on stability and technology. The model on *system burdens* involves evaluation of system burdens and their risks by means of the interpretation models "dependency" and "throughput." The evaluation model *irreversibility* is concerned with evaluating system behavior in terms of the irreversibility of processes, done directly from variables and cybernetic characteristics. In the case of the evaluation model *self-regulation*, attention is given to the tendency toward self-regulation in terms of the degree of cross-linkage and the loop relationships with the interpretation model "feedback." The *stability factor* model evaluates stability from the relationship between degree of cross-linkage, diversity, and throughput, and also utilizes the "cybernetic scale."

The evaluation models are handled as follows: The first model is manually processed, followed by three computerized models. Allowance is made for mutual influences of these three models in the evaluation model "stability factor." This model involves a computer run and relative weighting of risks. A total of seven risk factors from the several models allows for the estimation of the most urgently needed safeguards. It also allows for estimation of first guidelines for assessment of existing planning devices.

It must be possible to check the results of one evaluation model against the results from the others. The last three evaluation models are related to the interpretation models and simulation level by linked computer programs. It is important, then, that automatically coupled results be checked by manual examination methods and evaluation. The developers of the sensitivity model stress that this is especially the case with the cybernetic scale and basic bio-cybernetic laws that were not integrated into the computer program.

The reader is here reminded that at the level of analysis and evaluation, matters of cybernetic interpretation according to structure and distribution, numerical ratios, existence or nonexistence of cybernetic characteristics, and the like are of major concern. The procedure results in a macroscopic view of general system behavior, and this forms the basis for all further strategies.

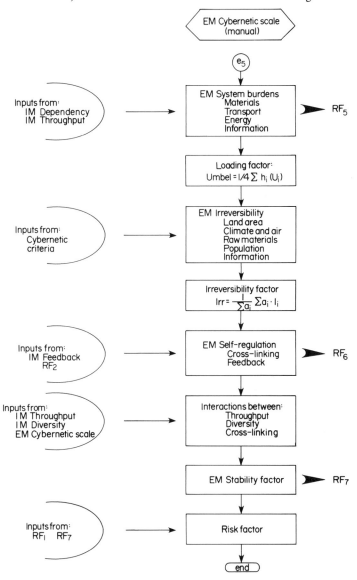

Figure 3-29. Evaluation models (computer flowchart). From Sensitivitäts-modell, Regionale Planungsgemeinschaft Untermain, Frankfurt.

An overview of the computer flowchart of the evaluation is found in Figure 3-29. Note that IM stands for interpretation model and EM for evaluation model.

It shows the computer and manual programs starting at the simulation model and going through to the strategy level. An open computer program, under development, is intended to be of special assistance in moving through the several stages and procedures minus difficulties.

What is provided to the planner by the evaluation and interpretation models is information enabling him to recognize actual system-relevant roles of various systems components in context. As examples, he finds revealed whether road-building acts as a control element, whether it stimulates or suppresses negative feedbacks, whether development of a system of rapid transit is an element that controls or buffers. Also shown are how role change or introduction of new elements affects system behavior relative to its self-regulation, sensitivity to developments, energetic efficiency, and quality of life, among others.

Reaching the Strategy Stage

The application in planning and decision making at the interpretation stage allows for the planner's becoming familiar with the cybernetic role of system elements, and makes it possible to give weighted significance to variables having seemingly equal importance. Analysis and evaluation are later carried out, and cybernetic system criteria are worked into the form of biocybernetic laws by use of mathematical criteria for system stability, analogies with viable biological systems, and thermodynamic criteria. Checking individual planning actions is then feasible by utilizing the criterion of system behavior (this complements criteria already taken into account—public opinion, finances, etc.). It also re-places system-irrelevant criteria, such as growth of the gross national product.

Final statements of the sensitivity model are geared to providing evidence of expectable system reactions to specific development, with the intent of providing system-relevant decision-making aids and alternative operators.

At the stage of strategy, one is aware that "indications of a biocybernetically sensible strategy either come directly from interpretations of cybernetic characteristics or from statements of individual interpretation and evaluation models or from the summary results of the last evaluation model" (Vester and von Hesler, 1980). Clarification must be reached on the type of strategy, the type of operator, and the type of objective.

The possible specific strategies, based on laws of the biosphere, are now addressed. Questions are raised about how negative feedback, regulating limit values, and control elements can be found, and how self-regulation or developments can be established, using a minimum amount of control energy. Possibilities exist for using the sensitivity model as a decision aid to compare, assess, and choose possible strategies. It also allows for pointing to new strategies from the point of view of relevance within the system.

Evaluation of allocation of new housing areas, building of a highway, provision of subsidies (all of these representing a course of development) is simulated in an aggregated model. Results indicate where control cycles are broken. Also shown is where throughput is displaced toward higher risks or where cybernetic organization was worsened. Having this at hand enables the cataloging of prohibited interventions or zones in which certain interventions must be forbidden. Based upon this, it is possible to leave the remainder of zoning work to self-regulating economic and natural forces. The strategy that results is known as *avoidance strategy*, where everything that is not prohibited is permitted.

Another way in which the sensitivity model can be employed is where the evaluation just described is executed, but where one finds that "a sufficiently well-justified and admissible statement on definable prohibited zones (or the background conditions for ecologically inadmissible interventions) is not possible. The consequences differ from those of the first case: with the available instruments, we can at best make statements concerning the possibly permissible interventions and zones. Everything else would involve unforeseeable risks from the ecological or systematic viewpoint and should therefore be prohibited or only permitted after very extensive and careful studies of the individual case" (Vester and von Hesler, 1980, p. 203). This results in the *permissive strategy*, where everything that is not permitted is prohibited.

It is important to note that the evaluation produces no detailed indications. It provides statements on the desirable degree of cross-linkage, overall stability, or increased significance of self-regulatory mechanisms. Most of these take effect with a lag in time. In some cases the best approach may be to do absolutely nothing, and to allow natural regulation to catch up with events. The resultant strategy here is a passive strategy (i.e., one of recognizing that the most intelligently planned intervention cannot substitute for the natural self-regulation in a system).

The manner of recording and describing system behavior also determines the character of measures that are to be taken. The fact is that the main information resulting from the evaluation models concerns criteria that are rather unusual from the standpoint of classical planning. Further, the necessary levers may prove to be not ministerial decrees nor amendments to laws but work in the public relations sphere to have the population become aware of certain relationships. An example provided is that of public rejection of washing machines that cannot be connected to the hot water system, regarded as a "special rejection of consumption that will give innovative impetus" (Vester and von Hesler, 1980).

In dealing with the operator (specific measure for the realization of a strategy), comparison is drawn between human illness and warning systems that are rendered inoperative or bypassed on the one hand, and the situation in a cross-linked overall system consisting in some cases of already badly damaged ecosystems. Humans would suffer much more sickness and injuries if not protected by built-in sensors and indicators, many of which work with automatic feedback. In total human ecosystems, sensors and behavioral repertoire are inadequate and too slow to react to massive and faster interferences with natural balance and important material cycles resulting from our energy-intensive technologies.

Operators of strategy resulting from the sensitivity model consist of establishing a catalog of system-oriented behavioral rules (comparable to conditioned and unconditioned reflexes), as well as planning guidelines, planning principles, and early warning indicators (social and biological). In the latter are included stress, anxiety, taboos, addiction, level of air pollution, and water quality. The catalog is meant to allow planning to correct ecologically harmful measures before the beginning of catastrophic and inhuman self-regulatory processes, such as a sudden sharp rise in the death rate, begin.

In terms of types of objective in strategy, it is concluded that there is no final objective in the viability context (just as there are not absolute but only relative entropy values). Thus, for strategic considerations, "Quantitative objectives, states that can be described in absolute terms, are not important but relative movements are. From the direction of these movements, such as can be visualized by means of the cybernetic scale of the first evaluation model, we can see which direction is the preferable one in every case, regardless of how far we can go in that direction. One can be sure than an improvement in respect of one basic law will lead to a series of other positive developments, because of the links between all basic laws and their communicative self-amplification" (Vester and von Hesler, 1980, p. 208).

The Untermain sensitivity model makes possible the development of a matrix that couples tendencies of system behavior with indications of control measures as decision-making aids. A *behavioral matrix* results. The process's use of a checklist of basic biocybernetic laws allows one to allocate a cybernetic function that will improve cybernetic behavior to any type of system behavior. Such a functional matrix enables a new type of information in the form of cybernetic control measures. This should be the introduction of a certain control element, the "damping" of a controller, or the coupling of separate control cycles to safeguard self-regulation below certain limit values.

But planners will find no practical value in this unless it is related to definite specific measures to realize strategy. Here, an *operator matrix* is introduced, consisting of an alternative list of possible modifications, measures, and control mechanisms allocated to the several cybernetic functions.

In concluding our consideration of the sensitivity model, we emphasize that it offers many widely differing operators for a given cybernetic strategy. Work on special system strategy and corresponding instruments remains for a further development project.

The sensitivity model's relevance to planning and decision making is clear. Much less clear, however, is the practical way of linking its approaches to those of currently existing mechanisms and approaches. Undoubtedly, production of greatly simplified descriptive training devices by the developers of the model (to expose planners and decision makers to the intent, perspectives, and procedures of the model) would help. The present publication of Vester and von Hesler (1980) is unduly burdened by wordiness, perhaps due to the translation from German, and by using technically heavy jargon and terminology. In short, the writing needs to be simplified if it is to reach that part of its intended audience resident in the English-speaking countries. Its value is clear; its contents need to be made clearer.

Environmental Impact Assessment
and Environmental Impact Analysis

The processes known as "environmental impact assessment" and "environmental impact analysis" are used in several parts of the world, especially in the United States, where federal and some state laws demand environmental impact statements for specific projects. Much of what is said in this chapter and particularly in relation to the sensitivity model, has relevance to these terms. The need for a holistic approach and for truly interdisciplinary (and not only multidisciplinary) teamwork is also essential for these procedures. The role of interdisciplinary teams in environmental impact assessments has been dealt with by Erickson (1979) in discussing principles and applications. Some of the critical problems of such environmental impact assessments developed in the United States and adapted to the specific needs and conditions of the United Kingdom were discussed in a special seminar in 1975 (O'Riordan and Hey, 1976). Further information can be found in the recent comprehensive handbook of environmental impact analysis edited by Rau and Wooten (1980). It should be mentioned, however, that in the Rau and Wooten handbook the term "landscape" is not mentioned, nor is landscape ecology or its concepts dealt with. The handbook's last chapter provides a summarization of environmental impacts dealing with aspects of land use, water and air quality, noise and vibrations, community transport, and service infrastructure and wildlife and vegetation. It thus contains many of the elements treated by Vester and von Hesler (1980), but it lacks the biocybernetic approach to enable a full holistic analysis and synthesis, and therefore, prediction.

Another important point not to be overlooked is the need for the linking of environmental impact statements with the decision-making and planning processes, something often discussed but not yet fully realized. Such an integration in the planning process has been attempted successfully in West Germany by the team of landscape ecologists and planners from the Technical University of Berlin (Bachfischer et al., 1977) in the urban–industrial agglomeration of Middle Franconia, with the help of an ecological risk analysis. On the basis of an analysis of present land-use demands, spatially differentiated impact sensibilities, impact intensities, and the risks of impacts for special natural resources were derived. Thereby three alternatives could be proposed for the future development of this region.

In Israel, such an ecological risk analysis is being carried out presently for the landscapes and their natural resources in the Mediterranean coastal strip as guidelines for future planning and management with special reference to recreational uses, by the Environmental Protection Service (Kaplan, 1981) with the help of S. Amir, a leading landscape planner from the Technion-Israel Institute of Technology.

A further important tool for the planning and decision-making process in land use, applied presently in West Germany, is the analysis of the benefit values ("Nutzwertanalyse"), defined by Zangemeister (1971) as the analysis of a great number of complex alternatives for action with the purpose of ordinating these in relation to a multidimensional goal system by stating the total benefit value of these alternatives. Bechmann (1978), one of the leading theoreticians of landscape planning on a cybernetic-system theoretical basis from the Technical University of Hannover has recently reviewed critically the application of this tool in the light of a general theory of evaluation. He proposed an improved "second generation" of methodology, which was applied in the above-mentioned comprehensive project of landscape planning for recreation of the Sauerland (Kiemstedt et al., 1975). He also discussed in depth the difficult problem of communicating these evaluation analyses to those taking part in the decision-making process, namely, the politician, the administrator, the public, and the advising scientist. In his opinion, in actual planning these benefit value analyses should serve as a communication link between science and politics and therefore their communicability would be improved if it were based more on everyday knowledge than on scientific perception.

Methodologies for Land-Use Capability Analysis and Regional Landscape Evaluation for Planning and Design

Historical Progression in Methodologies

The rapid proliferation over recent decades of methodologies for analysis of the capability of ecological systems to "accommodate" particular land-use needs of man has recently been accompanied by advances in predictive evaluation of regional landscapes that carry the process into areas of systems modeling.

Concern with land issues (including types of land uses, compatibility between such uses by man and ecosystems, consideration of special or critical resources and their identification and protection) has, in the last century and this one, been manifested by those in the landscape planning, design, and management fields.

In this chapter, we consider planning and design, and refer only tangentially to management and reclamation, which are covered in depth in Chapter 4.

As noted by Trowbridge (1979) in a brief review of ecologically based regional planning methods, *trial-and-error techniques,* in which planning decisions were made partially from common sense and experience and also through a series of related actions and reactions, were among the early clearly identifiable regional planning methods of the 20th century. Later, *inventory methods* were developed, characterized by systematic inventories of land uses that combine information from natural sciences with engineering classification of soils and aerial photographic interpretations of cultural land-use patterns.

Hand-drawn overlay methods were introduced to address the complex decisions to be dealt with at a regional level of planning. At this scale, a more comprehensive approach to land-use planning begins.

The progression from inventory mapping to the combination of related categories in hand-drawn overlay methods indicated both the use of finer distinctions between land uses and resources and the attempts to resolve conflicts by analyzing the relationships between different types of land use.

Data classification systems represent a subsequent stage. These began to emerge as knowledge and expertise in systems analysis were applied to computer systems and space technology. Information theory was employed as the organizing element in data classifications, this theory clarifying the relationships between data requirements, collection, storage, manipulation, and the users of that data.

Computer-based planning methods, including computer-aided assessment procedures (Fabos, 1978), enabled extensive use of quantitative weighting systems and the sophisticated techniques of derived weights and values, generating even more data that must all be stored and retrievable (Trowbridge, 1979).

Thus, as these developments show, rapid evolution may be said to have taken place in capabilities and approaches. A closer look at these developments is in order before specific consideration is given later in this chapter to some of the most developed, holistically based methodologies.

During the latter part of the last century, the work of Muir and of Pinchot in North America formed an early basis for ecologically based regional planning methods. The specific purposes of regional planning before 1900 were varied, but what is evident is a respect for available natural resources accompanied by much concern for the quality of the environment.

In this century in North America, trial-and-error techniques were first utilized, and perhaps described best, as Trowbridge (1979, p. 2) does, in allowing for decision making like a "chess game of moves and counter-moves . . . a series of successful or possibly unsuccessful activities. Since few replicable approaches existed it was difficult to specifically relate two complex series of decisions. This nonreplicable characteristic also made it extremely difficult to teach a land use planning approach or discuss strategies. Information with which decisions were made came in a variety of states. Early soil and physical resource data was often incomplete and published in literary form, at different scales which made comparison of data extremely difficult. *The Town Plan for Billerica, Massachusetts* (1912) and *The Graphic Regional Plan of New York and Its Environs* (1929) are early representative examples of documented land use planning."

Decision making based on trial and error is appropriate for certain projects, but although the results of personal and trial-and-error judgments may lead, through experience and personal knowledge, to the same decision as a more structural approach, such judgments must be carefully examined. There is, after all, a distinction between observation and analytical evidence. Partial or biased observations, and the coloring of judgments by an individual's experiences, among other factors, make for difficulties in arriving at agreement about the existence of the same environmental qualities in a particular region.

The development of inventory methods received a major boost from the great growth in knowledge about soils, geology, and hydrology (and ecology) during this century. Subsequently, remarkable innovations in the field of remote sensing, as well as the considerable refinement of classifications in soil taxonomy, and enhanced cartographic representations (in the United States, the continued work of the United States Geological Survey should be cited), all added to the greater availability of natural resource and land-use data. Further impetus for such inventory methods is discernible in the federal interest in statewide planning, which prompted funding of resource information at the state level. The organization in a classification system of existing resource data and the incorporation of new data derived from systematic aerial photographic interpretation were considered an accompaniment to statewide mapping.

Hand-drawn overlay methods involve maps of data developed at a common scale and overlaid in various combinations to create a new level of information, that of composite maps.

G. Angus Hills' plan for Ontario Province (1961) is an early North American example employing a sophisticated data overlay method. The work of Ian McHarg is about the best known. McHarg's studies, employing the overlay method, are well presented in the oft-cited *Design with Nature* (1969), which served as a "landmark exposition" of both McHarg's philosophy and his approach and technique. McHarg's method of "composite suitability mapping" is clearly summarized in that work, and case studies showing application of the method (a layered planning approach) include, among others, the study of Staten Island, New York; the *Plan for the Valleys* near Baltimore, Maryland; and planning for residential needs in the Washington, D.C., area. McHarg is regarded as a leading exponent of ecological determinism in regional planning. Although the specific hand-drawn overlay method has been surpassed by later developments, and McHarg himself has moved from his earlier concerns to more intricate concerns and approaches, his 1969 work remains a most important contribution to ecology-based planning literature in North America of the late 1960s. McHarg has headed the Department of Landscape Architecture and Regional Planning at the University of Pennsylvania at Philadelphia.

Philip Lewis, at the University of Wisconsin at Madison, is another individual active in land use and resource analysis utilizing land overlay training. His studies have focused on the landscapes of the midwestern United States, including development of an environmental corridor concept for the state of Wisconsin (Lewis, 1967). He investigated the consequences of rapid land-use change, and used local experts to identify indications of changes in land use. Mapping was then done of these indicators.

In considering the step to development of data classification systems, it is well to remember that the "progression from inventory mapping to combination of related categories in the methods using hand-drawn overlays pointed up both the use of finer distinctions between land uses and resources and the attempts to resolve conflicts by analyzing the relationships between categories of land use" (Trowbridge, 1979).

Geographic information systems have been developed by now that help serve these focuses, and are discussed in a publication of the International Geographical Union, Commission on Geographical Data Sensing and Processing (Calkins and Tomlinson, 1977), with investigations supported by UNESCO and by the Resource and Land Investigations (RALI) Program, Geology Survey of the United States Department of the Interior, as well as the Argonne National Laboratory, Energy Research and Development Administration. The support of United Nations organizations and governmental agencies in the United States has resulted in a text on geographic information systems for personnel who provide technical support to land and resource planners. Examples used in the text draw largely from United States state and regional systems, but reference is made to some West European, Canadian, and developing countries' systems. Description is provided of manual and computer-aided systems, equipment, the data considerations of quantity, reliability, accuracy, and resolution, costs, systems design, implementation strategies, institutional considerations, and data sources. The International Geographic Union publication is identified for the reader interested in the state of such spatial data systems as of 1977, and provides much detail of general and specific value about such systems. (See note on p. 255).

Computers may be used to produce and duplicate maps. Faster data sources have become more comprehensive, leading to more complex site analysis and land-use planning. The observation and recording of natural systems can be accomplished and economic and cultural data can be mapped as well. In the use of overlays, there occurs a recombination of each data category that must also be recorded and stored. A problem results: The drawing of a map for each category and the association of categories consumes much time. Adding in, as mentioned earlier, of quantitative weighting systems and derived weights and values makes for even more data to be stored and retrieved. Computers are most useful in just such situations, that is, when large amounts of data are to be collected, stored, or retrieved in graphic form. As a consequence, highly advanced landscape planning methods have been produced, which now include the utilization of computer technology allowing display of three-dimensional forms in perspective. As noted by Trowbridge, great progress occurred in computer programming in the 1960s, with the use of a cathode ray tube display screen to visualize the designs, and planning and design professionals have increasingly made use of these computer capabilities.

However, these advances are still widely scattered efforts, concentrated in certain universities and institutes around a few leading landscape and physical planners. They mostly lack a unified conceptual basis, which is provided to a great extent in Central Europe by landscape ecology, and their impact on actual decision making on land use over large areas is still very limited.

In discussing planning and landscape evaluation in a keynote address at the New Zealand Planning Institute Conference on Planning and Landscape, Julius Fabos (1978) delineated the University of Massachusetts Metland method for dealing with significant landscape problems and opportunities. He elaborated on the use of "landscape assessment and evaluation" procedures that can—through computer assistance—be meaningfully employed in regional planning. In discussing an approach that harnesses unfolding capabilities in achieving greater precision in definition of land, avoids subjectivity, and is quantitative, he noted the explosion in remote sensing and computer technology. "We are establishing better quality, larger data banks and developing data manipulation technology which was unthinkable a decade ago."

Thus it is evident that new capabilities exist, and are being activated by regional landscape planners in a major way, in highly developed countries and regions. As still higher levels of hardware become available, they will be sought after and introduced as planning tools. As discussed earlier in the section on remote sensing and the information-base needs of developing nations, lower levels of mechanization may be emphasized in acquisition of resource information in such situations. Attention should be given to the expenses involved and the professional expertise support base necessary for effective and efficient operation of very high-level mechanized systems for data manipulation. As noted by Hardy (1979), in determining the appropriate technology to use for inventories, it is necessary to determine whether to use a high or low level of mechanization. "The concept of using a great deal of hardware for resource inventories is common. At this time, the availability of machinery is very high and the incentive to use machinery is extremely strong." Beyond inventories alone, the question of degree of appropriate technology—mechanization in less-developed countries remains a deep concern. It is largely beyond the purpose of this book to address further in a detailed way these dilemmas of less-developed nations. The problems are real, and have been given much coverage in recent years (Bulfin and Weaver, 1977; National Research Council, 1977; Eckaus, 1977). Although both of this book's authors have much interest in and feel much concern for the landscape planning and management needs of developing nations [one of the authors (Lieberman, 1978) initiated and now serves as coordinator of an emerging international land-use planning training program for less-developed country participants, and the other worked several years on range and pasture development problems in northern Tanzania and Masailand (Naveh, 1966) and is a member of the International Union for Conservation of Nature and Natural Resources (IUCN) Commission on Ecology, being actively engaged in problems of conservation and rural development], our consideration of methodologies will, in the main, provide examples from highly and moderately developed nations in North America, Europe, southwest Asia, and Australia. Planning situations in less-developed countries will be dealt with in only a limited way.

Terminology of Land-Use Capability and Regional Landscape Evaluation

Before moving on to consideration of selected methods for land-use capability analysis and regional landscape evaluation, we must become more familiar with some of the perspectives and terminology involved.

Unless otherwise designated, the terms that appear here are based on a publication on assessing tropical forestlands of the East–West Center, Honolulu,

Hawaii (1980), the biogeographer Robert Orr Whyte's (1976) book on land and land appraisal, the Wildland Planning Glossary of the United States Forest Service (Schwarz et al., 1976), and a lecture by Julius Fabos (1978). These are abbreviated respectively as EWC, ROW, Schwarz et al. 1976, or JF after each entry. The entries are aggregated, and within each aggregation appear in the approximate sequence in which they are found in the planning process.

Inventory

1. The gathering of knowledge for future use. This involves the quantity or count of physical entities in the area and, in inventories in land-use planning, involves two basic types, existing and potential. Existing refers to measurement of what is actually "on the ground" (in terms of number) and potential considers the total identified undeveloped capability (Schwarz et al., 1976).
2. Enumeration of pertinent data, such as volume, increment, mortality of stands for assessment of productivity, and values on a certain time scale (Zumer-Linder, 1979).

Reconnaissance Survey

1. In the United States Forest Service usage, a preliminary survey usually executed rapidly and at a relatively low cost (Schwarz et al., 1976).
2. At the reconnaissance survey level, land systems are described in terms of their component land units which recur in association to form the land-system pattern. [Discussion by ROW of the Australian Land Research's approach, which recognizes these units in subdivision of landscape: site, land unit, and land system.]
3. Reconnaissance survey, more than 2000 square miles. May be as large as 100,000, having an objective of national or regional inventory, involving landscape analysis of land systems or higher categories, and in economic analysis, at the national or regional economy level. [ROW coverage of Land Resources Division, Overseas Development Administration, Great Britain: The principal types of survey (Baulkwill, 1972).]

Reconnaissance Scale

1. In classification procedures devised for assessing tropical forestlands, their suitability for sustainable uses, reconnaissance scale is 1:100,000 to 1: 1,000,000 (EWC).
2. Typical reconnaissance map scale is regarded as 1:250,000 and 1:500,000 (Baulkwill, 1972, appearing in coverage by ROW).

    Overview level refers to a land capability classification on a reconnaissance scale.

    Extensive level. Detailed inventory; broad assessment of agricultural potential. 1000–10,000 square miles. Typical map scale 1:100,000 to 1: 250,000 used in landscape analysis of land systems (Baulkwill, 1972, appearing in coverage by ROW).

Intensive level. Location and definition of development projects. 500–5,000 square miles. Typical map scale 1:25,000 to 1:100,000 landscape analysis of land systems and facets (Baulkwill, 1972, appearing in coverage by ROW).

Preproject surveys. (In general, these are detailed reconnaissance in nature.) The scale of aerial photographs suitable for project surveys depends partly on the nature of the project and partly on the natural complexity of the landscape. For example, where landscape patterns are simple, as in broad floodplains, or the projected scale of development is such as that in large-scale mechanized dryland farming, aerial photographs with the same scale as those used for reconnaissance surveys (1:40,000 to 1:80,000) should be adequate. For complex landscapes where intensive development or irrigated agriculture or forestry is contemplated, it is desirable to have new photographs taken. To avoid expense of possible further rephotography at the development survey stage, these photographs should be taken at a scale that will be appropriate for development surveys (i.e., in the range of 1:10,000 to 1:20,000). As with the scale of aerial photographs, the scale of maps produced in project surveys will vary, but generally they will fall within the range of 1:50,000 to 1:100,000 (Christian and Stewart, 1968).

Predevelopment surveys are required for the detailed planning of development following the approval of the project. As with preproject surveys, the nature of the resource studies will depend on the nature of the project. The survey of the appropriate resources is at detailed or semidetailed level. Because of the order of accuracy required, this work is normally carried out with large-scale aerial photography (1:10,000 to 1:20,000). A high proportion of field work is required to give the desired accuracy and aerial photograph interpretation is limited to direct and associative interpretation largely used in planning field sampling and using the aerial photograph as a convenient base map on which to work (Christian and Stewart, 1968).

Detail scale refers to large map scale (1:1,000 to 1:50,000).

Detailed level thus involves land capability classification at the detailed scale (EWC).

Development study. 1. Usually more than 50 square miles. Typical map scale 1:10,000 to 1:25,000. Objective of resource analysis and development planning. Landscape analysis of land facets and elements (Baulkwill, 1972, appearing in coverage by ROW).

Analysis

1. A detailed examination of anything complex in order to understand its nature or determine its essential features (Webster, 1963).
2. A separating or breaking up of any whole into its component parts for the purpose of examining their nature, function, relationship, and so on (Webster, 1963).

3. In mathematics and computer science, it pertains to solving problems (United States Forest Service, 1972).
4. A process by which the landscape is broken down into components (JF).

Assessment

1. A process of synthesis. Much like property assessment, where composite value is expressed based on size, age, quality, and location (JF).
2. See also *suitability assessment* below (EWC).

> *Valuation.* Procedures based on professional judgment, public preference, values of the elite, economic evaluation, and several others (JF).

Evaluation

1. A procedure that informs decision makers about the trade-offs among alternatives (JF).
2. An examination and judging concerning the worth, quality, significance, amount, degree, or condition of something (Webster, 1963).

Land capability. Capability classification is the description of a landscape unit unit in terms of its inherent capacity to produce a combination of plants and animals (EWC).

Capability (inherent capability, inherent carrying capacity, natural capability, natural carrying capacity, physical carrying capacity, resource bearing capacity, site capacity).

1. In land-use planning literature the terms "suitability" and "capability" (whether alone or accompanied by various modifiers) are often used interchangeably to refer to ratings based on two different evaluation procedures. One basic type of evaluation procedure is the rating of use or productivity potentials based on the present state of the resource. This type of rating is therefore an evaluation based on the resource's inherent, natural, or intrinsic ability to provide for use, and includes that existing ability which is the result of past alterations or current management practices. A second basic type of evaluation procedure rates the potential ability of a resource to produce goods or services on the basis of the maximum possible outputs for a given type and level of future alternative site or resource management inputs.

   G. Angus Hills was probably the first prominent wildland planner to recognize the importance of retaining the distinction between these and other types of evaluations in a process for land-use planning. Hills used the terms "use capability" to refer to evaluations based on a resource's inherent or present condition abilities, "use suitability" for evaluations based on potential management inputs, and "use feasibility" to refer to usability potential ratings based on an evaluation of offsite factors—such as accessibility, present and forecasted socioeconomic conditions, and technological developments (Belknap and Furtado, 1967; Hills, Love, and Lacate, 1970).

Perhaps even earlier than Hills, the United States Soil Conservation Service began widespread use of its *land capability classification* system (Wohletz and Dolder, 1952), which also uses "capability" evaluation in the sense of an inherent ability rating (Schwarz et al., 1976).

To avoid confusion, we recommend that in the future the Hills and Soil Conservation Service usages be recognized as having established the precedent for the proper use of these terms; thus "capability" should be used when referring to evaluations for usability based on the present state of the resource, and "suitability" should be used only for evaluations based on assumptions about potential usability or productivity if specified management alterations are to be made—such as drainage improvements, or added irrigation and/or fertilization.

Better yet, both terms should be accompanied by modifiers that make it absolutely clear just what type of an evaluation procedure is implied by the rating given. Thus the word "capability" should always be presented as "inherent capability," "intrinsic capability," or "natural capability." Similarly, "suitability" could be referred to as "managed suitability" (Schwarz et al., 1976).

2. The intrinsic ability (i.e., ability unaltered by any level of potential future human management activities or other type of alteration) to support a given use type, intensity, and quality on a sustained basis (i.e., without significant resource deterioration over the time span of renewing biogeochemical cycles) (Schwarz et al., 1976).

   Land suitability. Suitability assessment is the subsequent rating of a response of a landscape unit under a certain use, specifically whether that use on that unit will cause unacceptable degradation of productivity and therefore not be sustainable (EWC).

Landscape Unit. A mappable area, roughly homogeneous as to soil, topography, climate, and biological potential, whose margins are determined by change in one or more characteristics. A landscape unit is characterized as an ecosystem (i.e., the physical structure and relationships of soil, water, nutrients, energy, plants, and animals). Delineation of a geographic area as a landscape unit requires differentiation from adjacent units and recognition of similarity to the same type of unit recurring elsewhere (EWC).

Land System

1. A unit of physiographically related habitats as often contained in a watershed (EWC).
2. In the Canada Biophysical Land Classification System usage, an area of land with a recurring pattern of landform having homogeneous soil characteristics and supporting a particular type of vegetation (Bajzak, 1973).

   This is the second smallest of the four classification levels of that system.

The other three classes, in order of decreasing size, are land region, land district, and land type (Schwarz et al., 1976).

3. United States Forest Service usage. Typically used to categorize all the land in an area into each of various hierarchical levels by the application of increasingly more specific delineation criteria at successively lower (i.e., smaller units) levels in the system. Each set of delineation criteria is unique to its classification level but is also logically and systematically related to the other higher and/or lower levels in that classification system.

A hierarchical framework of land type units (after Nelson, 1975; Schwarz et al., 1976).

4. In the Australian Land Research approach, these units have been recognized in the subdivision of landscape: the site, land unit, and land system. The *land system* is the unit of mapping. At the reconnaissance survey level, land systems are described in terms of their component land units, which recur in association to form the landscape pattern. The site is the smaller identifiable unit. Land units may consist of a single site or a group of geographically associated sites (Christian and Stewart, 1968).

A *site* is a part of the land surface that is, for all practical purposes, uniform throughout its extent in landform, soil, and vegetation. A small amount of variability occurs within a site, but it is of such low order that at the level of the survey being made, the variation falls within the units of classification used in each discipline.

In practice, the *land unit* is usually a group of related sites that has a particular landform within the land system, and wherever it occurs again it would have the same association of sites. Land units are geomorphologically and geographically associated to form patterns that recur in the landscape. The boundary of the pattern generally coincides with that of some discernible geological or geomorphogenetic feature or process. Within the pattern the same land units recur. Where a different assemblage of land units begins, it is a different land system. Thus by subdividing a region into its land systems (patterns) and describing each in terms of its component land units and sites, it is possible to reduce the complexity of the region to a reasonable number of types which can be comprehended without having to resort to the time-consuming practice of actually mapping each occurrence of each land unit or site (Christian and Stewart, 1968).

5. An area or group of areas throughout which there is a recurring pattern of topography soils and vegetation (Basinski, 1978). This definition appears in an important study of the Commonwealth Scientific and Industrial Research Organization (CSIRO) on methods of acquiring and using information to analyze regional land-use options in the New South Wales area of Australia. Land unit is defined as a component element of a land system, equivalent to one or more facet combinations. *Facet* is an element of functional land unit differentiated on the basis of landform, geology, and terrain. *Facet combination* is a specific combination of facet, vegetation community, and soil association.

Land Units. In G. A. Hills' land evaluation usage, a portion of a landscape unit—
that is, groupings of physiographic site types and physiographic site phases into
16-square-mile minimum units with special significance for some specific use
(after Belknap and Furtado, 1967; Schwarz et al., 1976).

"Landtype" ("Watertype")

1. In G. A. Hills' land classification usage, units created by isolating areas of
   differing landform, geological composition, and water content within the site
   region.
      A "landtype" is a subdivision of a "site region" and is determined by cer-
   tain soil properties (such as texture, depth, mineral composition, and water
   content) and the nature of the parent material in large-scale land features
   (after Belknap and Furtado, 1967; Schwarz et al., 1976).
      The second largest unit in Hills's land classification system. The various
   levels, in order of decreasing size, are site region, "landtype," physiographic
   site class, physiographic site type, and physiographic site phase.
2. In United States Forest Service usage, visually identifiable unit areas resulting
   from homogeneous geomorphic and climatic processes and having defined
   patterns of soils and vegetative potentials.
      Landtype units range in size from about 0.1 to 1 square mile. Their size
   and composition depend on the significance of physical characteristics, which
   can be readily interpreted to identify hazard, capability, and productivity
   potentials that are reliable for land-use planning purposes.
      Landtype units generally have uniform management response characteris-
   tics and so can be used to identify areas for which location decisions can be
   made (after Wertz and Arnold, 1972; Schwarz et al., 1976).

Land Type. In the Canada Biophysical Land Classification System usage, an
area of land having a fairly homogeneous combination of soil features at a level
corresponding to the soil series; that is, a group of soils having horizons similar
in distinguishing characteristics and arrangement in profile and developed from
the same parent material and chronological sequence of vegetation.
   The land type is the basic biophysical land unit that can be interpreted to
provide land evaluation data useful for planning and development. The mapping
scale at the land type level ranges from 1:20,000 to 1:10,000. The dominant
ecological variables are soil moisture regime, physical and chemical soil properties, and microtopography (Wells and Roberts, 1973; Schwarz et al., 1976).

Land Region. In the Canada Biophysical Land Classification System usage, an
area of land with a certain type of vegetation that manifests a distinctive
climate (Bajzak, 1973). This is the largest of the four classes of that system. The
other three classes in order of decreasing size are land district, land system, and
land type (Schwarz et al., 1976).

Land Function. Designation of a way in which an area of land is useful to the community; includes what are normally called land uses and ways that do not involve purposive human intervention.

Landform. A topographic–geological unit, such as a beach deposit, dune area, lava flow, alluvial flat, or talus slope (EWC).

Land Use. The occupation or reservation of a land or water area for any human activity or any defined purpose. It also includes use of the air space above the land or water (California Council on Intergovernmental Relations, 1973).

Bioclimatic Region or Zone. The data ranges and threshold values are originally adjusted to certain spatial changes in the biota. The biotic characteristics (aspects of vegetation physiognomy) reflect macroclimatic differences (EWC).

Bioclimatic Zone. Refinement to include topography, geological substrate, land-form, and soil as additional descriptors (EWC).

Life Zone

1. A geographic area where the climatic conditions and vegetation are approximately uniform (Marsh, 1964).
2. An altitudinal or latitudinal biotic region or belt with distinctive faunal and floral characteristics (Hanson, 1962; Schwarz et al., 1976).
3. Areas of land classified into gross habitat types by a combination of elevational and climatic zones and other factors that determine the type of animal and plant life found in them—for example, the Upper Sonoran, Transition, Canadian, Hudsonian, and Alpine life zones (Schwarz et al., 1976).
4. Holdridge's (1947, 1967) life zones are bioclimatic zones based on particular combinations of mean annual rainfall and temperature. Such schemes have the advantage of rapid area mapping of macroclimates which may be assumed to convey a general level of biological homogeneity within each macroclimatic unit (Figure 3-30).

Plant Formation. A mid-sized vegetation unit distinguished by vegetation structure or architecture, such as closed forest versus open forest, or low-stature forest (EWC).

Forest-Cover Type. A vegetation unit usually distinguished by dominant species that cover mid-sized areas such as landforms or land systems.

Forestland Type. A specific habitat or site identified in relation to its current or potential forest cover; for large scale and intermediate scale.

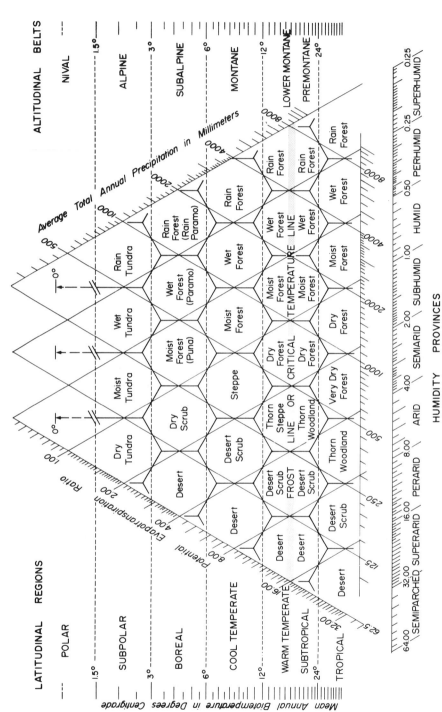

Figure 3-30. Diagram for the classification of world life zones or plant formations. Reproduced courtesy of L. R. Holdridge, Tropical Science Center, San Jose, Costa Rica.

Forest Ecosystem Type. Mid-sized unit in relation to a certain land system— a watershed, for example.

Forest Site Type. Small-area unit distinguished by a unique forest vegetation, soil–moisture regime, and habitat combination (e.g., alluvial bottomland, periodically flooded).

Forest Habitat Type. Definable physiographic segment of land, with all its particular environmental factors.

Sustainable Use. Continuing rational use of land without severe or permanent deterioration in the quality and quantity of the integrated ecosystem or landscape unit (EWC).

Procedures and Methodologies

Procedures and methodologies for landscape resource analysis have been greatly enhanced in recent years, and allow focus on the needs of decision makers at several levels of concern, addressing questions of environmental impact and assessment of ecological stability.

The Metland Procedure (Massachusetts, USA)

The perspective of one group of planning researchers in the United States introduces some very recent thinking that both serves as a contrast to earlier approaches and raises basic questions about the intentions and characteristics of landscape analysis, assessment, and evaluation. Challenges are raised to the approaches of the 1950s and 1960s, and it is contended that the parametric approach, what with its much-improved data gathering and manipulation capabilities, supersedes the landscape approach.

The Metland procedure of the University of Massachusetts landscape planning team (whose overall objective is "to facilitate land use decision making based on environmental factors for the general public good") incorporates scientific knowledge and advanced technology into a landscape planning model. It moves from the first phase of composite landscape assessment [addressed are landscape, ecological, and public service value profiles, dealing with special (or critical) resource components; hazard components, and development suitability components] through phase two, alternative plan formulation, to phase three, plan evaluation. Phase two involves the development of three alternative plans: a composite landscape value bias alternative, a status quo bias alternative, and a community preference bias alternative. This is based on the notion that some assumptions are needed in terms of land conversion or new land uses. The formulation of "alternative plans or alternative scenarios is regarded as an appropriate approach to obtain needed assumptions." Phase three utilizes an evaluation procedure that predicts the effect of proposed land uses and is intended to result in the selection of the alternative plan that satisfies the majority of the human objectives, but at the same time has less negative impact on the composite landscape, ecological, and public service values. Considered are the effects of alternative plans on the landscape value profile, the ecological compatibility profile, and the public service value profile (Figure 3-31).

Looking back to phase one, we note that the general aim of what are identified by the landscape planning team at the University of Massachusetts as environmental assessment procedures is to provide information concerning the identification of environmental resources or land-use constraints, examples being

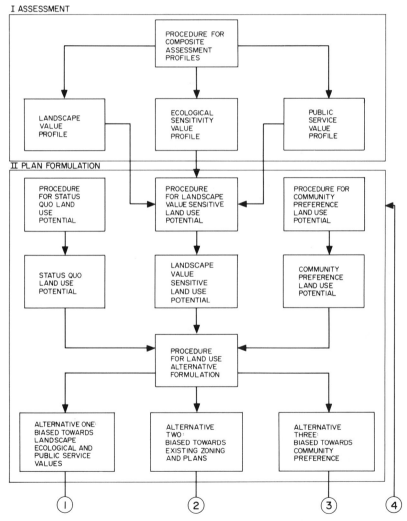

Figure 3-31. Metland landscape planning process flow diagram. Massachusetts Agricultural Experiment Station Research—METLAND.

water supply or floodplains; an indication of the quality and quantity of each resource or constraint, the location and distribution of each area of resource or constraint; and an estimate of the potential impact of alternative uses on each resource or constraint area (in this segment, we could note that suburban development over an aquifer area will probably pollute existing groundwater unless precautions are taken (Fabos, 1979). Further, these types of assessments, involving as they do environmental resources and hazard potential, provide information of both short-range and longer-range relevance as consideration is given to trade-offs to meet human objectives. Thus, in dealing with the "environmental assessment procedures," as viewed by the University of Massachusetts landscape planning research team, the processes of analysis and of assessment should be clearly distinguished from one another (other works often employ the terms "analysis" and "assessment" interchangeably). It is to be further noted that several landscape assessment and evaluation procedures have distinguished between problems and/or constraints on the one hand and opportunities on the other. In the procedures for environmental assessment of the Massachusetts team are included an assessment of landscape resources, an assessment of landscape hazards, and an assessment of development suitability, as well as an assessment of human impact on environment. What is regarded as synthesis, based on the

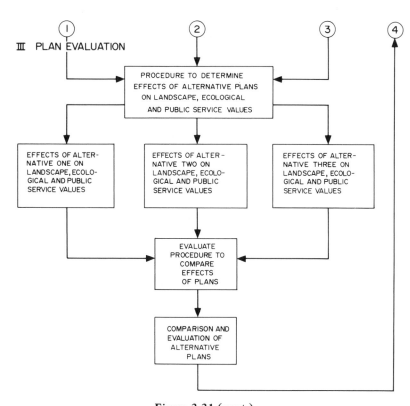

Figure 3-31 (cont.)

findings of the environmental assessment, appears in the landscape planning procedures (involving "comprehensive environmental planning procedures"), which
can be used in making several general types of planning decisions. Each of the
decisions deals with different (although sometimes overlapping) requirements.
Preservation, protection, development, reclamation, and management decisions
come into consideration. Fabos (1979) defines preservation decisions as an
attempt to maintain a piece of land in its existing state, whereas protection (or
conservation) decisions imply limited human use within specified guidelines.
Development decisions aim at transforming a rural landscape into an urban and
metropolitan landscape. Reclamation decisions are aimed at undoing the damage
caused by previous development or management decisions, ones that have in
some way damaged (impaired) an environmental resource. These four types of
decisions are accompanied by the fifth, management decisions, which specify
both the land-use activities allowable and the strategies for managing these
activities.

In the Metland procedure, an argument is presented that landscape evaluation
is different from the placing of values on landscape parameters while assessing
landscape. Rather, landscape evaluation is viewed as a process used to estimate
the effects of various alternative plans. In this procedure, then, landscape evaluation follows plan evaluation, while landscape assessment precedes it. This distinction should be borne in mind. It is proposed that the landscape evaluation be
done with a landscape bias, much as social evaluation is biased toward social
values. The developers of the Metland work portray their parametric approach
as replacing the earlier landscape approach, which evolved in the 1960s. The
landscape approach, in their words, "was a systematic way to determine land
use constraints and opportunities. Due to better data gathering and improved
manipulation capabilities, the parametric approach is seen as superseding the
landscape approach. Use of the computer is an integral part of the landscape
evaluation" (Fabos, 1978).

The matter of the landscape approach versus the parametric approach deserves further attention. The portrayal of the landscape approach as coming
earlier in time, and having been bypassed by more advanced approaches, the
example being that of the Metland procedure, can be regarded as an accurate
reflection of evolution in landscape planning procedures. Confusion can come
about if, on the other hand, a total integrated holistic landscape approach is
regarded as outdated; in this case, unfortunate retrogression in concept and practice can result. The fact is that advanced landscape approaches that make use of
holistic landscape-ecological perspectives, combined with the capabilities of
geographic information systems and appropriate use of computers, can make for
a most potent landscape planning procedure. It will be helpful to utilize concepts and terminology employed by various researchers and formulators of procedures to gain a sense of the trends in this fast-moving area. This will assist in
the incorporation of landscape-ecological knowledge into landscape planning.

It is also well to keep in mind that use of a landscape approach in the survey
of natural resources should in no way counter or preclude the use of the latest

tools and procedures known in landscape planning. It is the harnessing of these in the overall process that needs to be considered, and indeed in these regards much evolution has occurred in the period from the 1960s until now. True, the parametric approach "allows the values of various landscape attributes to be weighted differentially to account for their relative significance. These weightings are usually perceived economic, social, political, professional, or other values. Comparisons among the results of different weightings can also be readily made" (Fabos, 1979). But within the framework of a total landscape perspective it should be possible also to explore individual attributes.

We shall now explore some methodologies for capability analysis and regional landscape evaluation against this backdrop, remembering that here the term "evaluation" is used in its broader sense. Some of the methods are in reality the initial stages of a larger, more comprehensive physical planning process, whereas others represent the total process itself. The examples are drawn from The Netherlands, Australia, north-central New York State, and Germany. Although not elaborated on here, further development of important computer-focused landscape interpretation methods of Steinitz (1977) at Harvard University and their adaptation to the ecological, economic, and sociological conditions of Israel, has been accomplished by Professor Shaul Amir of the Technion-Israel Institute of Technology at Haifa. These methods are already in use in planning of the coastal region of Israel.

The Netherlands General Ecological Model and Environmental Survey

In Chapter 1 we stressed the outstanding position of The Netherlands in regard to the appreciation of nature, conservation, and landscape-ecological principles in the implementation of comprehensive landscape and physical planning on a nation-wide basis. There we also briefly discussed the leading role played by several vegetation scientists and plant ecologists, such as Westhoff, van der Maarel, van Leeuwen, and Zonneveld, in developing what could be called "The Netherland school of landscape ecology," which culminated in the organization of the first international congress of landscape ecology in April 1981 and the foundation of the International Association of Landscape Ecology (IALE) in 1983, its headquarters in The Netherlands. We illustrated this approach by several examples (Tables 1-3, 1-4, and 1-5 and Figure 1-4). Here we intend to present a more detailed description of the general ecological model (GEM) of the Ministry of Housing and Physical Planning. The GEM and the Netherlands Environmental Survey both serve to illustrate the degree of perception of—and commitment to utilize—such inputs from landscape ecology in planning (Ministry of Housing and Physical Planning, 1977).

The National Physical Planning Agency prepared study reports intended for professional physical planning circles and containing treatises geared to physical planning subjects at the national level. The series "general physical planning outline" includes the general ecological model, the result of four years of consideration and discussion by ecologists and planning experts. This group was in effect designated by the National Physical Planning Agency to study the GEM. The basic study, "Towards a General Ecological Model for Physical Planning in The Netherlands," and the study "National Environmental Survey" constitute the ecological basis of the process planning and the WERON ("Werkproces Ruimtelijke Ontwikkeling Nederland," Working Procedure Physical Development of the Netherlands) system, which designates the form for process planning by the National Physical Planning Agency.

In Chapter 1 we referred to the confusing use of the same "ecosystem" term for different levels of integrations and approaches by natural scientists and regional planners, a practice also adopted in the GEM report. We attempted to rectify this ambiguity by introducing the THE concepts and the distinction between different holons of bio- and technoecosystems. Despite this semantic criticism, however, the conceptual and epistemological basis of the GEM study has much in common with that outlined by us in Chapter 2, and the GEM can be considered one of the most comprehensive and valuable attempts at an ecological evaluation of physical development and a formulation of the options society is faced with regarding the natural environment. The main issue is the ecological foundation of WERON: The natural environment is regarded as a subsystem within the physical system, which is the central planning object of WERON. Thus the GEM study first describes the elements and processes within this subsystem. Also, indications are provided of relationships with other subsystems of the physical system. Further, "in physical planning practice it will be necessary to direct further attention to the development of physical planning evaluation techniques: techniques for finding optimal physical planning solutions (given a set of objectives and priorities of objectives). The GEM study makes use of the results of the National Environmental Survey, for which it provides the theoretical background at the same time."

In relation to WERON, and the position of the GEM in WERON, the question of the approach used becomes important. WERON follows a systems analysis approach, in which the aspects of reality relevant to physical planning are represented schematically. With the physical system in the central position as an object of planning, distinctions are made among the social system, involving social structures and processes between individuals and groups in society; the economic system, involving economic relationships in society; and the ecological system, in which is reflected the relationship between society and the natural environment. Separately taken up is the administrative system, which includes planning objectives, decision-making bodies, and procedures and planning instruments. The GEM study essentially is an attempt to provide a theoretical basis for the ecological system and for its methodological elaboration.

The GEM method includes four stages: (1) description of the functions of the natural environment for society (survey of functions); (2) description of the properties of the natural environment with respect to the fulfillment of functions (ecological evaluation); (3) description of the threat to function fulfillment emanating from side effects of social activities (ecological interaction analysis); (4) description of the interest groups and conflicts of interests with respect to

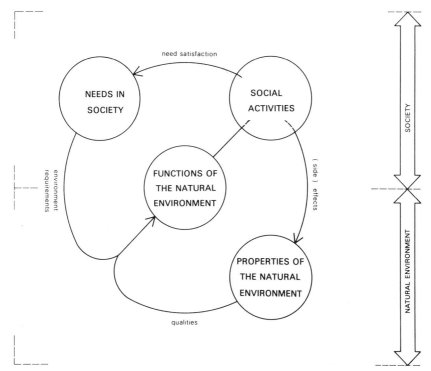

Figure 3-32. Outline of the General Ecological Model. Courtesy of the National Physical Planning Agency, The Netherlands.

function fulfillment (social evaluation and conflict analysis). Figures 3-32 and 3-33 illustrate the outline of and steps in the GEM method.

The GEM study tried to arrive at a main classification of groups of functions of the natural environment from a single point of view. Toward this end, a distinction was made between flow of energy and materials on the one hand and information flow on the other. In a survey of functions of the natural environment and their relationship with natural spheres of influence, a listing of produc-

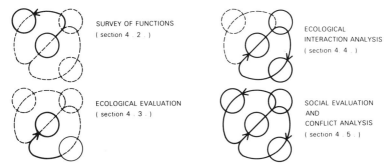

Figure 3-33. Steps in the GEM method. Courtesy of the National Physical Planning Agency, The Netherlands.

tion functions, carrier functions, information functions, and regulation functions was developed, and positive relationships, as well as possible or future relationships with the cosmosphere, atmosphere, hydrosphere, lithosphere, pedosphere, and biosphere were recorded in a matrix arrangement. (The functions are presented in Table 1-4.)

The summing up and classifying of the functions of the natural environment were intended to provide a broad base for ecological evaluation and planning. Ecological evaluation is defined in the GEM as the determination of the suitability or the capacity of the natural environment for the fulfillment of its various functions. In the summary report of the GEM study, the criteria for the information and regulation functions are elaborated on, these functions having been stressed because of "the underdeveloped state of knowledge concerning the fulfillment of the information and regulating functions; the highly complex structures and mechanisms in the natural environment associated with the fulfillment of these functions; and the vulnerability of these structures and mechanisms to large scale and dynamic spatial developments." Mention is made that the validity of various criteria is a matter of discussion among ecologists and planners, and further, that the measurability raises some questions, including the lack of requisite basic information ("for the time being we shall have to accept estimates for most of the criteria, based on superficial understanding, supplemented with results from representative field research"). The National Environmental Study in effect represents an initial attempt to map such data. Not only should basic inventories be carried out, it is felt, but interpretations of the basic material should follow (with the criteria as a guideline).

Some criteria for the fulfillment of information and regulation functions are listed as follows:

Function	*Criteria*
Information functions	
Orientation	Variety of vegetation structures and landscape elements
	Continuity of vegetation structures
	Age of trees
	Characteristic identity of landscape patterns
Education	Richness in remarkable species of plants and animals
	Variety of soil types
	Variety of ecosystems
	Richness in geological and geomorphologic phenomena
Indication	Biotic variation (variety of plants, animals, microorganisms, and biotic communities)
Scientific information	Richness in species
	Richness in ecosystems
	Variety of soil types

	Geological–geomorphologic variety
	Richness in fossils
	Rarity of species, ecosystems, soil types, geological phenomena, geomorphologic phenomena, fossils
Reservoir	Richness in species

Regulation functions

Regulation of factors from the cosmosphere	Ozone content of atmosphere, water vapor content of atmosphere
Regulation of factors from the atmosphere	Maturity and naturalness of ecosystems
Regulation of factors from the hydrosphere	Maturity, structural differentiation, and naturalness of ecosystems
Soil regulation	Naturalness of ecosystem
Biotic regulation	Maturity of ecosystems and richness in species

Purification functions

Noise absorption	Density of vegetation
Filtration of dust	Height and stratification of vegetation
Biological purification	Natural dynamics (absorption capacity for waste materials in the mineral cycle)

Stages in the GEM Method

Stage 1. Procedure of function description. In stage 1 of the GEM method, a study phase takes on the diagramming of energy, material, and information flows and considers the overall picture of needs in society, which results in function groups. Spheres of influence of the natural environment are noted, and along with the function groups lead to the development of subfunctions. In the planning phase, elaboration of objectives results, with influence, of course, from the policy area (main objectives of physical planning policy).

Stage 2. Procedure of ecological evaluation. In setting forth the procedure for ecological evaluation, a study phase moves from consideration of functions of natural environment and criteria for function fulfillment to coverage of properties of the natural environment [and the carrying out of inventory (basic maps)]. Attention is given to the capacity and/or suitability of the natural environment. This activity in the study phase leads to the planning phase of preparation of suitability maps (from the basic maps) and their division into compartments, eventually being influenced by norms and priorities of policy.

Stage 3. Procedure of ecological interaction analysis. A study phase at this stage addresses social activities and influences, or the susceptibility of the natural environment to activities, leading to the effects on properties of the natural environment. By combination with functions of natural environment, it is possible to note the degree of incompatibility of activities and functions. In planning, the preparation of susceptibility maps and the taking into account of potential areas of conflict (based on the degree of incompatibility of activities and functions) result in the preparation of both alternatives of allocation of activities and (spatial) restrictions with respect to activities. As in previous stages, norms and priorities of policy are utilized in the procedure of ecological interaction analysis.

Stage 4. Social evaluation and conflict analysis. As stressed in their summary of GEM,

> a proper foundation for ecological planning requires more than just insight into ecological relationships. Natural sciences may provide insight into the consequences of alternative physical developments, whereby the options are clarified. It cannot tell us which choice we must make. The concept of "ecological evaluation" explicitly excludes subject estimates of "values," because otherwise we would introduce a clouded mixture of scientific information and individual standards of value. Nor has it been attempted to develop an economic cost–benefit analysis in which the social value of the natural environment is determined. We assume that the weighing of (spatial) alternatives takes place on the basis of norms which are of a political nature, i.e., norms which are derived from specific social objectives. (Ministry of Housing and Physical Planning, The Netherlands, no date)

Social evaluation and conflict analysis attempts to define the interests of different groups in the fulfillment of functions and to confront them with one another. Emphasis here once again is given to interests associated with the fulfillment of information and regulation functions. But at the time that these interests get weighed against each other, it is also necessary to bring in those interests that are involved with the carrier and production functions. Claims on space (which are independent of functions of the natural environment) must also be considered.

The GEM is not regarded as having the task of indicating priorities and norms in a weighting problem of this kind. But the GEM does "search for information that can contribute to a more effective political weighting."

An example is provided in the form of a fictitious decision to build a road in a rural area. Different interest groups and sectors are identified as follows, in what is admittedly a very rough survey.

Function	Properties of Natural Environment	Interest Group or Sector
Carrier functions	Carrying capacity of soil for road body	Road user (taxpayers)

Production functions	Suitability of soil for agriculture	Farmers (consumers)
Information functions	Recognizability, familarity of natural elements	Recreationists, inhabitants
	Rare ecosystems	Education, scientific research
Regulation functions	Storage of water in soil, canals and lakes	Inhabitants of the area

It is recognized by the preparers of the GEM basic study that such summing up of social factors represents only the beginning of social evaluation and conflict analysis. What does result, however, is the placing of ecological evaluation and interaction analysis "in a social context that joins with other problem fields."

In the procedure of evaluation and conflict analysis, the study phase includes exploring functions of the environment and incompatibility of functions and social factors with interest sectors and groups, providing the identification of conflict of interests between functions. In the planning phase, the social description conflict area is brought into the picture and social description alternatives produced. Again in this procedure, norms and priorities of policy are introduced.

Some conclusions about the GEM are that (1) although it is regarded as theoretical, an effort has been made to render the model operational in physical planning. (2) The method is of a largely verbal and qualitative character, but a mathematical and quantitative elaboration is regarded as possible. "Environmental surveys will be able to make an important contribution to the quantitative elaboration of the method. These surveys have hitherto been concerned with the determination of suitability. Quantification and surveying of influences of social activities and of the susceptibility of ecosystems to certain activities could in a supplementary way lead to an elaboration of the ecological interaction analysis. As regards quantitative elaboration the social evaluation and conflict analysis is still the least developed one" (Ministry of Housing and Physical Planning, 1977).

A connection with other model techniques developed for physical planning is considered necessary. Linking the GEM with allocation and competition models used in the MIRAD study (Buchanan, 1976) was contemplated. The MIRAD models focus on land-use functions, emphasizing location of population in the form of residential areas. At the supraregional level, MIRAD uses a regional allocation model that produces a rough allocation of population and employment. The competition model of MIRAD is involved with urban–regional to regional level concerns. It is interesting to note that in both models, "suppositions concerning possible restrictions with respect to land use originating from natural environment play a part. The results of GEM can be used to produce further support for and to elaborate these suppositions."

The Netherlands Environmental Survey

The national environmental survey in The Netherlands was undertaken on be-half of WERON, which is an attempt to make long-term physical planning into a

> real and continuous element of the work of the National Physical Plan-ning Agency. What it amounts to is to put the ideas on planning into effect as a cyclic and continuous process. The cyclic character reveals the possibility of feed-back, whilst the process is taking its course, to previous phases, data or goals. The continuity implies that the topic studies are constantly being adapted to recent information, augmented with fresh information and that new methods and techniques are intro-duced all the time. This makes it possible to work with a constantly advancing time horizon and extract data at any time, for instance on behalf of the preparations for a policy report (Ministry of Housing and Physical Planning, no date given).

The Netherlands National Environmental Survey was carried out by staff re-cruited for the purpose, and the study team was accommodated at the National Institute for Nature Conservation. Guidance for the team was provided by a panel for national environmental surveying that included representatives of the Soil Survey Institute of Wageningen, the National Physical Planning Agency, and the Roman Catholic University of Nijmegen.

Over a long period of time basic material on its natural environment has been collected in The Netherlands. Early in the last decade, it was felt necessary to make these data suitable for use in behalf of national physical planning policy. Recent years have seen an increasing number of environmental surveys on a regional and local scale. Generally such surveys involved performing "a concen-trated study of natural elements in the area where specific planning develop-ments will take place. This then fits in with investigation of the physical system essential for planning, which is the system of artifacts and natural environment (altogether also referred to as the physical elements) and the activities and func-tions located there." Such data collection serves as the basis for physical plan-ning efforts, and allows for measurement of the consequences of various physical development possibilities.

The basic goal of national physical planning policy in The Netherlands is de-fined as "The promotion of such physical and ecological conditions as to ensure that: (a) the true efforts of individuals and groups in society are given the best possible chance; (b) the diversity, coherence and duration of the physical en-vironment are guaranteed in the best possible way."

The National Physical Planning Agency ordered a survey of The Netherlands natural environment to be executed on a national scale in mid-1972. The object of the survey was "The drawing up of an inventory, typology and survey of natural environment in The Netherlands, centered on physical planning at national level and in such a way that relationships can be established with activi-ties of groups within society and with the functions which the natural environ-ment fulfills to this end."

Such environmental survey is part of a larger study of an overall character needed in physical planning. The environmental survey is part of the landscape survey, the latter also including study of the scenery and the landscape's socio-historic characteristics. Also, a thorough study is needed, including the environmental load caused by pollution of air, soil, and water as well as by noise pollution.

For physical planning, an environmental survey that is properly executed may be regarded as "an instrument enabling the activities of society to be located in such a way that they can take place in the greatest possible harmony with the natural environment." Again, the point is made—as in landscape planning documents elsewhere in the world—of arranging for activities to take place in locations "where natural environment offers the best facilities for the activity concerned while on the other hand, such activities will have to be located in places where they do the least harm to the environment."

The survey in The Netherlands was intended to provide for physical planning at the national level, and it was decided that the basic data collection would take place systematically and be completed in about two years. A determination was made to represent the results of the national environmental survey in maps to a scale of 1:200,000. Little time was available for field work, and existing information was to be employed as far as possible.

As mentioned in Chapter 1, in the European landscape-ecological planning process much emphasis is placed on vegetation as a crucial characteristic of the physical environment. This is well reflected in the WERON survey, which regards vegetation occurring in any particular location as constituting a representation of the interplay between soil, climate, and human influences, while on the other hand characteristics of biotic communities are determined in large measure by the type of vegetation, its structure, and its variety. The vegetation map is regarded as "a sound basis for a landscape ecological survey."

A practical problem encountered in The Netherlands is that the actual vegetation now present is, indeed, "cut up." This tended to make the preparation of a map at 1:200,000 representing actual vegetation an impossible task. For this reason maps of natural potential vegetation of The Netherlands were prepared, along the model of those described in Chapter 1 for West Germany. A move was made to prepare instead a map of vegetation that would finally develop within a period of from 50 to 150 years if direct human influences were to cease there; the map would cover the potential natural vegetation of The Netherlands. Since the type of potential natural vegetation (p.n.v.) to be expected in any area depends on soil factors, the survey of p.n.v. was based on soil maps to a scale of 1:50,000, on detail maps, and for the remainder on maps scaled to 1:200,000.

In addition, ecological interpretation of the soil maps was done in collaboration with the Soil Survey Institute of Wageningen. Thus a map to a scale of 1:200,000 was obtained of areas that "as regards soil conditions are more or less homogeneous or possess a specific combination of soil characteristics. It was subsequently ascertained in respect of these areas what potential natural vegetation was to be expected there. The areas having one and same potential natural vegetation, are indicated as p.n.v. areas."

The potential natural vegetation was determined through investigation of the near-natural and related woodland vegetation still occurring there per map unit of the soil map. (These kinds of vegetation approach the potential natural vegetation.) In instances where these types of vegetation were not present, a study was made of derivate plant communities—such as plant communities in an earlier developmental stage—but also vegetation types that replace natural vegetation under various forms and different degrees of human influence.

"Groups of plant communities can be distinguished, which are associated with a given type of potential natural vegetation either because they belong to the natural development series of this type or because they can occur within the area of the p.n.v. type, where their form of appearance is determined by the degree of anthropogenic influence. As a separate group there are then also the Aquatics and marsh vegetations, which likewise exhibit a certain bond with a given type of potential natural vegetation" (Ministry of Housing and Physical Planning, no date).

These development, substitution, and accompanying vegetation communities belonging to a type of p.n.v. can be combined and arranged in a vegetation series, making it possible to portray the actual vegetation. In The Netherlands, seven main series were distinguished, namely:

1. Main series of woodlands belonging to the alliance of summer and winter oaks (*Quercion robori - petraeae*);
2. Main series of the woodlands of the Carpinion;
3. Main series of the woodlands of the Alno-Padion;
4. Main series of the woodlands of the Alnion;
5. Main series of the scrubs and woodlands of the *Salicion albae*;
6. Main series of the bogs;
7. Main series of the dune woodlands and scrubs.

Plant communities in the series are divided over five groups. These are (1) woodland vegetation; (2) scrub vegetation; (3) seminatural land vegetation (comprising the plant communities maintained by a slight, but constant, human influence; this kind of community includes borderline communities of woodlands, moors, extensively used grassland, and limestone grassland); (4) little or no natural land vegetation (includes the strongly man-influenced communities of arable land, intensively used grassland, and wasteland, as well as certain types of dike vegetation and trampling plant communities; (5) aquatic and marsh vegetation (includes aquatic and waterside communities of oligotrophic to eutrophic waters; the marsh vegetation of the bog and fen areas also belong to these groups).

For all of the p.n.v. types there is an indication of which plant communities may be encountered within the five main groups of the relevant vegetation series. In an inventory, the p.n.v. areas were looked at to ascertain the areas occupied percentagewise by the five different groups within the vegetation series. (Area percentages were expressed in code figures, with distinctions made for area, line, and point elements.) Data for this purpose were derived from topographical maps at a scale of 1:50,000. The eventual result of this activity was the preparation of a 1:200,000 map divided into 7000 sections.

Color is importantly utilized in this process [for an excellent discussion of the use of color in the mapping of vegetation see Kuchler (1967)]. It is employed to designate what p.n.v. prevails in a particular area; in addition, the actual vegetation of the map section can be derived by using a five-figure code with the aid of a table of the vegetation series. To understand this better, an example is provided of an area having a p.n.v. consisting of a complex of oak–birch woodland and beech–oak woodland on a hydromorphic podzol soil. The coding is as follows: 2 .... 25% woodland; . 1 ... few copses . . 3 . . 25–50% moorland; ... 4 . 50–75% cultivated land; .... 1 few ditches. The code 2134 is thus used here to indicate the actual vegetation.

It should be noted that gradient situations, with gradual transitions between p.n.v. areas occurring in such situations, represent a problem not covered in the vegetation survey, but there is recognition that such transition areas are important, partly because they may be characterized by a large variety of species and many rare species may be represented among them. The size of the transition areas made it impossible (because of the scale of the national environmental map) to present them as separate map units.

In the summary of The Netherlands Environmental Survey it is stressed that ideally a survey of complete biotic communities, including the fauna, should logically extend the vegetation survey. This was not found to be possible, because "the branch of science dealing with this has hardly come to development." However, data on occurring animal groups supplemented the vegetation data, but few inventories of animal groups suitable for survey needs were available. Effort was limited to making distribution maps of a number of animal species "for which national inventories had already been drawn up in such a way as to enable the data gathered to be used for the national environmental survey, either directly or after simple processing. Distribution maps of rare breeding birds, groups of certain migratory birds, ducks, geese, waders and grassland birds; all reptiles and a number of amphibians and butterflies. Also, as a supplement to the vegetation map, distribution of rare types of vegetation were also incorporated."

The Netherlands National Environmental Survey had as the purpose of its landscape-ecological survey the elaboration of theoretically demonstrated relationships between community and the natural environment for concrete situations. Thus, collected data were interpreted in terms of the following:

(a) indications of where and in what measure specific functions of the natural environment are fulfilled within a given planning area (ecological evaluation).

(b) indications of where functions of natural environment are likely to be affected by specific activities of human society or side-effects thereof. This is referred to as compatibility of functions and activities. Next, the effort is made to indicate how and where functions of the natural environment that are highly valued socially could be further developed (potentials or development possibilities). (Ministry of Housing and Physical Planning, no date).

As was shown in the extensive discussion of the GEM, careful analysis was undertaken in The Netherlands of the functions the natural environment is capable of fulfilling with regard to society or groups within society. This is followed by coverage of the way in which these functions can be rendered measurable, recognizing that it is not now possible to establish measuring criteria (for all functions) with ample certainty.

In considering the suitability of the natural environment in The Netherlands for fulfillment of ecological functions (on the basis of the inventory drawn up on behalf of the national environmental survey related in particular to the concept of information functions), a map of The Netherlands dealing with the capability of the natural environment to fulfill information functions was developed in addition to the potential natural vegetation map. Taken into account were (1) differences in completeness of the inventories drawn up from points of view of different biological disciplines (botany, zoology, ornithology, hydrobiology); (2) current theoretical problems related to the integration of inventories drawn up by various biological disciplines into a single evaluation.

As a consequence, the method chosen was the stepped suitability determination, meaning that first a suitability map was made on the basis of actual vegetation; then it was determined which areas had to be placed in another category on the basis of data provided by the other biological disciplines. The map that resulted showed a recording of the occurrence of natural elements of special significance. Categories were (a) area with almost-everywhere natural elements of national significance; (b) area with in large parts natural elements of national significance; (c) area with in many places natural elements of national significance; (d) area with in some places natural elements of national significance; (e) area with practically no natural elements of national significance; (f) open water with almost-everywhere natural elements of national significance; (g) open water with in large parts natural elements of national significance; (h) open water with in many places natural elements of national significance. Thus, five categories or areas were distinguished for the land, three for the water. One category was created for not-yet-classified areas. The map, "Survey Map of the Ecological Significance of the Natural Environment in The Netherlands," can be properly used only in combination with the vegetation map.

Concerning the determination of the (in)compatibility of functions and activities, the national environmental survey did not intend to produce maps of the vulnerability of the natural environment to activities and side effects thereof. In the case of vegetation, on the basic national environmental survey map, by using a code, eutrophication, trampling, and lowering of the water table and how great the chance is of being negatively affected, were estimated. In the p.n.v. areas, a figure was used to express the estimate. Several categories were delineated to convey the degree of vulnerability.

Social goals are important because they bear on the functions of the natural environment. In the case of The Netherlands vegetation map, once these goals were formulated, the table of vegetation series provided windows to see in what areas it is possible to realize these goals.

In The Netherlands survey, development possibilities are given for a number of vegetation types rare in that country. The information functions of the natural environment would thus be improved. Examples included vegetation of shifting sands, wet to moist hay meadows, flowering grasslands, and heathlands.

Returning to the matter of ecological evaluation, it was felt that this kind of evaluation is not really complete until it is followed by the "allocation of social weights." After the development of the ecological model, the matter of giving social weighing an explicit base has begun to take place. Two phases of the evaluation process are discernible, namely, that of (1) "determining suitability of the natural environment for fulfillment of ecological functions" and (2) social evaluation. Interestingly (and remembering also the concern of the Metland procedure in this regard), the terms "evaluation" and "valuable" are confined to the second phase. However, the terms "ecologically valuable areas" and "valuable" are still often used in certain documents for the first phase, undoubtedly creating a degree of confusion in terminology, the report notes.

The matter of application in physical planning of The Netherlands Environmental Survey bears on such issues as whether it will primarily be checking individual physical plans or comparing alternatives. The checking of alternatives seems to be the more productive route to follow, recognizing that norms concerned with quality of the environment will be subject to constant change, and this creates a problem. But the setting of conditions or tolerance limits is indeed possible, and in exploring individual plans it is possible to determine the loss or gain with respect to the natural environment and its functions for society, but "not whether as a result the plan becomes acceptable or not."

It is here that the WERON approach is recalled, since in it yet another application is seen. This approach gives the ecological system its "own autonomous entry in a very early planning stage. In this way the significance of a system of formulated quality requirements is actually manifested."

The Netherlands survey is also considered in terms of its application in a structure scheme, which is a "long-term plan relating to a sector having a strong physical planning accent." It lies at the intersection of facet and sector planning (the latter being preparations of policy for road construction, house building, electricity supply, public transport, drinking-water supply, educational facilities, agricultural policy, welfare facilities, etc.). "A structure scheme is a report with maps on the long-term policy to be conducted on behalf of specific provisions, which are relevant to the physical planning policy and for which the central government bears a considerable degree of responsibility." It presents an overall estimate for a period of 25-30 years of the extent and location of the provisions concerned. It is reviewed once every five years. Also, there exists a structural outline sketch, defined as a long-term physical facet plan.

Incidentally, the term "facet planning" has been in use since 1968, when a report of policy formation for future social structure was produced. Distinction is made among economic, physical, and sociocultural facet planning. Economic and physical planning already were represented through two facet planning agencies (Central Planning Agency and National Physical Planning Agency). Sociocultural planning was not so represented. In 1974, as a result of the report, the Socio-Cultural Planning Agency was created.

Facet planning checks the work of individual sector departments against economic, physical, and social–cultural criteria. Checking the facet policy occurs with the assistance of facet reports. An example is the Government Report for Physical Planning in The Netherlands.

With provisional results of The Netherlands National Environmental Survey, a number of structure schemes in an advanced stage of elaboration, such as the network of main roads and railways and the national network of pipeline tracks, were checked. In these schemes appear statements based on expected requirements concerning the nature and extent of the provisions concerned. In addition, indication is given in an initial way of the location or track for each provision. Rough insight is gained early in the planning process as to the amount of space, and where (which area) there is likely to be need for this provision in the future. The National Environmental Survey is employed to check these locations and tracks to see if, when the space is used, there will be damage to the natural environment.

The additional applications of the national environmental survey are evident in the following account by the survey investigators:

> A start was made with the application of the National Environmental Survey in the preparatory work for the Structural Outline Sketch for Urbanization. The Structural Outline Sketch for Urbanization contains statements concerning the physical structure of urbanization. Norms were established, depending on the significance of the goal with respect to the natural environment and on the chosen strategy with respect to the open space—i.e. greater emphasis on physical integration or segregation.
>
> For the various urbanization alternatives the conclusion was drawn that in Class A and B (and F and G) of the survey map, no urbanization should take place (these are areas which practically everywhere consist of large natural elements of national significance).
>
> Hence, the differences between the urbanization alternatives developed stand out particularly in the evaluation of the areas in the other classes.
>
> National rarity was not the only criterion that played a part in this latter evaluation. Regional parity and the ecological relationships between the p.n.v. areas have as far as possible also been introduced as criteria. Especially in the case of the urbanization strategies aiming at integration, the suitability aspect (for urban provisions) has also been introduced by relating the operations necessary on behalf of urbanization to the abiotic data (groundwater, soil). Finally, an effort was made to incorporate the influence of the side effects occurring as a result of increasing urbanization (such as increase of infrastructural facilities and

increase of recreational activities in the surrounding urban area) on the fulfillment of the functions of the natural environment.

The Netherlands general ecological model and The Netherlands National Environmental Survey, although not providing a total set of methodologies for decision making, do make an important contribution to the planning process, and insert ecological functions and evaluation into the larger planning picture. They show clearly how holistic landscape-ecological concepts and methods can be used as practical tools on a nationwide basis for decision making in land-use planning.

An Australian Approach to Gaining and Using Information for Analysing Regional Land-Use Options

The government of the state of New South Wales, located in southeastern Australia, which includes the major cities of Sydney and Canberra, called in March 1972 for the national Commonwealth Scientific and Industrial Research Organization (CSIRO) to join in a joint study of land use. The area involved was to be the south coast of New South Wales. CSIRO was invited to undertake a pilot survey of resources in the area with the objective of providing a "rational basis for planning decisions on a wide variety of land uses" (Basinski, 1978). The request for CSIRO assistance evolved from heightened concern about expanding conflicts (and differences of opinion) as to how the region's land should be used. As is true in many other developed societies, the region has experienced subdivision and loss of productive agricultural holdings; recreation areas have become more urbanized; and along with the extension of national parks, timber resources have been lost. The south coast has changed from a rural backwater to a popular resort area.

Beyond the matters of what land uses are to be permitted and where they should be permitted, major conflicts exist about how such uses should be managed. The intensity of use (an example being foreshore degradation and erosion) and managerial control (such as fire control in national parks) exemplify areas of contention.

By June of the same year, CSIRO's Division of Land Research (as it was then called) was requested to submit a research project proposal to deal with the problem. Since the problem related to how the Australian land base should be used, the project was considered an appropriate one for the division to handle. (At the time, the research goals of the division were to specify Australia-wide possibilities and limitations for a range of productive land uses and to identify and measure biological functions of nonurban land.) The then-followed approach involved developing more potent methods of land data acquisition and of information storage and retrieval, and developing mathematical models of land productivity and of biophysical processes suitable for broad areas.

However, on the New South Wales south coast, the land-use problem involved all functions of land (well beyond just primary productivity and biological function). The context was a very dynamic one. It also concerned "balancing existing and new ways of using land, as contrasted with new technologies of land use," in a specific, relatively small area in New South Wales.

It was recognized that the development of new frameworks, capabilities, and approaches would be necessary. Research needs were immediately identified as including (a) a philosophy and understanding of regional land-use planning; (b) both biophysical and socioeconomic input to the project; (c) the problem of integrating biophysical and socioeconomic research into one unified study; (d) selection based on appropriateness for regional planning studies; (e) methods for communicating resource information useful for regional planning.

In addressing the land-use problem and seeking approaches to its solution, questions were raised about the nature of the land-use problem confronting the New South Wales south coast and what feasible approaches exist for its solution. An exploration of the regional planning and land-use planning literature suggested four areas of options, namely, identification of the "optimal" pattern of land use; studies of the social context of planning the information flow–decision-making system within which land change is initiated; elaboration and evaluation of alternative patterns of land use that reflect different land-use ethics; and methods of acquiring basic planning information (Cocks and Austin, 1978).

It should be noted that in 1973 the CSIRO implemented a reorganization of divisions concerned with land. The Division of Land Research became the Division of Land Use Research, with a socioeconomic research group added to researchers already involved in studies of land and water potential (Whyte, 1976). This conveys recognition of the need to look beyond biophysical concerns toward a more inclusive approach.

With the reorganization of the division in 1973, and the participation of the Australian National University's Center for Resource and Environment Studies, the project was expanded, and ultimately involved in some way about half of the scientists in the division (Division of Land Use Research, 1978).

The initial request to CSIRO had been for information on "total resources" of the region. As noted by Cocks and Austin (1978), decisions on which facts to provide and how they are to be presented involve value judgments (a fact not always recognized by physical scientists). A purpose must be present to guide this exercise; this must be chosen from individual perceptions of the problems lying behind a request for assistance, they contend.

The Nature of the Land-Use Problem. One consideration is that the land-use problem in essence reflects conflict between different uses for limited land. Particularly is this true in areas that are developing, the south coast of New South Wales being an example. Another alternative consideration is that of "unacceptable interactions" between land uses, as is the case where subdivision influences the property rates of nearby agricultural lands. The CSIRO researchers correctly stress that such forms of conflict are "objections to an evolving pattern of land use, but conflict can also occur over the process of land use control as well as over the products of that process, i.e. over the legitimacy and acceptability of procedures for determining zoning regulations, etc." The general purpose of the program took a turn from the narrow provision of information useful for resolving land-use conflicts to that of providing information for facilitating an informed consensus within society on how land should be used and how land should be controlled. Information here refers to relevant data and to methods for its use.

Land use and control are variously viewed by institutions and individuals who implement and influence decisions. The ideas may deal with context-free principles (all erosion is bad) or specific ideas for a specific area (e.g., conserve Mount Dromedary). The ideas may be partial or comprehensive and emphasis might be devoted to control or function. "Such ideas of how land should be used and controlled can be thought of as a *land use ethic*." In the instance of government, land-use ethic can be regarded as its land-use policy.

Understanding of the historical progression of land-use planning methodologies (presented earlier in this chapter) provides a sense of evolution in these methodologies. In the New South Wales study, the underlying techniques of some of these methods were critically reviewed before proceeding. In addition, objectives of planning lie between goals and implementations. Rawls (1972) is cited as regarding the planner's aim is to come to a "reflective equilibrium in which all three levels are compatibly and sufficiently linked." Etzioni's (1968) perspectives view this as a

> mixed scanning procedure involving *iteration* (successive approximation narrowing the options until a solution is found); *review* (examining each step in terms of all the planning processes mentioned); *reticulation* (proceeding to that step which generates the most problems whether it be goal generation, control instruments or information acquisition); and *design* (the generation of relevant ideas). There is recurrence and interaction between the processes of expanding the set of potential objectives (by increasing awareness or understanding of the nature of the system being planned); narrowing the set of potential objectives (by evaluating or appraising possibilities against external criteria). (Etzioni, 1968)

Survey Techniques, Including Land Capability and Sieve Mapping. Survey techniques for expanding and narrowing particular options within a range of planning focuses include land capability and sieve mapping techniques, infrastructure-oriented techniques, and regional economics and planning techniques. We will take note here of the first type only. In the case of land capability techniques, several land capability classification systems have been developed and are being used. Land with certain characteristics is determined to be capable (or incapable) of certain uses. Development possibilities for agriculture or forestry have been considered by such land capability systems, one example being that of Lacate (1969). An oft-noted criticism of such systems is that they are divorced from economic, social, and technological considerations and "are limited to a sequential evaluation of unchanging objectives" (Cocks and Austin, 1978) and that they are based on a large intuitive element in the classification procedures leading to ambiguities. As one moves to sieve mapping (McHarg, 1969), the best location for a particular land use is identified by successively eliminating areas. Sieve mapping helps in choosing among options but, as we have already noted, does not represent the most highly developed methodology available today.

The South Coast Study Sequence. The study of the south coast of New South Wales was aimed at seeking the data that seemed necessary for comprehensive regional land-use planning and at examining the use of these data for planning. Attention was devoted to developing methods for acquiring pertinent bio-physical and socioeconomic data, for storing and displaying these data, and for analyzing and integrating the data for comprehensive planning. Particular consideration was devoted to allocation of land for different uses.

Concurrent studies were undertaken. Collection of information that could be used for comprehensive regional planning was carried out. The intent was to employ the information in the context of a planning methodology, supported by a computer data bank and/or mapping system to examine land-use possibilities in that portion of the south coast consisting of Eurobodalla Shire and its catchments. Specifically undertaken were biophysical studies—data on land attributes such as climate, geology, hydrology, terrain, soils, fauna, and vegetation; socioeconomic studies that dealt with demography, economics, legal aspects, individual attitudes, and area institutions; land-use studies—with the intention of gaining information on current land use and possible future land use, including urban activities, agriculture, forestry, apiculture, fisheries, minerals, water supply, recreation, conservation and transport. The development of methods of integrating the information acquired through the resource studies for planning purposes was emphasized.

A two-level system of data storage and presentation used previously in work by the division was employed, but two modifications were introduced: "(1) Functional units (FU's) were adopted as the basic mapping entities instead of land systems, in order to: (a) preserve locational information lost by grouping unique mapping areas (UMA) into land systems; (b) facilitate the integration of socio-economic and bio-physical information by using both socio-economic and bio-physical attributes to differentiate mapping entities" (Austin and Basinski, 1978). Functional units are subdivisions of unique mapping areas based on land tenure and zoning boundaries. They are to depict areas relatively homogeneous in their functional possibilities (these possibilities are dependent upon qualitative characteristics as well as on the size, shape, and distribution pattern of component elements-facets). (2) Elements of functional units that were described but not mapped were defined in terms of facets instead of *land units*.

For planning purposes, the study area was initially divided into functional units based on both biophysical and socioeconomic boundaries. We now look more closely at the procedures.

It should be reemphasized here that the area comprises rugged forested mountains and hills, and has been relatively undisturbed, but it is evident that coastal belt recreational pressures have introduced changes in land use. The character of the area in the future will depend heavily on the allocation of resources by planners and others—and also on the maintenance of balance between development and the conservation of a diversity of natural landscapes.

The land area was subdivided into 14 functional districts along the limits of the three main catchments and the broad belts of undulating hilly and mountainous terrain. These lie virtually parallel to the coast. The area was further subdivided into 1482 unique mapping areas, this being done on the basis of

> distinctive patterns in geology, terrain type, and vegetation identified in 1:50,000 scale aerial photographs. Many of the unique mapping areas have similar patterns but each occurrence is regarded as unique in terms of its size, shape and geographic location. They have been grouped according to their geology, terrain type, soil and vegetation into 64 *land systems.* Imposed over the unique mapping areas are various administrative and land use boundaries which provide a further breakdown into . . . *functional units* of which there are 3855. The boundaries of the functional units have been entered in the data bank and form the *map base.* Finally, the functional units are subdivided into smaller components termed *facets.* These are relief units based on position on slope and aspect in various terrain types and are too small to be mapped or consistently identified in the 1:50,000 scale aerial photographs, but their occurrence and extent within each functional unit have been estimated by reference to 1:25,000 scale contour maps." (Gunn, 1978, p. iii)

The landform entities termed "facets" differ from land units (the term previously used) in that they are differentiated on the basis of landform and aspect, both of which can be determined directly from air photographs and contour maps. In addition, they are smaller and more uniform in size than land units. "Each facet is identifiably different from adjacent facets, is assumed to have uniform land function possibilities throughout its extent, and corresponds to a minimum unit of practical concern for all but the most detailed land use planning." The proportion and distribution of facets can be different in individual functional units within the same UMA. Further, it should be noted that facet descriptions are held in a computer data bank in terms of paradigms or ideotypes specifying landform, lithology, aspect, rockiness, soils, and vegetation. Paradigms specifying the first four variables are consistent for all occurrences. In some facets, paradigms specifying vegetation differ for different occurrences.

> The bio-physical features of the study area have been described by relating paradigms of landforms, soils and vegetation to the set of facets in each functional unit. The various attributes of these and other features needed for the development of exclusion criteria in the function studies . . . have also been recorded and entered into a computer data bank. These bio-physical descriptions of the functional units form the

attribute base and are linked to the elements of the map base. The data bank system therefore enables maps showing various attributes, classifications and land use options to be generated by machine. (Gunn, 1978, p. iii)

As indicated above, the functional uses are regarded as sufficiently uniform to have the same land-use potentials and limitations throughout. The stages in the methodology commenced with identification for each functional unit of those land uses regarded as socially unacceptable because they would be "unprofitable" or "lead to irreparable damage"—as with erosion, or would be incompatible with the use to which some other functional unit, adjoining or nearby, is already committed. The unacceptable uses are identified through the use of "exclusion rules" that are applied to the resource data collected. The exclusion rules quantify biophysical and socioeconomic attributes that would make an area of land unsuitable for a particular use. As an example, in the exclusion rule for forestry, slope and potential log volume are considered. Among urbanization criteria, land with a median slope above 6° is excluded.

The exclusion approach eliminates unacceptable land uses from further consideration. It has parallels to sieve mapping. However, the Australian researchers do not believe the exclusion approach has been used previously in the way it was employed in the south coast study. By application of the exclusion rules, it is possible to identify "option space" (a listing for each functional unit of unexcluded land uses). Following this, "preferred" land-use patterns are defined. Thus is identified for each functional unit land uses that should be encouraged if a particular policy of land use is to be executed.

Development of a regional land-use policy represents the initial step. Many component policies are brought together, each having a particular stance on definition of preferred land use. As an example of this, one can cite the limitation of urbanization to the coastal area, or that farming should be given favor over other uses on land having high agricultural capability.

Next, weighting is given to each component policy. This is based on its importance (as seen by decision makers who are involved in planning process).

Following this, it is necessary to develop measures conveying a sense of the degree to which a policy would be realized if a land use were carried out in an area. Rating weights, as these are known, are derived from land-use "support rules," the latter serving to quantify the potential of an unexcluded area.

After these steps have been taken, a preferred land-use pattern is further advanced by building a computer linear program procedure. Importance and rating weights are entered. In this procedure, one unexcluded use is selected for each of the functional units. "The sum of all the policies, as measured by importance weights, over all the functional units, is maximized."

Lastly, it is possible to investigate the effects of different policies on the pattern of preferred land use generated by changing the weightings assigned. The final stage involves appraising, assembling objections to, and (should it be needed) revising the preferred pattern of land use (see Figures 3-34, 3-35, Division of Land Use Research, 1978).

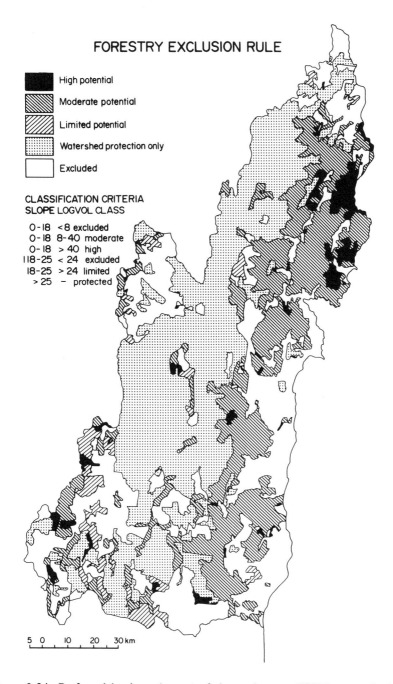

Figure 3-34. Preferred land use in part of the study area. CSIRO Australia Division of Land Use Research (1978).

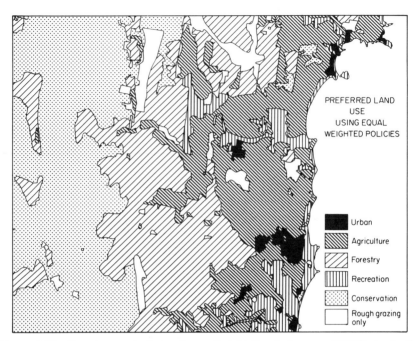

Figure 3-35. Forestry exclusion rule applied to the study area. CSIRO Australia Division of Land Use Research (1978).

The use of computers is strongly emphasized in the methodology utilized on the south coast. (Note the consideration for computer needs in systematizing land-use plans later in this chapter.) A system was developed that involved a structured data bank tied to a computer mapping system. This system makes possible rapid examination, by means of maps, of the implications of any set of exclusion criteria applied to resource data stored in the data bank. Further, should any objection be advanced to the selected exclusion criteria, substitution can be made of new criteria.

Products of the Study. As noted, results included biophysical studies, socio-economic studies, and land-use studies. In the last, land-use patterns and trends were examined in the south coast area. Identification was made of ten land-use categories, and the characteristics of land used for each purpose were summarized. Thus, much information was collected for each land use. The major product, from the standpoint of the overall project, was an exclusion map on agriculture, forestry, urbanization, recreation, apiculture, conservation, and residue assimilation (landfills and septic tank disposal). This showed areas from which the particular use in question should be excluded; it indicated areas that support rules indicated would have some additional potential for the land use.

A preferred land-use map was developed for the study area, with five land uses noted—agriculture, conservation, forestry, recreation, and urbanization—and a set of 36 policy statements was devised. From these, an acceptable policy might be developed. Several preferences were covered, such as giving an option to agriculture on predominantly freehold land, preference to recreation as an option in areas where demand is high because of ease of access from the city of Canberra, or preferring land uses that "imply a low change in current development status" (Division of Land Use Research, 1978). (See Figure 3-34.)

Judging the South Coast Project's Methodology. In attempting to judge the south coast project's methodology, one is first impressed by the researchers' evident contrasting of the current biophysical survey techniques with the division's earlier surveys. In those earlier surveys, there "existed close integration of specialist studies, extensive use of air photographs, minimum ground sampling and multi-attribute land classification" (Division of Land Use Research, 1978). The Land System surveys assumed that characteristics of terrain (and other attributes) identifiable on aerial photographs are utilizable for prediction and for extrapolation of spatial distribution of those attributes determinable only by examination on location. It further assumed that it is possible via multiattribute land classification to reach a meaningful generalization of land properties. Still further, the survey was based on the premise that every land use is constrained by combined and interacting effects of several land attributes. The implication is that balanced information on a range of relevant attributes is needed to assess land suitability for any land use and that multiattribute classification is necessary—or quite desirable—in evaluation (because the attributes cannot be considered in isolation). As a result of this type of thinking, the conclusion can be reached that the same resource data and the same multiattribute classification can be used to evaluate land for a range of uses, avoiding duplication of data collection, storage, and analysis. In addition, it is implied that closely integrated specialist studies, concurrent in nature, are required. The Land System approach surveys have been carried out in areas lacking a major human population and human impact; focus has been for early assessing of agricultural potentials.

Much attention is devoted to the limitations of the Land System surveys by Austin and Basinski (1978), as pertains to correlations, classifications, and extrapolation. They recognize that interattribute correlations do exist, but stress that usually the correlations are very complex and involve several attributes and their characteristics, and that even relatively simple correlations produced from survey data should be used with considerable caution for purposes both of classification and spatial extrapolation. An example is provided where correlations between lithology, landforms, and vegetation observed in one land system (or in one region) are not necessarily valid for other land systems or regions where different vegetation types occur on the same lithology and landform (or where the same vegetation types are found on different lithology and landform). Further, the qualitative methods of examining survey data are criticized as not enabling detection of valid and useful interattribute correlations, while numerical methods of exploratory data analysis are regarded as aiding in identifying correlations that are not readily apparent. In addition, intuitive thinking about causal relationships and/or existing land characteristics have been relied on in many studies. The need for a process model (which generally has not been developed for such a purpose) is identified for actual field studies.

Reconnaissance level surveys can result in the establishment of correlations that have limited utility for land classifications and spatial extrapolations. Also, subjective approaches and intuitive ideas of different survey teams have been involved in the grouping of mapping units into land systems. Another problem is that sampling schemes have not been sufficiently explicit for other researchers to repeat the procedure; also, different people's determining the sample pattern for different areas of the survey region is problematic.

In the south coast study, by contrast, stress was placed on developing explicit procedures for boundary delineation, grouping of mapping units into land systems, and extrapolating site data to unsampled areas. Survey techniques included (1) "explicit and consistent stratified sampling procedure based on preliminary mapping devised to ensure adequate coverage of all types of landscape and at the same time to provide more precise and reliable information for the more important parts of the region"; (2) the development of the concept of "facets" to allow more precise land description; (3) the use of a computer data bank and mapping system for data storage and analysis.

The south coast study represents a progression in utilizing integrated surveys for planning purposes. It builds upon earlier Australian experiences, recognizes some of the deficiencies in the earlier work, and attempts to rectify those shortcomings. The sequence followed could have been improved by changing the order of steps taken (Division of Land Use Research, 1978). The project should commence with policy, not with functional land definition. A more sequential project based on prior definition of the policies must be identified as a needed change for future efforts. Nonetheless, one must concur with the conclusion that the "main achievement is an integrated, comprehensive scheme for regional land use allocation and this appears to be relatively innovative" (Division of Land Use Research, 1978).

Recording of Visual Resources; Recreation Resources and Planning

Keeping in mind the focus of this book, it is not our intention to concentrate on techniques for visual resource recording, beyond coverage of the application of the Jacobs cataloging technique in Wayne County, New York. However, it is highly desirable to be aware of the great amount of activity and progress now occurring in this arena. As but one example, a major conference on applied techniques for analysis and management of the visual resource took place at Lake Tahoe in Nevada in 1979 (Elsner and Smardon, 1979) and covered technology available to solve landscape problems (including descriptive approaches to landscape analysis, computerized and quantitative approaches, technical standards for visual assessment procedures, psychometric and social science approaches, and evaluation of visual assessment methods), appropriate combinations of technology for solving landscape management problems (emphasizing surface mining and reclamation, urbanization, highway development, recreational development, rural and agricultural development, utility corridors and siting of power plants, timber management, water resource development, outer continental shelf and coastal energy development), landscape management systems, legal and policy tools available to use in solving landscape management problems (including litigation and landscape aesthetics, interface of legal and aesthetic considerations, landscape values in public decisions, visual quality testimony in adversary setting), new dimensions of visual landscape assessment, and future directions for research and management. The reader is encouraged to consult the proceedings of the conference, entitled "Our National Landscape."

In West Germany, Professor Hans Kiemstedt, now of the Technical University, Hannover, developed a method for evaluation of landscape resources and recreation planning. The method was intended to show that quantification of nature as a main resource for outdoor recreation is possible. The method has proved itself in a large-scale evaluation of recreation in the Sauerland area (investigation was carried out by a planning team from the Technical University of Berlin headed by Kiemstedt, then at that institution). The work was commissioned by Nordrheim–Westfalen Abteilung Landesplanung.

In Israel, the Kiemstedt method has been employed by a research team of the Center for Urban and Regional Studies of the Technion-Israel Institute of Technology, Haifa. They explored public evaluation of recreation resources in the planning process in the region of Mount Carmel (Schechter et al., 1979). The approach basically involves the integration of users' preferences regarding physical and aesthetic characteristics into an operational, ordinal measure of quality. The Israeli researchers also report on considerations involved in describing a model for evaluation of landscape resources in providing outdoor recreation services. It should be noted that the team in Israel has closely collaborated with Hans Kiemstedt, and their research formed one part of an overall research project in which he was the principal investigator.

The Kiemstedt method involves the "V-Value" for different landscapes, "V" standing for diversity. A basic assumption in the method is that natural conditions of a landscape play an important role for recreation, and that recreational effects can be reduced to a few essential factors for which measurement criteria can be found.

Holistic Preservation Planning, Decision Making, and Public Education

As noted in the section on remote sensing as a tool, a very useful methodology developed by the late Professor Stephen Jacobs of Cornell University (Jacobs, 1979) was applied in cataloging the visible resources of a rural area of New York State (Wayne County, near Lake Ontario). Emphasis lay in surveying readily observable features of the physical environment. The system that was used is readily stored and evaluated. The great majority of the items that were classified are architectural in character, but note is taken that the balance is adjustable, so that other features observable in the landscape of Wayne County could be favored. Involved in the Wayne County effort is a comprehensive, integrating approach, useful for planning and decision-making purposes and extremely helpful as well for public education needs. The product is an effective dovetailing of natural patterns, social developments, and economic development of the area.

Jacobs pointedly notes that few communities have successfully surveyed and recorded the totality of their man-made environmental heritage. Even when the will to survey was present, a satisfactory framework for identifying and classifying significant sites and structures was lacking. Further, "the will has tended to falter as people fail to agree on suitable standards for determining significance." Nonetheless, urban preservationists continue in their efforts to resist "mindless demolition working in the guise of social and economic progress."

There is increasing organization on the part of urban preservationists to realize their objectives. Jacobs stresses that outside the cities, the picture relative to preservation is quite different. As he states,

> Pickets publicize the impending doom of a Penn Station concourse but the crumbling of the ingenious canal lock or abandoned granary on which the economic structure of a whole rural community once rested goes unnoticed. Those who are not familiar with it, and many of those who have always known it, think of the rural world as relatively immutable—slow to degrade and more sheltered from thoughtless exploitation than the fragile wilderness of the dynamic urban area. They are wrong. The industrialization of agriculture and the abandonment of vast stretches of marginal farmland painfully cleared a century ago are producing drastic and rapid alterations. There is a real danger that the rural scene, with all its many evidences of symbiotic interaction between man and nature, may disappear without a trace—in a sense without ever having known it—because we have not looked at it carefully or attempted to record its manifold charms and quirks . . . . If works of genius or original aesthetic expressions are not generally expected there, all the elements of aesthetic experience are of course present. In the rural as well as the urban environment, we can be stimulated by the surface qualities of materials, by forms or combinations of forms, and by suggestions of association with other forms and expressions. Reflections of ingenious tradition based on available materials and local skills will be present, and when they are repeated often enough or give rise to forms that are naive or curious or unique, we can recognize these vernacular features as forging the local character. The number of visual stimuli in the rural scene is vast, and the range of items that can command our attention for an environmental survey on historic or aesthetic grounds is enormous. But the task is manageable if we can catalog features—providing a file classification or "pigeonhole" for every item of potential interest which may turn up.

One quickly recognizes, along with the sponsors, that his work concerns itself less with one county, although it deals only with features of that place, than with the general question of recognizing and preserving what is visually significant everywhere. The sponsor considers Jacobs's product as virtually a "how to" book for the would-be environmental taxonomist, "one that shows in particular how architectural data—in the form of photographs, drawings, structural and historical information—might be treated within a taxonomic classification . . . that was developed for surveying the visible resources of a rural area like Wayne County" (Jacobs, 1979).

Research for the Jacobs' work, which carries as part of its title *Catalog for the Environment* (in Wayne County) commenced more than a decade before publication occurred. Questions were inevitably raised about publishing materials about forms that were noted once as significant but were less than recognizable later on. Others were gone from the scene entirely. It was decided that since few environments were static, observation should be regarded as a continuous process. Further, it was concluded that each file compartment will contain differing and even conflicting reports on the same phenomenon as seen at different moments and by different persons.

The cataloging of Wayne County does not purport to be a complete documentation of the area's significant features. Rather, it surveys readily observable features of the Wayne County physical environment. The following classifications were established:

Natural Patterns

    Major surface features
      Hills
      Bodies of water
      River and stream drainage
    Geological character
      Soils and their distribution
      Underlying rock formations
    Aging
      Modification of area by glaciation
      Modification of area by erosion
    Ecological character
      Original flora and changes
      Original fauna and changes

Such subcategories as drumlins, mucklands and sandbars might be added as additional subcategories (to allow for closer analysis of the area's natural environment). However, the basic classifications are intended to provide a framework for documenting and comprehending the activities and material accomplishments of the county's first settlers and later inhabitants.

Social Development

    Aboriginal
    Frontier land distribution
    Early settlements
      Port
      Crossroads community
      Mill community

Township center
Ideal community
Cultural center
Planned community
Transport centers
Canal port
Railroad community
Highway junction village
Summer resort

Economic Development

Extractive industry
Agriculture (general, dairy, fruit, vegetable farm facilities; farmyard group-
ings), forestry, and fishing
Mining (limestone, iron, gravel)
Manufacturing and contract construction
Food processing (fruit, vinegar, malt, flour, sugar beets, vegetables, meat,
milk)
Agriculture related (ladder, basket, crate, steam engine factories)
Transportation, communication, and utilities
Water (lake, river, and canal facilities)
Road (dirt, county, highway; bridges)
Rail (freight and passenger stations, bridges, crossing guards)
Air (fields, towers, hangers, terminals)
Wire (telephone, telegraph)
Wireless (radio, television transmitters, studios)
Electric (substations, transformers, transmission lines)
Gas
Sanitary (waterworks, sewer plants)
Trade
Wholesale (wood and coal yards, cold storage, grain and fertilizer elevators)
Retail (isolated store, storefront unit, row building, business block, group,
shopping center)
Finance, insurance, real estate
Banks
Agencies (broker, insurance, real estate)
Services
Lodging (hotels, motels, boarding and tourist homes)
Personal (laundry, cleaning, barber, repair, photography, etc.)
Automotive (service station)
Medical and health (hospital and clinic)
Recreation (parks, boating, skiing, etc.)
Professional services
Education (one-room school, academy, consolidated school, college)

Other services
  Museum and garden (historical society, rose garden)
  Religious and secular organizations (churches; Grange, Elks' and Masons' halls)
  Cemeteries (layout; gates, chapel, vaults)
Government
  Federal (custom house, post office, mailboxes)
  State (school)
  County (courthouse, jail, poorhouse)
  Town (office)
  Village (firehouse)
Housing
  Single family (cabin, masonry or wood, story-and-a-half cottage, federal, Greek revival, or Italianate house, later farm and village or suburban houses)
  Multifamily (row, apartment, migrant housing)

The *Catalog for the Environment* is designed to provide a sense of the aesthetic heritage of a rural area. Thus it details much of the aboriginal historical evolution of land distribution and of settlements under social development. In considering economic development, it explores, among many other factors, the location close by Lake Ontario, and the construction and impact of the canal built across New York State, officially inaugurated in 1825. Throughout the *Catalog* there is an intensive review of multiple elements and factors involved in the transformation of the Wayne County landscape. Therefore, although certainly not devised as a landscape decision-making process piece per se, it has inestimable value to the physical, social, and economic planner as a compendium of the visible features and the aggregate aesthetic heritage of an area.

In dealing with the canal, it states that

a continuous waterway from Albany on the Hudson River westward into the interior of New York State was proposed in colonial times. Because of their familiarity with canal building in Holland and their experience as fur traders in the Iroquois country, merchants of Dutch descent were aware of the potential of an improved waterway along the Mohawk River, Wood Creek, Oneida Lake, and the Oswego River to Lake Ontario. George Washington investigated this route after the Revolutionary War (in 1783). Much interested in canal development, particularly along the Potomac River, Washington communicated his enthusiasm to Elkanah Watson when the latter visited him at Mount Vernon in 1785.

Watson had just returned from The Netherlands. He was intrigued by canal travel and became chief advocate of a New York canal. By 1787 the New York State legislature became involved, intent on making the Mohawk River navigable, and if practicable, extending it to Lake Erie. Private companies were chartered in 1792 to build canals linking the Hudson with Lake Champlain, and the Mohawk with Wood Creek. The objective was to divert the future trade of Lake Ontario and the Great Lakes above, from Alexandria (Virginia) and Quebec to Albany and New York.

Early travelers were impressed by the efficient canal operation, it is related, as well as the speed with which boats traversed the 83 locks and along the 18 aqueducts. However, traffic jams did occur, especially on the main line east of the city of Syracuse. In 1834 the enlargement of the whole Erie Canal was authorized by the legislature. Work began in 1837, but stopped five years later (as a result of financial and political complications). Unfinished structures were left along the route. In 1847, work started again. Only in 1854, however, after an amendment to the state constitution allowed appropriations for canal work, was it possible to progress rapidly with the work. Enlargement was finished in 1862.

Early in the present century, the legislature of New York State appropriated funds to modernize the Erie Canal and make it the New York State Barge Canal. The appropriation was made in 1903 and in 1905 work was begun. "The builders of the Erie, fearing the destructive effect of freshets and the danger of low water in midsummer, had attempted to avoid the use of riverbeds. The Barge Canal reverted to the natural drainage channels. In 1916, the Clyde River was dredged to combine it with the canal through Clyde Village." In other Wayne County villages—Macedon, Palmyra, and Lyons—the old canal-side commercial and wholesale facilities were cut off from access to the Barge Canal by new embankments. In Newark, largely a railroad town, the canal remains accessible.

In 1919, the Barge Canal system was completed. Since 1935 the federal government has cooperated with state authorities to assure its upkeep. Lyons continues to be a center of canal maintenance activity. West of the village, a basin has the canal's largest drydock and its home base for maintenance vessels of several types, including derricks, old steam dredges, and a floating dormitory boat for the workmen.

The canal today continues with its traditional functions. Shipment of bulk materials, mostly oil, in barges takes place on this watercourse. Controlled drainage of the surrounding area (particularly Montezuma Swamp and the Finger Lakes) is permitted. Pleasure boats are increasingly appearing on the canal. Marinas and parks are being established along its route. In 1976, special notation was made that canal features are visible in six Wayne County parks, and that "outstanding are provisions for viewing the Palmyra aqueduct and the old locks at Lock Berlin."

The section of the canal in Wayne County finds the most impressive relics of the Erie Canal of the 1800s, dating from the enlargement completed in 1862. The canal still depended on streams and lakes along its route. "Mud Creek waters supplied the section east of Palmyra, and feeder canals dug through Montezuma Swamp helped drain the swampy southeastern portion of the county. From the beginning, dams had been built either side of the canal to raise the waters intersecting streams to the level of the canal, and the sites were made available to waterpower industry. Guard locks were built either side and a special tow path bridge constructed. For the enlarged canal low level cut stone culverts with waste weirs were installed to control the influx of water and debris." (One is visible between Lyons and Clyde on the north of highway route 31.)

The intertwining of historical background with natural patterns and economic development, characteristic of the cataloging already alluded to, is further evident in a cataloging of roads in Wayne County. It provides yet an additional example of the comprehensive recording of heritage in the holistic orientation to be found in the Jacobs approach and will serve to conclude our inspection of his effort.

The aesthetic interest of a community or an area, he notes, is evident only to those who experience it. In our time, this means mainly those who use the roads.

> In the days of the Erie canal packet, much was done to charm or attract the passenger and the growing use of the Barge Canal and of the Lake Ontario bays and shore front by pleasure boats suggests that observation from the water is still significant. For a hundred years railroad trains offered unparalleled opportunities for observing both the countryside and the industrial and warehouse sectors of larger centers. Now that the railroad passenger train is nearly extinct, this kind of aesthetic experience, which involved travel along a fixed path and with relatively little interference with the view from surrounding traffic, is a thing of the past.

(Here Jacobs refers to the decline of passenger railroad travel in the years up to 1979. Since then, increasing use of rail and road mass transportation systems has become increasingly noticeable in several parts of the United States, albeit coinciding with removal of certain infrequently used train routes from the larger railroad network.)

> Air travel offers unparalleled possibilities for studying the ecology and cultural development of an area, as well as for appreciating the dynamic visual patterns of day and night movement about motor centers. However, the routes and altitudes are not fixed, visibility is often poor, and relatively little is done to enhance the effect of ground features for airborne travelers. For most people the public way is the principle path of travel, and the basis of the environment. In built-up areas, pedestrian activity is still found, although rarely encouraged. Except for a relatively limited group on bicycles, the motor vehicle—automobile, tractor, truck, motorcycle or bus—is the usual conveyance. This conditions our experience and, particularly, our response to places new to us . . . . Public appreciation of items of cultural or aesthetic interest will necessarily depend upon public awareness of them. While knowledge of some can be conveyed verbally, or in pictures, a physical confrontation is necessary for a full appreciation of a physical heritage . . . . All items visible from the public way are in the public domain and should be considered when a community's artistic and cultural resources are being inventoried. As more and more people have the leisure and resources

for travel, the value, both social and economic, of this approach will be clearer.

In prehistoric times in Wayne County, only pedestrians used the roads—then merely trails along the streams or over higher ground in marshy parts. For the Iroquois Indians, the Ridge Road, one of the east–west "paths of peace," was the main thoroughfare. North–south "paths of war" related to defensible positions on commanding hills. Routes established by the Indians are used by main highways today. When pioneers arrived, new trails were blazed. One such trail was laid out between the present-day Phelps and Palmyra. It is noted that subsequently they were improved by a process known as underbrushing, that is, the removing of brambles, bushes, and small trees so that horses could come through. They were then, of course, full of stumps and snags, and likely to be quite wet and muddy in the wake of spring thawing of the ground. To enable travel along a road for wagons and sleighs necessitated more effort. "Trees could be felled, sawed into lengths the width of the road, and then laid across it to form a log causeway, or corduroy road. Primitive log bridges were needed to cross streams."

A Summary of Contexts, Purposes, and Agency Orientation and Mission

Consideration has been given to the Untermain Sensitivity Model, and to several methodologies for land-use capability analysis and regional landscape evaluation for planning and design. It is now important to emphasize that these have each been developed in very different locational contexts and serve varying purposes (under context, the matter of scale exists as a subissue).

In addition, orientation and mission of the initiating agency (or sponsor) must be noted.

> Is the agency involved with land-use policy planning, concerned with the direct regulation or influence upon the use of land not owned or directly controlled by the planning government?
> Or, is land management planning involved, in which the land is under the ownership or direct control of the planning agency?
> Or, is the concern with land development planning in which a particular physical product is necessary?

Obviously, depending on the circumstances, any combination of these may be engaged.

The Untermain Sensitivity Model, the Metland Procedure, The Netherlands General Ecological Model and Environmental Survey, CSIRO's approach for analyzing land-use options, and the Wayne County, New York work of Stephen Jacobs reflect different focuses and scales. Because of this realization, no attempt has been made to adjudge any of these as having universal application, nor is any viewed as preferred to the others. Rather, it is recognized that each has its own relevance in addressing different emphases.

Conclusion

We hope that this chapter has fulfilled its purpose by showing that there is no dearth of valuable tools for holistic normative and purposeful landscape planning. Such availability of tools applicable to the critical interface of land use furthers the functional and structural integration of bio- and technoecosystems as the major goal of landscape ecology. The crucial question is whether their application will remain only isolated events, and therefore have only limited impact on planning and management, or whether they will become widely used as a normal procedure in the decision-making process of holistic landscape-ecological determinism.

As stated in Chapter 2, landscape ecology and the new paradigms of human ecosystemology can play an important role in this integration by supplying not only scientific information but also educational–cultural information (as cognitive and affective feedback loops for the education of a new generation of enlightened decision makers). One of the most important tasks will be to decode the information obtained by means of these tools into the symbols and language of everyday knowledge without altering its intrinsic values, and at the same time to show cogently how attainment of these values can serve the main purpose of creating better landscapes for our postindustrial total human ecosystem.

References

Agrios, G. N. 1969. Plant Pathology. Academic Press, New York.

Alexander, L., et al. 1974. Remote Sensing Environmental and Geotechnical Applications. The State of the Art. Engineering Bulletin 45. Dames and Moore, Los Angeles, California.

Allan, J. A. 1980. Remote sensing in land and land use studies. Geography, Vol. 65, Part I, No. 286, pp. 35–44.

Anderson, J. R. 1971. Land use classification schemes used in selected recent geographic applications of remote sensing. Photogrammetric Engineering 37 (4):379–387.

Anderson, J. R., et al. 1976. A Land Use and Land Cover Classification System for Use with Remote Sensor Data. Geological Survey Professional Paper 964. U.S. Govt. Printing Office, Washington, D.C.

Austin, M. P., and J. J. Basinski. 1978. Bio-physical survey techniques. In: Land Use on the South Coast of New South Wales, Vol. 1, J. J. Basinski (Ed.), General Report. Commonwealth Scientific and Industrial Research Organization, Canberra.

Austin, M. P., and K. D. Cocks (Series Eds.). 1978. Land Use on the South Coast of New South Wales. A Study in Methods of Acquiring and Using Information to Analyze Regional Land Use Options, 4 vols. Commonwealth Scientific and Industrial Research Organization, Melbourne, Australia.

Avery, T. E. 1970. Photo Interpretation for Land Managers. Eastman Kodak Co. Publication M-76. Rochester, New York.

Avery, T. E. 1977. Interpretation of Aerial Photographs. Burgess, Minneapolis, Minnesota.

Bachfischer, R., J. David, H. Kiemstedt, and G. Aulig. 1977. The ecological risk-analysis as tool for regional planning and decisions in the industrial region Middle Franconia. Landschaft + Stadt, 9:145–161 (German with English summary).

Bajzak, D. 1973. Bio-physical land classification of the Lake Tahoe Basin, California-Nevada. A Guide to Planning. U.S. Forest Service, South Lake Tahoe, California.

Basinski, J. J. (Ed.). 1978. General Report (Land Use on the South Coast of New South Wales, Vol. 1). Commonwealth Scientific and Industrial Research Organization, Canberra.

Bauer, H. J. 1973. Ecological aerial photo interpretation for revegetation in the Cologne lignite district. In: Ecology and Reclamation of Devastated Land. Gordon and Breach, New York.

Baulkwill, W. J. 1972. The land resources division of the Overseas Development Administration. Tropical Sci. 14:305–322.

Bechmann, A. 1978. Nutzwertanalyse, Bewertungstheorie und Plannung. Verlag Paul Haupt, Bern und Stuttgart.

Belcher, D. J., et al. 1967. Potential Benefits to Be Derived from Applications of Remote Sensing of Agricultural, Forest and Range Resources. Center for Aerial Photographic Studies, Cornell University, Ithaca, New York.

Belknap, R. K., and J. G. Furtado. 1967. Three Approaches to Environmental Resource Analysis. Conservation Foundation, Washington, D.C.

Berry, L., and R. B. Ford. 1977. Recommendations for a System to Monitor Critical Indicators in Areas Prone to Desertification. Clark University, Worcester, Massachusetts.

Buchanan, C., et al. 1976. Midden Randstad Study. Part II. Final Report. Adviersbureau Arnhem B.V., Gronmij N.V.

Bulfin, R. L., and H. L. Weaver. 1977. Appropriate Technology for Natural Resources Development: An Overview. Annotated Bibliography, and a Guide to Sources of Information. University of Arizona, Tucson.

California Council on Intergovernmental Relations. 1973. General Plan Guidelines. State of California Governor's Office, Sacramento, California.

Calkins, H. W., and R. F. Tomlinson. 1977. Geographic Information Systems, Methods and Equipment for Land Use Planning. 1977. Performed by International Geographic Union Commission on Geographical Data Sensing and Processing, Ottawa, Ontario, Canada.

Carpenter, R. A. (Ed.). 1981. Assessing Tropical Forest Lands: Their Suitability for Sustainable Uses. Proceedings of the Conference on Forest Land Assessment and Management for Sustainable Uses, June 19–28, 1979, Honolulu, Hawaii. Tycooly International Publishing Ltd., Dublin.

Chambers, A. 1979. Toward a synthesis of mountains, people, and institutions. Landscape Planning 6:109–126.

Christian, C. S. 1963. The use and abuse of land and water. In: The Population Crisis and the Use of World Resources, Vol. 2. World Academy of Art and Science. Dr. W. Junk, The Hague, pp. 387–406.

Christian, C. S., and G. A. Stewart. 1947. North Australia Regional Survey, 1946. Katherine–Darwin Region. General report on land classification and development of land industries. Commonwealth Scientific and Industrial Research Organization, Melbourne, Australia.

Christian, C. S., and G. A. Stewart. 1953. General Report on Survey of Katherine–Darwin Region, 1946. Commonwealth Scientific and Industrial Research Organization, Australian Land Research Series no. 1.

Christian, C. S., and G. A. Stewart. 1968. Methodology of integrated surveys. Aerial Surveys and Integrated Studies: Proc. Toulouse Conf., UNESCO, Paris, pp. 233–280.

Cocks, K. D., and M. P. Austin. 1978. The land use problem and approaches to its solution. *In:* Land Use on the South Coast of New South Wales, Vol. 1, J. J. Basinski (Ed.), General Report. Commonwealth Scientific and Industrial Research Organization, Canberra.

Division of Land Use Research. 1978. Research Report 1973–77. Commonwealth Scientific and Industrial Research Organization, Canberra.

Dorney, R. S., and D. W. Hoffman. 1979. Development of landscape planning concepts and management strategies for an urbanizing agricultural region. Landscape Planning 6:151–177.

Eckaus, R. S. 1977. Appropriate Technologies for Developing Countries. National Academy of Sciences, Washington, D.C.

Elachi, C. 1982. Radar images of the earth from space. Sci. Am. 247(6):54–61.

Elsner, G. H., and R. C. Smardon (Technical Coordinators). 1979. Proceedings of Our National Landscape, A Conference on Applied Techniques for Analysis and Management of the Visual Resource. April 23–25, 1979, Incline Village, Nevada.

Erickson, P. A. 1979. Environmental Impact: Principles and Applications. Academic Press, New York.

Etzioni, A. 1968. The Active Society: A Theory of Societal and Political Processes. MacMillan, London.

Fabos, J. G. 1978. Planning and landscape evaluation. Keynote address to New Zealand Planning Institute Conference on Planning and Landscape, Dunedin.

Fabos, J. G. 1979. Planning the Total Landscape: A Guide to Intelligent Land Use. Westview Press, Boulder, Colorado.

Fleming, P. M., and J. Stokes (Eds.). 1978. Land Function Studies. (Land Use on the South Coast of New South Wales, Vol. 4). Commonwealth Scientific and Industrial Research Organization, Canberra.

Food and Agriculture Organization. 1974. Remote Sensing and Its Use in Agriculture and Forestry. Leaflet. United Nations Food and Agriculture Organization, Rome, Italy.

Godby, E. A., and J. Otterman (Eds.). 1978. The Contribution of Space Observations to Global Food Information Systems. Proceedings of the Twentieth Plenary Meeting of COSPAR, Tel Aviv, Israel, 7–18 June 1977. Pergamon Press, New York.

Gunn, R. H. (Ed.). 1978. Bio-physical Background Studies (Land Use on the South Coast of New South Wales, Vol. 2). Commonwealth Scientific and Industrial Research Organization, Canberra.

Grey, A. L., et al. 1973. Applications of remote sensing in urban and regional planning. Proc. Symp. Remote Sensing, Sioux Falls, South Dakota, October 29–November 1, 1973. American Society of Photogrammetry, pp. 115–116.

Hammerschlag, R. S., and W. J. Sopstyle. 1975. Investigations of remote sensing techniques for early detection of Dutch elm disease. Proc. 4th Ann. Remote Sensing of Earth Resources Conf., Tullahoma, Tennessee.

Hanson, H. C. 1962. Dictionary of Ecology. Philosophical Library, New York.

Hardy, E. E. 1970. Inventorying New York's land use and natural resources. Food and Life Sciences Quarterly 3(4):4-7.

Hardy, E. E. 1979. The design of natural resource inventories. Resource Information Laboratory, New York State Cooperative Extension, Ithaca, New York.

Hills, G. A. 1961. The Ecological Basis of Land Use Planning. Research Report 46, Ontario Department of Lands and Forests, Toronto, Ontario, Canada.

Hills, G. A. 1976. An integrated interpretive holistic approach to ecosystem classification. In: Ecological (Biophysical) Land Classification in Canada. pp. 73-98.

Hills, G. A., D. V. Love, and D. G. Lacate. 1970. Developing a Better Environment. Ecological Land-use Planning in Ontario—A Study of Methodology in the Development of Regional Plans. Ontario Economic Council, Toronto.

Holdridge, L. R. 1947. Determination of world plant formations from sample climatic data. Science 105(2727):367-368.

Holdridge, L. R. 1967. Life Zone Ecology. Tropical Science Centre, San Jose, Costa Rica.

Jacobs, P. 1979a. Landscape planning in Canada. Landscape Planning 6:95-100.

Jacobs, P. 1979b. George Angus Hills: A Canadian tribute. Landscape Planning 6:101-107.

Jacobs, S. W. 1979. A Catalog for the Environment. Wayne County: The Aesthetic Heritage of a Rural Area. Wayne County New York Historical Society.

Johannsen, C. J., and J. L. Sanders. 1982. Remote Sensing and Resource Management. Soil Conservation Society of America, Ankeny, Iowa.

Kaplan, G. A. 1981. Identification of impacts on development of recreation and tourism on coastal resources. Habiospherah 9-10:24-31 (Hebrew).

Kiemstedt, H. A. 1975. Landschaftsbewertung fur die Erhohlung im Sauerland. Dortmund.

Kuchler, A. W. 1967. Vegetation Mapping. Ronald Press Co., New York.

Lacate, D. S. 1969. Guidelines for Biophysical Land Classification. Canadian Forestry Service Publication No. 1264. Department of Fisheries and Forestry, Ottawa, Ontario, Canada.

Landgrebe, D. 1972. Systems approach to the use of remote sensing. In: International Workshop on Earth Resources Survey Systems, Vol. 1. NASA, Washington, D.C.

Lands Directorate. Canada Land Inventory. Environment Canada, Ottawa, Ontario, Canada.

Lands Directorate. Canada Land Data System. Environment Canada, Ottawa, Ontario, Canada.

Lewis, P. 1967. Regional Design for Human Impact. U.S. Department of the Interior, National Park Service, Northeast Region, Madison, Wisconsin.

Lieberman, A. S. 1978. The Need, Potential and Feasibility for Utilization of Remote Sensing in Land Use Planning and Natural Resources Management in Israel. A report to the Director of the Center for Urban and Regional Studies, Technion-Israel Institute of Technology, Haifa, Israel.

Lillesand, T. M., et al. 1978. Quantification of Urban Tree Stress through Microdensitometric Analysis of Aerial Photography. State University of New York College of Environmental Science and Forestry, Syracuse, New York.

Lillesand, T. M., and R. W. Kiefer. 1979. Remote Sensing and Image Interpretation. Wiley, New York.

Lund, H. G., and E. McNutt. 1979. Integrating Inventories, An Annotated Bibliography. Bureau of Land Management, U.S. Department of the Interior, Denver, Colorado.

Marble, D. F. (Ed.). Computer Software for Spatial Data Handling, Vol. 1: Full Geographic Information Systems. Prepared by the International Geographic Union, Commission on Geographical Data Sensing and Processing, Ottawa, Ontario, for the U.S. Department of the Interior Geological Survey.

Marsh, W. L. 1964. Landscape Vocabulary. Miramar, Los Angeles, California.

McCauley, J. F., et al. 1982. Subsurface valleys and geoarchaeology of the eastern Sahara revealed by shuttle radar. Science 218(4576):1004–1019.

McHarg, I. L. 1969. Design with Nature. Natural History Press, Garden City, New York.

Ministry of Housing and Physical Planning. 1977. Summary General Ecological Model. Study report 5.3.B. National Physical Planning Agency, The Hague.

Ministry of Housing and Physical Planning. Summary of The Netherlands Environmental Survey. Study report 5.3.A. National Physical Planning Agency, The Hague.

National Academy of Sciences. 1977. Resource Sensing from Space: Prospects for Developing Countries. National Academy of Sciences, Washington, D.C.

Naveh, Z. 1966. The development of Tanzania Masailand—a sociological and ecological challenge. African Soils 10,1/3:449–539.

Nelson, D. 1975. Land systems, some basic concepts. U.S. Forest Service, Region 9, Milwaukee, Wisconsin. 16 p.

Nir, Y. 1976. Shovrei-galim M'nutakim, Dorbanot, V'gufim M'lachutiyim Aherim Behof Ha-yam Hatichon, Vehashpaatam al Mivneh Hehof Shel Israel (Detached breakwaters, groins and other artificial bodies on the Mediterranean coast, and their impact on the structure of the coast of Israel). Hebrew Report No. Gimel-Yud/2/76. Department of Marine Geology, Geological Survey of Israel, Ministry of Commerce and Industry, Jerusalem.

Noi-Meir, I., and N. G. Seligman. 1979. Management of semi-arid ecosystems in Israel. In: B. H. Walker (Ed.), Management of Semi-Arid Ecosystems. Elsevier, Amsterdam, Chapter 4.

O'Riordan, T., and Hey, R. D. 1977. Environmental Impact Assessment. Saxon House, Lexington Books, Lexington, Maine.

Otterman, J., et al. 1975. Western Negev and Sinai ecosystems: Comparative study of vegetation, albedo, and temperatures. Agro-ecosystems 2:47–60.

Otterman, J., et al. 1978. Rangeland recovery in an exclosure: Satellite and ground observations. In: The Contribution of Space Observations to Global Food Information Systems, Vol. 2. Advances in Space Exploration, COSPAR Symposium Series. Pergamon Press, New York.

Paul, C. K. 1978. Internationalization of remote sensing technology. In: Photogrammetric Engineering and Remote Sensing 44(5):625–632.

Qureshi, A. H., et al. 1980. Assessing Tropical Forest Lands: Their Suitability for Sustainable Uses. A Report on the Conference on Forest Land Assessment and Management for Sustainable Uses, June 19–28, 1979, Honolulu, Hawaii. East-West Environment and Policy Institute, East-West Center, Honolulu, Hawaii.

Rau, J. G., and Wooten, D. C. 1980. Environmental Impact Analysis Handbook, McGraw-Hill, New York.

Rawls, J. 1972. A Theory of Justice. Oxford University Press, London.

Schechter, M., R. Enis, B. Reiser, and Y. Tzamir. 1981. Evaluation of landscape resources for recreation planning. Regional Studies 15:323–392.

Schwarz, C. F., et al. 1976. Wildland Planning Glossary. USDA FS Gen. Tech. Rep. PSW-13/1976. Pacific Southwest Forest and Range Experiment Station, Berkeley, California.

Senykoff, R. S. 1978. Transferring Satellite Remote Sensing to Kenya: Application of Appropriate Technology to Spatial Characteristics of Population. Ph.D. Thesis. Cornell University, Ithaca, New York.

Steinitz, C., et al. 1977. Managing Suburban Growth: A Modeling Approach/ Summary. Landscape Architecture Research Office, Graduate School of Design, Harvard University, Cambridge, Massachusetts.

Trowbridge, P. J. 1979. Ecologically-based regional planning methods. Lecture notes for landscape architecture class, Cornell University, Ithaca, New York.

United Nations. 1978a. Desertification: Its Causes and Consequences. Pergamon Press, New York, p. 10.

United Nations. 1978b. United Nations Conference on Desertification. 29 August–9 September 1977. Round-up, plan of action and resolutions. United Nations, New York.

U.S. Agency for International Development. 1980. Remote Sensing: A Technology for Economic Development. U.S. Agency for International Development, Washington, D.C.

U.S. Army Engineer Division, North Central. 1970. Upper Mississippi River. Comprehensive Basin Study. UMRCBS Coordinating Committee.

U.S. Forest Service. 1972. Planning and Programming Glossary. Unpublished manuscript. Planning Glossary Task Force, Washington, D.C.

Van Genderen, J. L., et al. 1978. Guidelines for using Landsat data for rural land use surveys in developing countries. Invited paper, ITC International Symposium 1976. ITC Journal 1978(1):30–45.

Verstappen, H. T. 1966. The use of aerial photographs in delta studies. *In:* Scientific Problems of the Humid Tropical Zone Deltas and Their Implications. Proceedings of the Dacca Symposium. UNESCO, Paris.

Verstappen, H. T. 1969. Aerial survey and rural development in South Asia. *In:* Problems of Land Use in South Asia. U. Schweinfurth and M. Domrös (Eds.), Yearbook of South Asia Institute. Heidelberg University, Otto Harrassowitz, Weisbaden.

Vester, F., and A. von Hesler. 1980. Sensitivitätsmodell. Regionale Planungsgemeinschaft Untermain, Frankfurt am Main.

Walker, P. A., and J. F. Angus (Eds.). 1978. Socio-economic Background Studies (Land Use on the South Coast of New South Wales, Vol. 3). Commonwealth Scientific and Industrial Research Organization, Canberra.

Walter, H. 1979. Vegetation of the Earth and Ecological Systems of the Geobiosphere, 2nd edition. Springer-Verlag, New York.

Watkins, T. 1978. The economics of remote sensing. Photogrammetric Engineering and Remote Sensing 44(9):1167–1172.

[Webster] 1963. Webster's Third New International Dictionary of the English Language. Unabridged. G & C Merriam Co., Springfield, Massachusetts.

Wells, R. E., and B. A. Roberts. 1973. Bio-physical Survey of the Badger-Diversion Lake Area, Newfoundland. Canada Forestry Service, Newfoundland Forest Research Center, St. Johns, Newfoundland, Canada.

Whittaker, R. H. 1975. Communities and Ecosystems, 2nd edition. Macmillan, New York.

Whyte, R. O. 1976. Land and Land Appraisal. Dr. W. Junk, The Hague.

Wilford, J. N. 1982. Spacecraft detects Sahara's buried past. New York Times, November 26, 1982.

Zangemeister, C. 1971. Nutzwertanalyse in der Systemtechnik. München.

Zonneveld, I. 1979. Land evaluation and land(scape) science. In: Use of Aerial Photographs in Geography and Geomorphology. ITC Textbook of Photo-interpretation. Vol. VII. ITC, Enschede, The Netherlands.

Zonneveld, I. S. 1978. A critical review of survey methods for range management in the third world with special emphasis on remote sensing. Proc. First International Rangeland Congress, Denver, Colorado, pp. 510–513.

Zumer-Linder, M. 1979. Environmental Word-List. Swedish University of Agricultural Sciences. International Rural Development Centre, Uppsala.

Note Added in Proof

The *Cornell Remote Sensing Newsletter* reported the following on the various Landsat activities in the September–October 1983 edition (the Newsletter is published by the Remote Sensing Program, begun in 1972 with a grant from the National Aeronautic and Space Administration to the Cornell University School of Civil and Environmental Engineering).

Landsat 4. The landsat-4 multispectral scanner (MSS) is operating normally and data are routinely available. Although acquisition of thematic mapper (TM) data was halted last February by the failure of the satellite's X-band transmitter, limited TM data have been acquired since mid-August through the Tracking and Data Relay Satellite, TDRS-A. The TDRS-A was deployed from the 6th Space Shuttle last April but did not achieve proper orbit. NASA controllers were able to stabilize the satellite in its correct geosynchronous orbit in July.

Landsat 4 is operating on approximately 50% power due to the failure of power supply cables from two of its four solar arrays. TM data acquisition will most likely be restricted given the additional power needed for the TDRS antenna system and the Ku-band communication link through the TDRS. The operational MSS system would have priority over the experimental TM system should the power be further limited. Notably, Landsats 2 and 3 are no longer acquiring data and are scheduled for permanent retirement on 30 September 1983. For information on data availability, contact: NOAA Landsat Customer Services, Mundt Federal Bldg., Sioux Falls, South Dakota 57198 (tel. 605-594-6151).

In addition to the Calkins and Tomlinson, 1977 reference, the following latter work should be identified:

Calkins, H. (Coord.), and D. F. Marble (Ed.) 1980. Computer Software for Spatial Data Handling: Volume 1, Full Geographic Information Systems. Performed by International Geographic Union Commission on Geographical Data Sensing and Processing, Ottawa, Ontario, Canada. (Volumes 2 and 3, by other Coordinators, with D. F. Marble as Editor, cover, respectively, Data Manipulation Programs, and Cartography and Graphics.)

Remote Sensing, Geographic Information Systems, and Landscape Ecology

Since the original edition of this book appeared, rapid developments have taken place in remote sensing and its applications. Increased spatial and spectral resolution capabilities have become available. Some technical elements described in our initial edition have been superseded.

It is beyond the scope of this edition to detail the numerous advances and changes in remote sensing technology and information-processing techniques. Readers may consult Lillesand and Kiefer (1987), who provide thorough technical coverage of these areas, as well as Quattrochi and Pelletier (1991). The latter deal with remote sensing for quantitative analysis of landscapes.

In a parallel way, advances in geographic information systems (GIS) have moved at a very fast pace. GIS is a computerized mapping system for capture, storage, management, analysis, and display of spatial and descriptive data. Burrough (1988) provides a comprehensive review of GIS principles and uses for land resources assessment. The March 1990 issue of *Landscape Ecology* addresses spatial and temporal analysis using GIS, and contributes to landscape-ecological investigations. The work of Davis and Goetz is among several relevant articles to be found in it.

The following will serve to illustrate a few of the notable developments in these fields as they bear on landscape ecology and applications in landscape-ecology-based planning.

The fields of remote sensing and information science have a significant role to play in holistic landscape evaluation. They are of vital relevance in dealing with issues of total human ecosystems, where cultural and natural interactions need to be identified and clarified. One example of strides in the last decade has been the use of GIS for conservation and development planning (Olson and Lieberman, 1985). Also, application has been made of previous findings to explorations of specific regional landscapes, such as the Great Barrier Reef off Australia (CSIRO, 1984).

As covered in Chapter 3, remote sensing is regarded as an integral tool for landscape-ecology-based planning. In the remote sensing community, more participants have appeared, among them the French SPOT satellite. SPOT and Landsats 4 and 5 provide much finer resolution than was available with Landsats 1, 2, and 3. However, as noted by Roller and Colwell (1986), some investigators have explored the interesting and difficult proposition that data with low resolution will be useful for applications in inventory, monitoring, and biophysical characterization of very large areas—much larger than the 34,000 km^2 in a single Landsat image. Data from satellites such as those of the U.S. National Oceanic and Atmospheric Administration (NOAA) can be useful for such applications. Roller and Colwell conclude that such NOAA satellite data should prove valuable to ecologists because they provide a chance to measure directly (and map) ecological parameters at regional and global scales, and allow for a frequent coverage on a key date.

In the case of the high-resolution capability aboard landsat MSS/TM mentioned in Chapter 3, the spatial resolution is 30 m, while the French SPOT's HRV (high resolution visible) multispectral mode yields 20-m resolution, and its panchromatic (black-and-white) photographic mode gives 10-m resolution. Greegor (1986) notes that the multispectral band position and width were chosen to provide specific ecological information (e.g., for discriminating vegetation and analyzing water surfaces). Greegor stresses that both low- and high-resolution sensors offer specific applications for many ecological systems.

Another example of the intersection of these fields lies in the work of Davis and Goetz (1990), who employed GIS to test the value of predictive vegetation mapping in regional vegetation analysis. In their work, weighting and overlay was performed on digital maps of geological and topographical variables to predict distribution of a coast live oak forest in a 72-km^2 region near Lompoc, California. They compared the predicted pattern to the actual distribution of oak forest mapped from thematic mapper simulation (TMS) data. They concluded that GIS-based spatial modeling can contribute to analyzing vegetation of other landscape variables in several ways, especially for extrapolating multidimensional ecological processes that can only be measured at points or in small areas. Davis and Goetz feel "the types of cartographic analyses conducted here complement traditional field survey methods by measuring associations or testing field results with many more random samples and at larger spatial scales than can practically be collected in the field, facilitating the analysis of larger heterogeneous landscapes."

Another work with inportant international implications is that issued in 1990 by the German National Committee for the UNESCO Program "Man and the Biosphere" (Ashdown and Schaller, 1990). It discusses GIS and its application in MAB projects, ecosystem research, and environmental monitoring. It focuses on GIS creation and application, trends in GIS software, and applications of GIS in ecosystem research and monitoring. It has been applied by a number of researchers, including Haber (1990), in his study on planning and management in Germany, mentioned in the supplement to Chapter 2.

Intelligent Geographic Information Systems and the Management of Natural Resources

Management of natural resources involves integration and interpretation of such forms of knowledge as simulation results, historical data bases, technical reports, and heuristic information. GIS serves as a key technology for investigation of landscapes. Beyond the representation of landscape features, GIS is useful in (1) predicting consequences of an action being considered, (2) evaluating results of actions that have already been taken, and (3) comparing alternative actions. Conventional GIS can be greatly expanded by adding artificial intelligence (AI) methodologies (see remarks in the supplement to Chapter 2). The product which results is an intelligent GIS, or IGIS (Coulson et al., 1991).

A fine illustrated discussion of GIS and AI may be found in Coulson et al. (1991). They point out that partnerships between landscape ecologists, natural resource managers, geographers, and GIS developers are increasing. AI, as an interdisciplinary subject, draws from serveral academic specialties, and AI research addresses inference, knowledge representation search, and pattern matching. In terms of natural resource management, AI enhances the function of GIS in three ways: automation of interpreting relations within and among landscape data themes, selection of appropriate analytical solutions for natural resource management and landscape-ecological problems, and guidance in the use of the system.

A natural resource manager must integrate both quantitative and qualitative information. GIS permits analysis and representation of quantitative spatial and tabular data, while AI provides the means for using qualitative information. Thus, the IGIS, already mentioned in the supplement to Chapter 2 in connection with fuzzy sets and expert systems, blends methods for representation, analysis, and interpretation of quantitative data with knowledge of experts. Coulson et al. (1991) stress that it is especially useful for interpreting relations within and among landscape data themes and for selecting appropriate methods of analysis. A cited application in problem-solving and decision-making concerns the United States Environmental Protection Agency's Endangered Species Protection Program. Its goal was to protect endangered species from further adverse effects from pesticides. The ecological impact of this program cannot use traditional analytical tools. Because a great deal of information available on the subject exists as the qualitative knowledge of experts, decision-making is largely judgmental. IGIS methodology provides a means to organize, integrate, and interpret qualitative and quantitative information on the impact of pesticide use on endangered species. The IGIS approach solves a complex landscape-level problem for which spatially referenced data, tabular data, and heuristic knowledge must be integrated.

Ecology-Based Planning and Management for Coastal Lands

Development pressures to alter and build up coastal zones are being experienced worldwide, especially in the Mediterranean (see supplement to Chapter 4). The Australian approach (cited in Chapter 3) for analyzing regional land-use options along the New South Wales south coast, deals with conflict between different uses for limited land (see pp. 231–240). This coast, like many other locations in both developed and developing countries, changed from a rural backwater to a popular resort area. In just such circumstances, the use of ecology-based coastal land and resources planning is essential.

Several organizations have prepared reliable, easily comprehended literature to help decision-makers, planners, land managers, and the general public understand ecological considerations for management of coastal zones. These include the Conservation Foundation in the United States (Clark, 1974) and the Sea Grant Extension Program of New York State (O'Neill, 1985, 1986; Lieberman and O'Neill, 1988). Clark (1974), focuses on ecological considerations, environmental disturbances, resource evaluation, and protection and constraints on specific uses. He later produced an enlarged technical manual (Clark et al., 1977). O'Neill's two works respectively cover coastal erosion processes and the issues and consequences (as mentioned later in this supplement in the case of artificial offshore bodies along Israel's coast) of building structures to control coastal erosion. The work of Lieberman and O'Neill explores vegetation use in coastal ecosystems, including holistic approaches to the utilization of vegetation to reduce shore erosion on marine and interior seashores.

Shaul Amir of the Technion-Israel Institute of Technology, whose work on the Mediterranean Coastal Strip is cited in Chapter 3, has produced tools for coastal resource management that involve multiple objectives. These include resource description and assessment tools to identify, classify, and evaluate coastal resources; to simulate their interrelationships; and to forecast and monitor development impacts. Unlike a typical land-use policy designed to answer the demand for a given growth in activities, the policy advanced for the coast in Israel is "supply oriented." Accordingly, the amount of activity and its location on the coast will be decided largely on the basis of environmental characteristics and conditions of each section of the coast. A separate review and analysis of Israel's coastal policy and its resource management program and the potential challenges to their full implementation, appeared in the *Coastal Zone Management Journal* (Amir, 1984).

Monitoring the dynamics of coastal systems is becoming ever more relevant in light of the anticipated rise in sea level and changes of shorelines and coastal areas resulting from global climatic changes.

A special issue of *Landscape Ecology* (1991) dealt with the impact of climatic change on coastal dune landscapes of Europe. It should be consulted for articles dealing with the consequences for the abiotic and biotic environments, and on the use of GIS in assessing the impacts of sea level rise on natural conservation along the Dutch coast (van der Meulen et al., 1991).

Monitoring and Characterization of Coastal Development Impacts along Israel's Mediterranean Coast

An important example of coastal monitoring and characterization is that by Nir. In a consideration of twenty-five years of development along Israel's Mediterranean coast, aided by ongoing aerial photographic monitoring, Nir and Elimelech (1990) address goals and achievements. He continues his investigation of the effect of offshore structures built for recreational purposes (see pp. 165–167, and Figures 3-13 to 3-18, for Nir's earlier documentation and photographs from his earlier work of the impact of breakwaters and other bodies on the structure of the coast).

It is important to note that the portion of the Mediterranean coast which forms the southeastern part of the Levantine basin is smoothly curved, creating a large arc that originates in Northern Sinai and continues toward the point where Mount Carmel projects into the Mediterranean Sea. Nir (1990) observes that these beaches belong to the Nile littoral cell that commences at the Nile River outlets at Rashid (Rosetta) and Dumyat (Damietta). The main net sand transport in this cell is from the Nile Delta to North Sinai and the Israeli beaches, terminating at Akko, at the northern end of Haifa Bay, in Northern Israel. Sand quantities are known to gradually decrease from the sources to the edge of the cell. Quartz is the main component of the sand, along with calcium carbonate from broken shells of local origin and a wide range of heavy minerals.

The Israeli shoreline, as may be seen in the map in Figure 3-13, has no pronounced bays (except for the Haifa bay, which resulted from the promontory formed by Mount Carmel) that can provide shelter and quiet water. Consequently, Nir points out, the 240-km-long beach is exposed during the summer to westerly and southwesterly waves originating from the long fetch of the Mediterranean Sea. Thus, the sea is relatively rough during the height of the swimming season, representing a danger to humans. To counteract this problem, various types of structures have been built to create shelter and quiet water for bathers. While these structures have resulted in the development of larger beaches and some areas of quiet water, they have also resulted in serious problem for beaches and bathers.

Since 1965, about 30 structures have been built for recreational purposes, mainly next to large urban centers. These include detached breakwaters, groins, sea walls, marinas, and small ports. The early detached breakwaters were very long in relation to their distance from the original shoreline. This conflicted with the empirical rule relating to the structure's distance-to-length relations. As a result, huge amounts of sand accumulated between the beach and the breakwater. In most cases, they connected the beach and the breakwater with a "tombolo" sand-body which is trapezoidal or triangular-shaped. The bathing area was very small and produced only a slight improvement in bathing conditions. As noted by Nir (Nir 1982a,b), tombolos mature over an average period of four to five years. Later structures, built with the help of knowledge gained about the need to avoid large sand accumulations, were located further offshore or were shorter in length, resulting in smaller accumulations of sand and smaller areas of quiet water.

Such offshore structures, constructed at any orientation to the original shoreline, interfere with longshore sand transport, which essentially has a net northerly trend in the cell considered along the Israeli coast as far north as Akko. Nir observes that sand usually accumulates on the "upcurrent" side of the transport direction, while pronounced erosion occurs on the "downcurrent" side of the structure. The latter phenomenon results from both a sand deficit and wave refraction on the northern edge of the breakwater. Thus, beaches become narrower, the natural beach and backshore protection is minimized, and the whole system experiences accelerated erosion.

On the basis of his observations, Nir notes that some of the declared goals of offshore detached breakwaters have been achieved (beach areas were expanded, quiet water largely exists in the bathing season, and the size of sand-beaches in relation to the length of original shoreline has been extended). However, severe problems have shown up (the stone and concrete structure represent a danger to swimmers in both the sheltered area and the open sea; fishermen using the stony platform area are at risk during stormy days). Turbulent currents, holes in the ground, and the impression of a safe shelter in the "protected" regions can result in an increase in drownings relative to the era before the structures were built.

Nir concludes in his 1990 piece that "the totals for the entire Israeli coast, show practically no improvement in beaches with sheltered areas, as compared with beaches without any offshore structure. There is therefore no doubt that construction of offshore structures should be planned only in regions where conditions are very rough and no other method can be implemented. In other regions, structures should either be improved or not constructed at all."

CLEARS: A Focus on Environmental Applications of Remote Sensing

The Cornell Laboratory for Environmental Applications of Remote Sensing (CLEARS) provides an example of interdisciplinary activity that harnesses coordinated knowledge bases in several fields which can be linked to landscape-ecological-type research. CLEARS, located at Cornell University, is the focal point for that institution's activities in remote sensing, GIS, and resource inventory. Among other foci, CLEARS is involved in cooperative research, technology transfer, instruction, and communication.

CLEARS is a part of the land-grant university framework. In the United States, land-grant universities have an educational outreach program called Cooperative Extension, and are in a unique position to work with both formal and informal educators. Cooperative Extension extends research-based knowledge of over 70 land-grant universities across the United States.

In its research program, CLEARS seeks to optimize the use of remote sensing and GIS to inventory and monitor environmental resources, particularly land and water. Current projects include inventory and assessment of crops and natural resources, characterization of land and water systems, GIS-based spatial process modeling, automated image classification, and remote laser spectroscopy.

A Community Science Education Program through CLEARS

A training example from CLEARS with implications for environmental education elsewhere is its Community Science Education Program. The Resource Inventory Program of CLEARS (Barnaba, 1992) collects and summarizes information on the type and location of land use and cover, as well as on renewable and nonrenewable resources. Its objectives are to design, administer, and coordinate regional and statewide education programs and projects in resource inventory and remote sensing. The audience includes Cornell University faculty, students, and staff; other academic institutions; Cooperative Extension field staff and clienteles; government agencies at all levels; attorneys, planners, and environmental consultants; environmental commissions; and secondary school educators and students.

Eugenia Barnaba of CLEARS and Marianne Krasny of the Department of Natural Resources at Cornell are involved in a major project entitled Remote Sensing and Map Interpretation: A Community Science Education Program. The three-years project is intended to reach over 500,000 people to improve their knowledge of the basic sciences and mathematics, and to teach integration and application of these disciplines through analysis of local environmental problems using remote sensing imagery and maps. This application serves as an effort across disciplinary lines, and can be further enhanced by the bridging concepts and methods of problem-solving-oriented landscape ecology.

CLEARS is also engaged in consideration of the use of aerial photographs in county inventories of waste disposal sites (Barnaba et al., 1991). In collaboration with the National Estuarine Research Reserves (NERR) in Ohio and New York and with NASA, educational materials are being developed to integrate the use of aerial photographs and map resources into estuarine education programs for NERR staff, resource managers, students, and teachers.

Environmental Impact Assessment and Evaluation of Landscape Functions

The process of Environmental Impact Assessment, discussed in Chapter 3 (pp. 199–200), has become the most widely used—and unfortunately also misused—tool for the evaluation of predicted ecological alterations arising from changes in land use through development. Among the numerous publications on this subject the books by Westman (1985) and de Groot (1992) deserve special attention. They deal with natural and semi-natural landscapes not only in the narrow sense of impact assessments, but also consider their broader roles in environmental management and decision-making.

Ecology, Impact Assessment, and Environmental Planning (Westman, 1985) is a comprehensive attempt to assess the impact of human actions on ecosystems and landscapes by drawing together the perspectives of three distinct subdisciplines: applied ecology, environmental planning, and ecological impact assessment.

Of special importance for landscape ecologists (who generally are not well-versed in this subject) is Westman's review of environmental law, public policy, and environmental decision-making. He covers a broad field in a condensed and lucid way, using a holistic approach to planning and management. The same is true for his discussion of economic approaches to impact evaluations, where he reviews some of the relevant economic theories and methods, dealing with the evaluation of environmental methods by "shadow pricing" of natural resources and free services which are "not for sale." He also emphasizes the severe limitations of these economic evaluation methods for pricing natural, nonmarket goods and the valuation of human life. He concludes:

"The ever-present danger with any evaluation is that decision makers will accept numbers as an objective rationale for a decision, when such numbers merely reflect a quantification of particular human values" (Westman 1985, p. 193).

For predicting impacts on the physical environment and biota he makes use of the state of the art in the relevant fields of community and ecosystem ecology, landscape ecology, air and water pollution, and ecotoxicology. The high quality of his discussion of species and landscape diversity, and succession and resilience of ecosystems matches that of the best ecology textbooks. His definitions of inertia and resilience are now widely applied and have great relevance for landscape ecology and restoration ecology.

In the foreword to Westman's book, Woodwell states, "we need such a book for the synthesis it brings to the challenges of measuring and interpreting the values of resources that are commonly discarded and ignored."

This challenge has been taken up by R. de Groot (1992) in his *Functions of Nature: Evaluation of Nature in Environmental Planning, Management, and Decision-Making*. He developed further the functional evaluation method of the General Ecological Model (GEM) discussed in Chapter 3 (pp. 221–223). After a detailed description of these regulation, carrier, production, and information functions, he discusses their socioeconomic value. Socio-economic value indicates that these values go beyond the narrow interpretations of economic theory limited to monetarized market economics; they include also nonmonetary values of goods and services contributing to human welfare, which we have called "soft values." Thus, the environmental functions of ecological conservation and existence values can often only be described in qualitative terms. Social health and option values may be quantified by setting standards for minimum requirements for the availability of a given function, such as air quality. Only such economic values of consumptive use, productive use, and employment can be expressed directly in monetary units. Thus, integrated economic assessment of the environmental functions, providing goods and services that satisfy human needs in a sustainable manner, serves as a common indicator for environmental quality and quantity of life.

This quantification of the socioeconomic benefits of natural areas and wildlife in monetary units must be seen, in de Groot's opinion, as an addition to—and not a replacement of—their many intrinsic and intangible values, which we called "noneconomic richnesses." There are, of course, great differences in the cultural perceptions and in the interpretation of the implications of intrinsic and ethical values placed on the natural environment. Thus, it may be unwise to use these as the only argument for conservation and sustainable use of nature and natural resources. Many other arguments and values should be considered.

De Groot included "option" values as social values, measured by the importance people place on the future availability of a given amenity, good, or service and of the so-called "serendipity value" of the potential benefits to human society of natural processes, components, and species that have not yet been discovered. Since the future is uncertain, all types of optimum values may be used to assigning value to risk aversion (McNeilly, 1988).

The discussion of these problems could bring, in de Groot's opinion, ecologists and economists closer in their approaches to solving environmental problems. He concludes:

A change in attitude of economic and political decision-makers in favour of long-term sustainability is probably more important than constructing an artificial yardstick for measuring all economic benefits of environmental functions. We may never be able to quantify the spiritual experience of nature, but at least we should attempt to consider all values of the natural environment in the economic planning and decision-making process. (de Groot, 1992, p. 140.)

The usefulness of this approach is demonstrated in three case studies.

Table S3-1 is an example of the functions used for the socioeconomic evaluation of the Galapagos National Park. The estimation of the total monetary return from those environmental functions, such as conservation values, productive use value, and employment value amounted to US $120/ha/year. At an interest rate of 5%, this amounts to a capitalized value of almost US $2.8 billion of the net present value for the entire study area. It showed clearly that in addition to their great ecological and

Table S3-1. Functions of the Galapagos National Park

1. Regulation functions	**2. Carrier functions**
Climate regulation	Aquaculture
Coastal protection and flood prevention	Recreation and tourism
Water catchment and erosion prevention	Nature protection
Bio-energy fixation	
Storage and recycling of human waste	
Providing biological control	
Migration habitat and nursery function	
Maintenance of biological diversity	
3. Production functions	**4. Information functions**
Food/nutrition (edible plants & animals)	Aesthetic information
Genetic resources	Spiritual/ethical information
Raw materials for building/construction	Historical information
Biochemicals (e.g., Orchilla)	Cultural/artistic inspiration
Energy (fuel wood, solar energy, etc.)	Scientific & educational information
Ornamental resources (e.g., black coral)	

Source: de Groot (1992, p. 220).

intrinsic value, the Galapagos Islands, in their present largely natural state, represent also a considerable economic value.

The annual monetary value of sustainable use of all the functions of a natural tropical moist forest in Panama were estimated at least as high as US $500/ha, and their net capitalized value as US $10,000/ha. From all functions of the tidal wetlands of the Dutch Wadden Sea, the combined potential annual return is at least US $6,200/ha/year, representing a capitalized value of these functions for the whole wetland of about US $124,000/ha.

Unfortunately, present-day market economics do not recognize the monetary value of these free natural services to society. Even if their exact amount may be disputable, such calculations create greater awareness of these functional values and thereby contribute to their preservation and sustainable utilization.

Table S3-2. Framework for an Expanded Cost Benefit Analysis: Costs and/or Benefits of Each Project-Alternative[a]

Effects	Costs (ie: negative effects)	Benefits (ie: positive effects)	Neutral (no effects)	Monetary effects[b] (sum of costs and benefits)
Environmental Effects				c
(on functions & hazards)				c
regulation funct.				c
carrier functions				c
production funct.				c
information funct.				c
Socio-Economic Effects				c
Land use				c
Employment				c
Health				c
etc.				c
Cultural Effects				c
Social structure				c
Cultural identity				c
etc.				c

[a] Effects should be quantified in their "natural" dimensions as much as possible. If quantification is not possible, at least a qualitative indication of the expected effect should be included in the analysis.

[b] For some changes in availability of environmental functions (and occurrence of natural hazards) and the effects on human welfare (quality of life) it is possible to calculate a monetary value. Some methods to calculate market values and shadow prices for both man-made and natural goods and services are discussed in DeGroot, chapter 3.

[c] Sum may be positive, neutral, or negative.

Source: de Groot (1992, p. 252).

Table S3-2 represents the proposed framework for such an expanded cost-benefit analysis.

In the concluding chapter, de Groot discusses the role of functional evaluation in environmental assessment and cost-benefit analysis. He argues for providing a systematic checklist of the many functions and socioeconomic values of sustainable use of natural ecosystem complexes (we would have called these "Landscape units"). This will enable more balanced decision-making, guided by the concept of sustainability. The shortcomings of traditional cost-benefit analysis limited to economic financial trade-offs could be overcome by a better accounting for non-monetary social and environmental values. It is important also to improve the determination of economic values of environmental goods and services by these functional evaluation methods.

In conclusion, these two books are in many respects complementary. They could be of even greater value for holistic landscape planning and management if their insights and methods could be expanded by recent advances in landscape ecology and in computerized knowledge engineering by fuzzy sets, discussed in the supplement to Chapter 4.

Holistic Land Management Efforts in the United States

1. A Strategy to Protect Texas Hill Country as an Entire Inhabited Landscape: Treating Humans as Part of the Ecosystem

As part of a novel strategy in the hill country of the state of Texas in the southern United States, an experimental project which departs from traditional conservation is being put into practice. It focuses on the entire inhabited landscape, rather than on individual species and habitats.

The area concerned consists of deep wooded canyons, and includes grassland, desert, and forests, as well as "vast honeycombed formations of limestone where strange creatures whose kind have not seen the light of day for millenia live in subterranean rivers, caves, and flooded crevices" (Stevens, 1992). It is an area twice as large as the state of New Jersey, has a human population of 2.5 million, and includes the cities of San Antonio and Austin. The area has become a magnet for urban expansion, retirees, and tourists: the population of 26 counties of the conservation project grew by 28% from 1980 to 1990.

The new approach combines three elements in an attempt to integrate human activity with nature. The strategy assumes that with planning for an entire ecological region, and preserving the most essential tracts of land, it becomes less necessary to protect every scrap of habitat. "The hope is that fewer knock-down fights will develop between environmentalists and economic interests, and that fewer wild species will wind up on the endangered list" (Stevens, 1992). The strategy, involving the nature conservancy; local, state and federal authorities; environmentalists, and landowners, is as follows: (1) Set aside highly protected and carefully managed "core" reserves for native fauna and flora, in addition to existing park lands. (2) Surround the core reserves with "ecologically friendly" buffer zones. Ranches that maintain wild habitat for tourists as well as undeveloped areas of military installations are examples of such buffer zones. (3) Create "ecologically healthy" islands of wildlife habitat on the outskirts of cities while allowing development to proceed around them.

In the first category, four to six core reserves are to be established to protect and manage natural habitats. In the case of "eco-ranching", large ranches are encouraged to bring ecological perspectives to their money-making plans, and some of these cater to nature study, bird-watching, and like activities. With regard to habitats near cities, governments, environmentalists, and landowners in the Austin area are attempting to put together healthy islands of wildlife habitat. A conservation plan for the Austin area, developed by a committee composed of parties on all sides of the issue, aims to carve 60,000 acres of the least fragmented habitat within the metropolitan area and establish it as a system of preserves. The most important habitat chunks, many contiguous, will be set aside as a viable preserve system around which development can occur.

The results of these Texas efforts, and the question of how successful they will actually prove to be, will bear close watching. The innovative character of the strategy exemplifies what has been termed "New Conservation". It reflects a turning away from the traditional conservation, in which humans were historically ignored or excluded, as somehow separate from and antagonistic to the rest of nature. By contrast, the Texas hill country effort aims to "stitch together a quilt of healthy natural habitat and settled areas, saving as many species as possible while preserving economic vitality" (Stevens, 1992).

2. A Center for Holistic Resource Management

In his major work on holistic resource management, Savory (1988) presented a comprehensive planning model focusing on land management, which deals with people and their environment as a whole. Savory conveyed an innovative approach which has been further articulated at the Center for Holistic Resource Management in Albuquerque, New Mexico.

The center is a not-for-profit organization established in 1984 by a group of farmers, ranchers, and environmentalists to serve as a focal point for the exchange and dissemination of knowledge on holistic management. In holistic resource management, humans, their landbase, and their wealth are treated as resources and managed as one indivisible unit. The concept derives from Savory's experiences in rangeland science and takes a very different and controversial position in attempting to prevent desertification: he claims that overgrazed and desertifying areas could be restored by dramatically increasing herds of domestic livestock. The proposal is derived from observation of the positive impact of wild herbivores on grasslands in Africa. Savory has asked why land thrived under the heavy impact of vast wild animal herds but turned to desert on year-round thinly stocked cattle ranches. Was overgrazing really a function of animal numbers; were herd behavior and timing of herd moves the most critical factors; and might not grass which had evolved with grazing animals perhaps also depend on them?

The wild herds, Savory observed, grazed, fertilized the soil with concentrated dung and urine, trampled the ground, and moved on, not to return until the grass grew fresh again. By contrast, domestic stock tended to stay in one place, grazing some plants again and again. Thus, time and the thundering herd appeared to Savory to be the tools of choice for restoration of drylands with seasonal rainfall. Savory has by now considered holistic management on public lands and in multiple-use situations.

One project is entitled the Navajo Mine Holistic Resource Management. It includes the people who live near the mine, along with government agencies with interests in the reclamation of the mined land. The project involves long-term planning for return of the mined land to the Navajo nation and individual grazing permittees.

The holistic resource management approach deserves close attention. It has important implications for integrated land management strategies, and it is now being applied by more and more ranchers and others in the western United States, who realize the practical benefit of this holistic approach.

3. The Greater Yellowstone Coalition's Environmental Profile of the Greater Yellowstone Ecosystem

Another example of a nongovernmental conservation advocacy effort is that of the Greater Yellowstone Coalition (GYC) in the western United States. Located in Bozeman, Montana, this group has served to mobilize a growing network of grassroots conservation organizations. It is involved in charting a course for an area that encompasses nearly 18 million acres (7,200,000 hectares), and includes Yellowstone and Grand Teton National Parks. It also contains seven National Forests and three National Wildlife Refuges. Greater Yellowstone is surrounded by some of the richest and most diversified semi-natural landscapes of the temperate zone. It is the location for the most extensive array of geothermal features in the world, the largest concentration of elk and bison in North America, and populations of grizzly bear and other species reduced in their former range.

GYC is working to create an alternative to the projected grim future for the wildlands of the region implied by current development trends. GYC has been involved in the development and publication of an "Environmental Profile of the Greater Yellowstone Ecosystem" (Glick et al., 1991), which has been disseminated via a community outreach program.

The Greater Yellowstone Tomorrow Project is "proactively planning for the future protection of the Greater Yellowstone Ecosystem" with three principal goals: development of a blueprint for action to provide steps for long-term ecosystem protection in Greater Yellowstone; organization of an informal and motivated constituency broad enough to ensure that recommended actions are implemented; and creation of a catalyst for implementation of the blueprint by the year 2000.

Greater Yellowstone is surrounded by development and is already isolated from other major wildland systems. The Greater Yellowstone Ecosystem includes state and private lands, and contains nearly all of the living organisms found in pre-Columbian times (though generally not in the same numbers). At the heart of the ecosystem is Yellowstone National Park, a high volcanic plateau with an average elevation of 8,000 feet (2,440 meters), surrounded by a series of spectacular mountain ranges. In Yellowstone Park may be found the world's largest petrified forest.

Despite the ecosystem's size, its biodiversity is in jeopardy. A matter of serious concern is the fragmentation of habitats such that plant and animal populations become isolated from one another and are cut off from processes essential for survival. The coalition's environmental profile considers potential effects of development on Greater Yellowstone's natural systems, for example, timber harvesting, oil and gas development, forest roading, hard rock mining, developed recreation sites, and rural subdivisions. Questions of human lifestyles and cultural attributes are addressed, as are regional economic trends and resource management. Steps toward ecosystem management include defining the ecosystem, including its ecological components, processes, and its problems; identification of unified management goals and programs; coordination of resource management and protection programs; and identification of information needs, filling in of these gaps, and effective utilization of this knowledge.

The Greater Yellowstone Coalition's activities convey an intention to think and act comprehensively about an entire landscape, and to engage governmental and nongovernmental participants in an attempt to deal with the concept of the Greater Yellowstone ecosystem. It is an exciting and important undertaking that appears to reflect an integrated landscape-ecological approach. It stresses that to achieve true sustainability there must be recognition of ecological limits to material growth.

Toward a European Ecological Network

A report of the Institute for European Environment Policy (Bennett, 1991) covers the first attempt to elaborate the concept of a European ecological network. An important goal of the European Community's (EC) Habitat Directive is the realization of a European ecological network, or EECONET, based on the notion that the conservation of Europe's natural heritage is a common responsibility. Each member state is responsible for a specific part of the European network. The European ecological network involves a framework to conserve the ecological nature of Europe and complex interrelationships which can extend over thousands of kilometers. Thus, it is a means to develop international priorities for action in a European context that will serve to guide practical conservation measures at national and local levels.

The report explores biogeographical structure of habitats in western Europe, including coastal areas; inland wetlands and rivers; forest and scrub; grassland; bogs; and inland rocks; cliffs, and dunes. It considers the decline of western European habitats and species, as well as the development of an ecological network. Three related shifts in emphasis representing a major realignment of nature conservation policy are discussed: shifts from species to habitats, from sites to ecosystems, and from national to international measures. Interrelationships between habitats, habitat change, and ecological corridors, are considered. The EC Corine program, adopted in 1985, has the objective of collecting and coordinating information on the state of the EC environment and improving its consistency. Color maps in the report show areas of European nature conservation, biogeographical zones, and animal migration routes.

Case studies dealing with Spain and The Netherlands (including coverage of the Dutch national ecological network) provide insight on environmental policy within the European Community. The report also looks at EECONET and EC policy:

"Nature conservation has not been a major priority for EC policy in the past. Policies for species protection and the establishment of national parks, nature reserves and other protected areas have developed mainly at a regional or national level and there remain substantial differences in the approach adopted in different parts of Europe. Nonetheless, the international framework for these diverse policies is growing in substance and influence and the Community's own role is expanding from a primary focus on birds, to the wider area of habitat conservation. (Bennett, 1991.)

EECONET is characterized not as a straitjacket obliging national authorities to comply with a detailed EC master plan for a network, but rather a planning framework built on scientific principles for guiding and coordinating the nature conservation efforts of a wide range of different authorities. The report examines creation and implementation of EECONET and its extension beyond the EC, and looks at EECONET in relationship to EC policies on agriculture, regional development, transport and energy, and fisheries. EECONET may serve not only to prevent actions which would cause further damage to existing habitats, but may also promote enhancement of ecological systems.

As noted by Naveh in writing on biodiversity and landscape management (1993), "probably of greatest relevance for conservation management is the determination of connectivity" (see also the Chapter 1 supplement). Citing Merriam (1984), he describes connectivity as a functional parameter of connectedness (by structural links or corridors between landscape units, described from mappable elements such as woodland patches, hedgerows, roads, etc.) As explained in detail by Merriam (1989), connectivity has led to new insights into the effect of man-made corridors and barriers on meta population demography and genetics. EECONET is based on this concept of connectivity.

Warwickshire Landscapes Project: Assessment and Conservation of Landscape Character

In England and Wales rural landscape conservation is practiced on a large scale under the administration of the Countryside Commission (1991). One of the examples of this integrated approach in the United Kingdom is on open landscape planning and management to provide a systematic and thorough analysis of the Warwickshire landscape, together with practical guidelines for its conservation and enhancement.

The Warwickshire effort embraces landscape classification, description, and evaluation, producing a step-by-step approach to landscape assessment that can be adapted to the needs of the user. Close cooperation and discussion with organizations involved in countryside matters (farming and landowning representatives, foresters, county and district planners, highway and river engineers, landscape architects, ecologists, and archaeologists) has been an important part of the project.

Further case studies illustrating the concept of protected and managed landscapes have been provided by F. H. C. Lucas (1992) in his book on protected landscapes. Based on his international experience as chair of the IUCN Commission of National Parks and Protected Areas and as Fellow of the Environment and Policy Institute at the East-West Center in Honolulu, this book is a valuable guide for planners and decision-makers. It deals with the benefits of protected landscapes, biosphere reserves, and world heritage sites by examining the process of their selection and legal measures for their establishment and management. It is a practical guide to the implementation of holistic landscape planning and conservation under a range of socioeconomic and cultural conditions.

Planning and Design Implications of Landscape Ecology

A special issue of *Landscape and Urban Planning* (Cook and Hirschman 1991) considered planning and design implications of landscape ecology. The goals of the symposium from which the papers arose were to "help solidify theoretical methodological and applied aspects of landscape ecology from a planning/ design perspective and to expand the planning/ design participation in landscape ecology" (Cook and Hirschman, 1991). Articles addressed three questions: (1) How can landscape-ecological principles be fitted into traditional planning/design frameworks? (2) How can landscape ecological research be used to create new or more appropriate planning/design processes? (3) What is our responsibility in planning and design to incorporate landscape ecological principles and data?

We here focus on two of the papers. Golley (at the Institute of Ecology and School of Environmental Design at the University of Georgia, U.S.A.) and Bellot (at Institute of Mediterranean Agronomy at Zaragoza, Spain), deal with the interaction of landscape ecology, planning, and design (Golley and Bellot, 1991). They note that landscape ecology provides planners and designers with information which depicts the structure of the physical and biological environment at a scale practical for humans, describes dynamic processes in time and space, and explores ways in which structural shape is processed. This understanding of processes allows for planning and predicting the consequences of design. The program of the Institute of Mediterranean Agronomy at Zaragoza, Spain is discussed in which a connection between landscape ecology and design is achieved in a location devoted to advanced postgraduate training for agriculturally oriented professionals in developing countries. The method of landscape analysis of an irrigated landscape at La Violada is used to reinforce the authors' points about interaction of landscape ecology, planning, and design.

Lyle (1991) deals with semi-formal models in ecological planning, and stresses that formal mathematical models have proven difficult to use in landscape planning and design. Lyle, however, recognizes that models are quite useful as a means of making invisible processes tangible. He urges adoption of an adaptive and pragmatic approach to modeling that will enable creative use of models in planning and design processes. In "semiformal modeling", he refers to models which have definite form and consistency without the rigid formality of strict quantification. Lyle

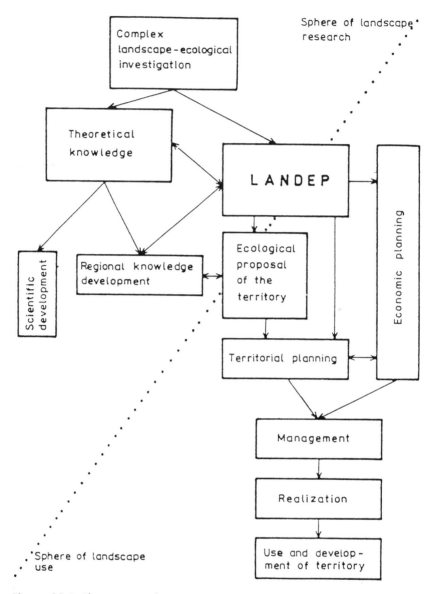

Figure S3-1. The Position of LANDEP, a Comprehensive Set of Methods for Land Planning and Optimization in Science and Practice. (Source: Ruzicka and Miklos 1990.)

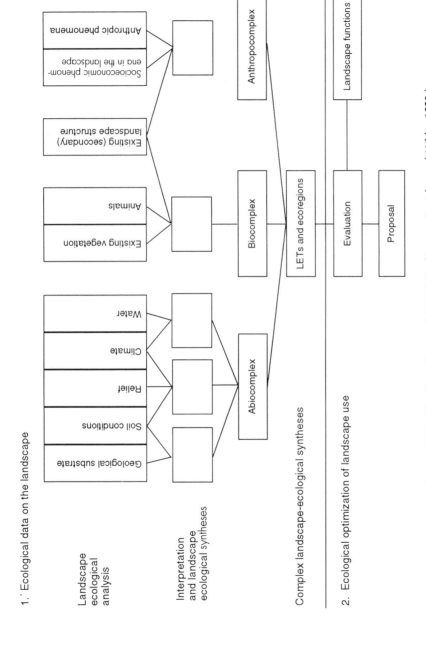

Figure S3-2. The Main Steps of LANDEP. (Source: Ruzicka and Miklos 1990.)

is the author of the acclaimed book *Design for Human Ecosystems: Landscape, Land Use and Natural Resources* (1985).

Stabilizing Spatial and Functional Relationships through Landscape-Ecological Planning in Czechoslovakia

The close integration of landscape ecology with land-use planning and decision-making in Europe has been noted in the supplement to Chapter 1. As cited there, the work of Ruzicka and Miklos (1990) in Czechoslovakia deserve special mention for their development of one of the most significant and practically applied integrated landscape ecological planning methods to date.

Landscape Ecological Planning (LANDEP) is described by Ruzicka and Miklos (1990) as a system which includes a comprehensive landscape-ecological analysis, a synthesis component, a landscape-ecological evaluation of the territory, and a proposal for optimum land uses. LANDEP contains two basic parts (see Figures S3-1 and S3-2).

1. *Landscape ecology data.* These data comprise inventory and assessment of the abiotic and biotic components, contemporary landscape structure, ecological phenomena and processes, and effects and consequences of human activities upon the landscape.
2. *Ecological optimization of the landscape use.* Optimization relies on the landscape ecology data, particularly for ecologically homogeneous spatial units. Spatial units are compared with development needs for a particular territory. Following the evaluation of the degree of appropriateness of each spatial unit for a particular human activity or land use, a proposal is made on the most suitable locations of the activity in the landscape based on landscape ecology criteria.

Maria Kozova and Tatiana Hrnciarova of the Center of Biological and Ecological Sciences of the Slovak Academy of Sciences, observe that "landscape-ecological planning is recognized as a new progressive form of planning which has gained its important position within the frame of the decision making processes in territorial planning" (Kozova and Hrnciarova, 1988). The LANDEP method of landscape-ecological planning focuses on elaborating solutions for ecologically optimum utilization of landscape. For each area, optimum utilization should be consistent with its ecological prerequisites. Consequently, in the long-term an environmentally sound proposal also becomes economically profitable.

LANDEP is a system of methods intended to ensure long-term stability of the landscape, linking up to all utilizable stabilizing elements and mechanisms. It has been used to address the needs of agriculture, territorial planning, nature conservation and management, and social practice.

Clearly, LANDEP requires interdisciplinary cooperation among scientists.

Jurko (1987) defines ecological stability as a continuation of an ecological system at its "unstressed" state during stress situations, and its ability to return to the original state after slight deformations. Landscape stability may be studied and evaluated at different levels—through the stability of natural cycles, stability of communities and ecosystems, up to the stability of global processes.

Within LANDEP, procedures to evaluate ecological stability at the community and landscape levels have been elaborated with regard to constancy (stability against stress situations ranked to the "normal" living regime) and resistance (dynamic stability, or stability against external perturbations).

LANDEP is involved with (1) proposals to ameliorate the negative consequences of disturbed landscape resulting from development (industrial production, raw material exploitation, transportation, agricultural production, etc.); (2) proposals for economically optimum functional landscape organization in which societal requirements and demands consistent with landscape and ecological conditions of the territory are considered; and (3) proposals for a network of stable elements in the landscape.

LANDEP has been applied in areas with different degrees of disturbance by industrial production, traffic, urbanization, agricultural and forest production, etc. These include (1) proposals for environmental regeneration and rational utilization of territory after disturbance by magnesite processing, (2) ecological evaluation of prerequisites for developing agriculture and recreation on land of the Klenovec water reservoir, and (3) ecological study of the Danube River water works site (see supplement to Chapter 1). In this area, LANDEP proposes ecologically optimized land use, considers the structure and location of the forest, proposes a territorial system of ecological stability which preserves the ecological role of the local fauna and flora, and divides the area into functional zones.

LANDEP has a wide range of potentially applications and deserves to be thoroughly examined by landscape ecology planners worldwide.

Conclusions

Employment of holistic landscape ecology is increasing in land planning and resource management, in both industrialized and developing countries. It is being used as a basis for decision-making for creation of standards and policies, and for implementation at scales from subregional to global.

A challenge to landscape-ecological researchers has been presented by landscape planners and resource managers to convey comprehensible information in formats that can be easily utilized. Researchers involved in holistic landscape-ecological research appear to be aware of the practical needs of those who must apply research results to actual needs in specific locations. This has meant that holistic landscape-ecological planning and management is gaining credence, adherents, and adoption (as documented in Chapter 3). This is even occurring in circles which questioned or actively opposed the resolution of conservation and development issues through insights gained via holistic landscape ecological research.

The increased capability of remote sensing and geographical information systems are undoubtedly contributing factors. These new methods for land planning and management have heightened acceptance of holistic landscape ecology by those involved in the "real world."

The growing interaction between public officials, physical/social/economic planners, private and public land managers, and motivated citizenry and environmental educators is necessary and desirable. This interaction will help bring about rational, sustainable, and ecologically consonant results.

References

Ahern, J. 1991. Planning for an extensive open space system: linking landscape structure and function. Landscape and Urban Planning 21(1,2):131–145.

Amir, S. 1984. Israel's Coastal Program: Resource protection through management of land use. Coastal Zone Mgt. J. 12(2/3):189–223.

Ashdown, M., and J. Schaller. 1990. Geographic Information Systems and Their Application in MAB Projects, Ecosystem Research and Environmental Monitoring. German National Committee for the UNESCO Program, Man and the Biosphere (MAB), Bonn, Germany.

Barnaba, E. M. 1992. CLEARS Resource Inventory Program. Annual Report, Center for the Environment, Cornell University, Ithaca, New York.

Barnaba, E. M., W. R. Phillipson, A. W. Ingram, and J. Pim. 1991. The use of aerial photographs in county inventories of waste disposal sites. Photogrammetric Engineering and Remote Sensing 57(10):1289–1296.

Bennett, G. (Ed.) 1991. Towards a European Ecological Network. Institute for European Environment Policy, Arnhem, The Netherlands.

Burrough, P. A. 1988. Principles of Geographic Information Systems for Land Resource Assessment. Clarendon Press, Oxford.

Center for Holistic Resource Management. 1990. Holistic Management on Public Lands and in Multiple Use Situations. Center for Holistic Resource Management, Albuquerque, New Mexico.

Clark, J. 1974. Coastal Ecosystems: Ecological Considerations for Management of the Coastal Zone. Fourth printing, September 1989. The Conservation Foundation, Washington, D.C.

Clark, J. et al. 1977. Coastal Ecosystems Management. A Technical Manual for the Conservation of Coastal Zone Resources. John Wiley and Sons, New York.

Cook, E. A., and J. Hirschman. 1991. Guest editor's introduction. Special Issue on Landscape Ecology of Landscape and Urban Planning 21(1,2):1–2.

Countryside Commission. 1991. Assessment and Conservation of Landscape Character. The Warwickshire Landscapes Project Approach. Countryside Commission, Cheltenham.

Coulson, R. N., et al. 1991. Intelligent geographic information systems for natural resource management. *In*: Quantitative Methods in Landscape Ecology. M. G. Turner and R. H. Gardner (Eds.), Ecological Studies 82. Springer-Verlag, New York.

CSIRO. 1984. Collected workshop and conference papers from the Great Barrier Reef Marine Park Project. Technical Memorandum 84/8. CSIRO Institute of Biological Resource. Division of Land and Water Resources, Canberra, Australia.

Davis, F. W., and S. Goetz. 1990. Modeling vegetation-pattern using digital terrain data. Landscape Ecology 4(1): 69–80.

de Groot, R. S. 1992. Functions of Nature. Evaluation of Nature in Environmental Planning, Management and Decision Making. Wolters-Noordhoff, Amsterdam.

Glick, D., et al. (Eds). 1991. An Environmental Profile of the Greater Yellowstone Ecosystem. Executive Summary. Greater Yellowstone Coalition, Bozeman, Montana.

Golley, F. B., and J. Bellot. 1991. Interactions of landscape ecology, planning and design. Landscape and Urban Planning 21(1,2): 3–11.

Greegor, D. H., Jr. 1986. Ecology from space. Bio Science 36(7):429–432.

Haber, W. 1990. Using landscape ecology in planning and management *In:* I. S. Zonneveld and R. T. T. Forman (Eds.), Changing Landscapes: An Ecological Perspective. Springer-Verlag, New York, pp. 217–232.

Hall, D. L. 1991. Landscape planning: functionalism as a motivating concept from landscape ecology and human ecology. Landscape and Urban Planning 21(1,2): 13–19.

Johnston, C. A. (Ed.) 1990. Spatial and Temporal Analysis Using Geographic Information Systems. Special Issue of Landscape Ecology 4(1):3–80.

Jurko, A. 1987. Constancy and resistance as basic components of stability in landscape ecology: a case study on vegetation in East Slovakian Lowlands. Ecology (CSSR) 6(4):417–438.

Kozova, M., and T. Hrnciarova. 1988. Stabilizing of spatial and functional relationships in landscape-ecological planning LANDEP. *In:* Proceedings of VIIIth International Symposium of Landscape Ecology Research, Zemplinska Sirava, October 3–7, 1988, pp. 39–50.

Lieberman, A. S., and C. R. O'Neill, Jr. 1988. Vegetation Use in Coastal Ecosystems. Information Bulletin 198. Cornell Cooperative Extension, Cornell University, Ithaca, New York.

Lillesand, T. M., and R. W. Kiefer. 1987. Remote Sensing and Image Interpretation, 2nd Edition. John Wiley and Sons, New York.

Lyle, J. T. 1985. Design for Human Ecosystems: Landscape, Land Use and Natural Resources. Van-Nostrand Reinhold, New York.

Lyle, J. T. 1991. The utility of semi-formal models in ecological planning. Landscape and Urban Planning 21(1,2):47–60.

Lucas, P. H. C. 1992. Protected Landscapes. A Guide for Policy-Makers and Planners. Chapman and Hall, London.

Merriam G. 1984. Connectivity: a fundamental ecological characteristic of landscape patterns. *In:* J. Brandt and A. P. Agger (Eds.), Methodology in Landscape Ecological Research and Planning. Proceedings of First International Seminar IALE, Roskilde University Centre, Oct. 15, 1984, Vol. I, pp. 5–15. Roskilde Universitatsforlag, Roskilde, Denmark.

Merriam G., M. Kozakiewicz, Tsuchiya and K. Hawley. 1989. Barriers as boundaries for metapopulations and demes of *Permyscus leucopus* in farm landscapes. Landscape Ecology 2:227–236.

Mc Neilly, J. A. 1988. Economics and Biological Diversity: Developing and Using Economic Incentives to Conserve Biological Resources. IUCN, Gland, Switzerland.

Milne, B. T. 1991. The utility of fractal geometry in landscape design. Landscape and Urban Planning 21(1,2):81–90.

Naveh, Z. 1993. Biodiversity and landscape management. *In:* K. C. Kim (Ed.), Biodiversity and landscapes: A Paradox of Humanity. Cambridge University Press, New York (in press).

Nir, Y. 1982a. Offshore Artificial Structures and Their Influence on the Israel and Sinai Mediterranean Beaches. Report No. MGG/4/82. Marine Geology and Geomathematics Division, Geological Survey of Israel, Ministry of Energy and Infrastructure, Jerusalem.

Nir, Y. 1982b. Offshore artificial structures and their influence on the Israel and Sinai Mediterranean beaches. *In:* Proceedings of the Eighteenth Coastal Engineering Conference, ASCE, Capetown, South Africa, November 14–19, 1982, pp. 1837–1856.

Nir, Y., and A. Elimelech. 1990. Twenty five years of development along the Israeli Mediterranean coast: goals and achievements. *In:* P. Fabbri (Ed.), Recreational Uses of Coastal Areas. Kluwer Academic Publishers, The Netherlands, pp. 211–218.

Olson, G. W., and A. S. Lieberman (Eds.). 1985. Proceedings of International Symposium on Geographic Information for Conservation and Development Planning. International Land Use Planning Program, Cornell University, Ithaca, New York.

O'Neill, C. R., Jr. 1985. A Guide to Coastal Erosion Processes. Information Bulletin 199. Cornell Cooperative Extension, Cornell University, Ithaca, New York.

O'Neill, C. R., Jr. 1986. Structural Methods for Controlling Coastal Erosion. Information Bulletin 200. Cornell Cooperative Extension, Cornell University, Ithaca, New York.

Quattrochi, D. A., and R. E. Pelletier. 1991. Remote sensing for analysis of landscapes: an introduction. *In:* M. G. Turner and R. H. Gardner (Eds.), Quantitative Methods in Landscape Ecology. Ecological Studies 82. Springer-Verlag, New York, pp. 51–76.

Roller, N. E. G. and J. E. Colwell. 1986. Coarse-resolution satellite data for ecological surveys. Bioscience 36(7):468–475.

Ruzicka, M. and Miklos, L. 1990. Basic premises and methods in landscape ecological planning and optimization. *In:* I. S. Zonneveld and R. T. T. Forman. (Eds.) Changing Landscapes: An Ecological Perspective. Springer-Verlag New York, pp. 233–260.

Stevens, W. K. 1992. Novel strategy puts people at heart of Texas preserve. New York Times Science Times section, March 31, 1992, pp. C1, C8.

Savory, A. 1988. Holistic Resource Management. Island Press, Washington, D.C.

Van der Meulen, F., J. V. Witter, and W. Ritchie (Eds.). 1991. Impact of Climatic Change on Coastal Dune Landscapes of Europe. Issue of Landscape Ecology 6(1/2): 3–113.

Westman, W. E. 1985. Ecology, Impact Assessment, and Environmental Planning. John Wiley & Sons, New York.

4

Dynamic Conservation Management of Mediterranean Landscapes

The Evolution of Mediterranean Landscapes

Bioclimate Delineation of the Sclerophyll Forest Zone

The sclerophyll forest zone (SFZ) of mediterranean climates covers all those regions that exhibit similar climatic characteristics of warm to hot dry summers, with high solar irradiation and high rates of evaporation, and mild to cool wet winters with low solar irradiation and low rates of evaporation. In these conditions, broad-leaved and mostly evergreen trees and shrubs with thick, but mostly small, leathery leaves, reach their optimum development and distribution. Forests dominated by such plants are considered the zonal vegetation. Köppen (1923) called this the *olive climate* because around the Mediterranean Basin the distribution of the (cultivated) olive tree—a typical broad-leaved evergreen sclerophyll tree—corresponds quite well with this climate type.

In addition to the Mediterranean, four other widely separated regions of the world, chiefly between 30° and 40° north latitude and south latitude on the west coasts of continents have similar bioclimates with comparable sclerophyll vegetation types; these regions are California, central Chile, southwestern and part of southern Australia, and the Cape Region of South Africa. This zone makes up roughly 1% of the world's terrestrial vegetation, about half of this concentrated around the Mediterranean Sea.

In France it is called maquis, in Italy macchia, in Spain and Chile mattoral, in Greece xerovoni, in Israel choresh, in California chaparral, in South Africa fynbos, in Australia mallee, heath, and scrub. The terminology of the lower and more xeromorphic vegetation types is even more confusing, and sometimes the same term is used in different countries for different formations.

As shown in Figure 4-1, Ashmann (1973), on the basis of the minimum and maximum growth requirements of these sclerophylls, restricted the sclerophyll forest zone in bioclimatic terms to regions having not more than 900 mm annual precipitation, with not less than about 250 mm in coastal regions and 350 mm in warmer interior regions; at least 65% of precipitation is concentrated in the winter half. But he also emphasized the great climatic diversity and peculiarity of each of these regions, which is hard to express in generalized climatic classifications.

Such a classification was attempted by UNESCO (1963). In it, the SFZ was subdivided into a drier "xerothermic" part that merges into the subdesert, (see also Meigs, 1964, for a subdivision of semiarid Mediterranean climates) and into a wetter *accentuated thermomediterranean* part, merging into the *attenuated thermomediterranean,* in which tall conifer and mixed broad-leaved summer-deciduous forests with higher moisture demands and greater cold resistance become the zonal vegetation. The SFZ is thereby confined chiefly to the xerothermic climatic index of 125–200 biological dry days.[1] This is the "true mediterranean fire climate" (Naveh, 1973), in which acute fire hazard prevails for dense sclerophylls and conifers, as well as for dry herbaceous vegetation from four to eight months.

In this zone, the difficulties of defining ecological homoclimes by single indices can be overcome only by closer comparisons of patterns of temperatures and their diurnal amplitudes, and of rainfall distributions and their reliability during the main growth season. For such comparisons valuable climatic diagrams have been used by Walter (1973) in his lucid ecophysiological description of this zone, by Horvat et al. (1974) in their perceptive description of this zone in southeastern Europe, and by Zohary (1973) in his broad geobotanical overview of the Middle East. They were also very illustrative in a detailed climatographic comparison of mediterranean ecosystems in California and Chile (di Castri, 1973), as well as in other comparisons between these countries.

The Evolution of Natural Mediterranean Landscapes

As has been described in more detail in a special volume on the structure and origin of mediterranean ecosystems (di Castri and Mooney, 1973), the Mediterranean Basin, California, and central Chile are relatively young orographic systems that gained their present high, sharp, folded and faulted landforms of mountain and hills, often rising close to the coast, by violent uplifting in the late Tertiary and early Quaternary Periods. Being highly dissected, complex, and partly unstable, with many steep slopes and shallow rocky soils, they are very vulnerable to sheet and gully erosion if their protective natural vegetation

---

[1]Following Gaussen, a xerothermic index (x) is calculated from the sum of monthly indices of dry months in which total precipitation $(P$ mm$)$ is equal to or lower than half the temperatures $(T°C)$: $P \leqslant 2T$. For this purpose, all biological dry days without rain, mist, dew, or fog (or half days with) are multiplied by a coefficient of relative humidity H, ranging from 1 $(H = 40\%)$ to 0.5 $(H = 100)$.

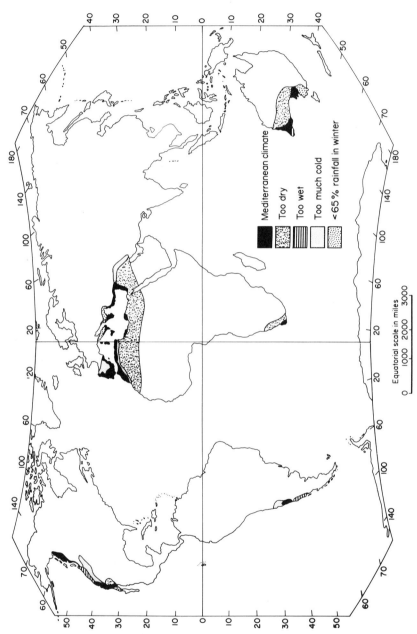

Figure 4-1a. World map of the areas with mediterranean climates (Ashmann, 1973).

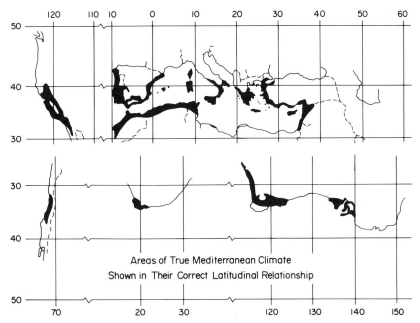

Figure 4-1b. Areas with true mediterranean climate shown in their correct latitudinal relationship (Ashmann, 1973).

canopy is denuded and their shallow soil mantle is exposed to desiccation in the dry summer and to torrential rains in the winter (see Figure 4-2).

The geomorphological similarity is greatest between central Chile and California. In both regions a coastal mountain range to the west is separated by a central valley from high mountain masses—the Andes and Sierra Nevada, respectively—to the east, and the SFZ is confined to chiefly nontillable foothills and uplands in the lower elevations. However, in Chile the narrow mediterranean strip gives way very soon to an extreme rainless desert to the north and to higher-rainfall areas to the south, whereas in California this zone widens in the south and merges gradually with the drier semidesert scrub. To the north the SFZ is bordered by more mesic conifer forests.

The Levant and the south Australian mediterranean landscape are rifted shields, and the South African and southwest Australian ones are ancient basement complexes. They all lack high relief and, with exception of the Cape Table mountain, have a subdued topography. The Australian and South African SFZ are of much more limited extent, bordered by the termination of these continents to the south, and their soils are extremely poor.

The plants of the Northern hemisphere SFZ evolved since Cretaceous times in the borderlines between temperate and tropical floras and as mixture of both predecessors, but in the Southern Hemisphere they developed entirely from tropical ancestors. The sclerophyll woody species were apparently preadapted to the mediterranean climate patterns of increasing summer drought and lowering winter temperatures that developed during the Pleistocene, but most herbaceous

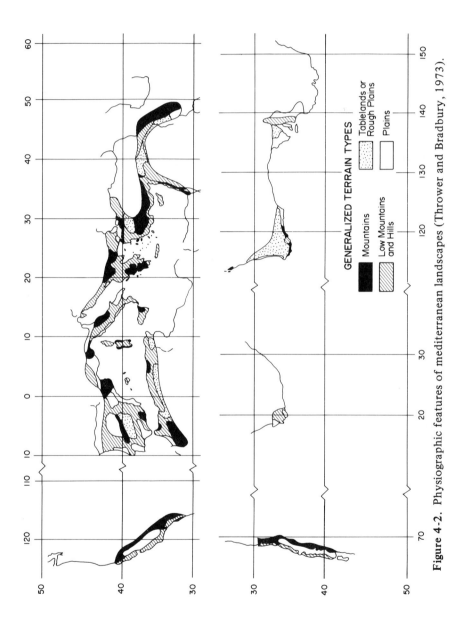

Figure 4-2. Physiographic features of mediterranean landscapes (Thrower and Bradbury, 1973).

species evolved in this period and became very rich in the Mediterranean, now constituting up to 50% of all species. In the Northern Hemisphere zones, 10% of all the plant genera and at least 40% of these species are endemic; but in the Southern Hemisphere these figures are much higher. This is especially the case in the Cape and southwest Australia, where these floras are derivatives of the ancient Gondowan heathlike communities. These floras were destroyed in central Chile by mountain upliftings, glacial periods, and extreme climatic changes, as well as by rejuvenation of soils, similar to those occurring in California. Therefore, both Southern Hemisphere Gondowan floras are older and richer, being adapted to very old, mineral-poor soils, which may also have induced their sclerophyllous features. For these reasons, as Naveh and Whittaker (1979a) pointed out, both Gondowan sclerophyll ecosystems are profoundly different in structure and soil and nutrient relationships and in their greater richness in woody plants from the other three more recent ones. They also differ entirely in present-day problems of conservation, and will therefore not be included in this discussion.

The Evolution of Cultural Mediterranean Landscapes

The Mediterranean Basin. The final shaping of the other three "true" mediterranean landscapes took place during the Quaternary in highly dynamic situations of climatic fluctuations, tectonic and vulcanic activities, and increasing diversification of local site conditions. In addition to drought, fire (and, at least in the Mediterranean, grazing pressure) became an important selection force. Here the evolutionary process of speciation and ecotypic variation was continued and gradually intensified during the long period of human occupation, and can be traced back to the middle Pleistocene in the fire-swept volcanic soils in the ecotones between semiarid tropical and mediterranean forests and savannas such as the woodlands of the Jordan Valley. Here, stone tools, artifacts, and fractured bones of 80 different species, chiefly land and water vertebrates of both tropical and temperate origins (Stekelis, 1966; Haas, 1966) have been found. This region was also one of the first cradles of cereal and stock farming about 10,000 years ago, and its drought-, fire-, and grazing-resistant annual grasses and legumes were among the first domesticated plants (Zohary, 1969). Further north, in the heart of the SFZ, the use of fire by paleolithic hunter-gatherers has been proven for at least 500,000 years (Naveh, 1977).

These evolutionary processes during geological times can be described by semiformal, multivariate, ecotope state factor equations of biogenic and pedogenic functions based on Jenny's (1961) functional–factoral approach and explained in Chapter 2:

$$E_{s,v} = f(P,R,Cl_{dr,fi}O_{gr} \ldots T < 1,000,000) \qquad (4\text{-}1)$$

where $E_{s,v}$ represents the dependent ecotope variables of soil and vegetation, P and R are the initial site condition of soil parent material and relief, respectively; Cl and O are the driving fluxes of climate and organisms, respectively, with their most important evolutionary forces of drought (dr), fire (fi), and grazing (gr); and T is the time function of the cycle.

With the emergence of Paleolithic hunter-gatherer economies in the Upper Pleistocene, these original biofunctions were gradually converted into an *anthropogenic* biofunction. This addition of psychogenic–noospheric forces and of cultural information marked the evolution of the semi-natural mediterranean landscapes:

$$E_{s,v} = f(H_{bu,hu\text{-}ga}, P, R, Cl_{dr,fi} O_{gr} \dots T < 100,000) \qquad (4\text{-}2)$$

where H is the driving flux of the human state factor through burning (bu) and hunting-gathering (hu-ga).

After the neolithic revolution, with the replacement of most wild ungulates by domesticated ones during the agricultural transformation, the speed of evolution of a diverse herbaceous flora of drought-, fire-, and grazing-adapted species and biotypes increased rapidly. It was reinforced by many invaders from adjacent, drier steppe regions. First the richest forests in the fertile lowlands were cleared for cultivation, and later on, with the evolving of dense pastoral-agrarian populations, all arable slopes were cleared, terraced, or patchcultivated, and the remaining sclerophyll forests and scrub thickets were opened periodically by fire (in the same way as by the Paleolithic hunter-gathers) to increase edible herbs, bulbs, and fruits and palatable pasture plants, and by woodcutting and coppicing for fuel, tools, and construction. The grazing (by goats, sheep, cattle, donkeys, etc.) of the cultivated uplands after harvesting in late spring facilitated the transfer of seeds to and from untillable "natural" pastures and created ideal conditions for introgression and spontaneous hybridization of wild and cultivated plants, as shown in the case of *Hordeum spontaneum* and *H. vulgare* (Zohary, 1960). Many of these plants, especially grasses and legumes, became the most valuable pasture plants in mediterranean and temperate climates. Others became widespread weeds, and many from both groups, pre-adapted to these combined pyric–pastoral–agricultural ecosystem modifications, invaded the SFZ thousands of years later in Chile, California, and Australia. Crowding out many indigenous plants, they created annual grasslands, that are strikingly similar in structure, productivity, and response to grazing.

During biblical and classical times, the densely wooded natural landscape was finally transformed into a more open and much richer cultural one. The natural vegetation was retained only on the least accessible rocky and mountainous sites. On untillable slopes and the edges of terraced fields, the man-modified semi-natural mixed woody and herbaceous plant communities were closely interwoven with terraced and well-managed (sometimes also irrigated and ferti-lized) field crops, vineyards, olive and fruit groves. The care, conservation, and management of this landscape is well documented in biblical, Talmudic, Greek, and Roman sources, and it can be considered the cradle of Western civilization. It became the granary of the Roman Empire, but after its downfall and the Moslem conquest, the onslaught of nomadic pastoralists marked the beginning of a long period of population decline, agricultural decay, landscape desiccation, and desertification. The denudation of the slopes and especially the destruction of terraces, water channels, and aqueducts, had far-reaching catastrophic geomorphological consequences leading to soil and water erosion, siltation, riverbed sedimentation, and the conversion of the fertile lowlands into badlands and malaria-ridden swamps. However, as pointed out by Naveh and Dan (1973), because of the decreasing population pressures in the last centuries and the great vegetative recuperative powers of the sclerophyll trees and shrubs, a new dynamic man-maintained equilibrium has been established on those nonarable and abandoned upland ecosystems that were neither overgrazed and heavily coppiced or too frequently burned nor completely protected. This man-maintained equilibrium has shaped the Mediterranean plant communities into a "mosaic of innumerable variants of different degradation and regeneration stages" (Walter, 1968). Thus, as a result of the combination of ecological heterogeneity, biotic richness, and human modification, a very attractive and diverse seminatural landscape has evolved. As will be shown later, this is now threatened by an unfortunate combination of excessive traditional and accelerating neotechnological land uses.

This evolution of the cultural landscape in the Holocene can be described as agropastoral biofunctions:

$$E_{s,v,a} = fH_{ag-pa}(P,R,Cl \ldots T < 100) \qquad (4\text{-}3)$$

where E_a is the dependent variable of man-made artifacts in cultural ecotopes and H_{ag-pa} represents the agropastoral human state factors of burning, clearing, cutting, coppicing, terracing, cultivating, grazing, browsing, constructing, and the like. These were various aggradation and degradation land-use cycles lasting thousands of years during which "Homo faber," through his noospheric inputs of agrotechnological energy (including fire and his own and his domestic animals' muscular energy) and material (including his agricultural crops and animals and their products), coupled with positive feedbacks of cultural information, gradually became the dominant state factor, changing all others and adding cultural artifacts as dependent variables, and thereby rural and urban ecosystems.

In the present neotechnological degradation cycles:

$$E_{s,v,a} = fH_{ne,ag-pa\ int.,ab,af,pr}(P,R,Cl \ldots T < 10) \qquad (4\text{-}4)$$

in which $H_{ne,ag-pa\ int.,ab,af,pr}$ represents the combined impact of modern human state factors of neotechnological interventions and intensified agropastoral land uses, coupled with abandonment, afforestation, and complete protection. In these anthropogenic biofunctions, "Homo industrialis" is accelerating and expanding the replacement of biosystems and distorting the above-mentioned dynamic flow equilibrium of the traditional agropastoral function.

As was mentioned in Chapter 2 and has been described in more detail elsewhere (Naveh and Whittaker, 1979a; Naveh, 1982), environmental and cultural negative feedback loops governing Equation 4-3 have now been replaced by positive feedback loops between energy/material production, conversion, and consumption and cultural information. These combined, and even synergistic, modern noospheric impacts, symbolized by semantic signs in these equations, have been partly quantified in seminatural and agropastoral ecotopes by using structural and species diversity as dependent vegetation variables of different biofunctions.

These major phases of evolution and degradation of the cultural mediterranean landscape have been summarized graphically in Figure 4-3 by an isomorphic model of the structural and spatial changes through time as effected by inputs of energy, material, and information from the geospheric–biospheric–noospheric Total Human Ecosystem (THE).

The above-described long process of human landscape modification has been condensed in Chile to 400 years and in California to less than 200 years. Although in Chile human impact started 11,000 years ago and slash-and-burn farming preceded European contacts, the latter very soon led to increasingly heavy defoliation pressures due to sheep and goat grazing; woodcutting for fuel, construction, and (later on) silver and copper mining; and careless crop cultivation on the upland slopes. Thus, at the end of the last century, not only were sclerophyll forests of the fertile lowlands destroyed, but most of the shrublands of the coastal ranges and Andes foothills were also decimated and severely degraded. Their remnants are left in scattered stands on the slopes, along riverbeds, in inaccessible mountain sites, and in some larger estates. As mentioned earlier, the indigenous herbaceous vegetation succumbed to more aggressive, preadapted mediterranean invaders, chiefly annual grasses, which colonized denuded and eroded slopes and now dominate the derived Chile grasslands (Ashmann, 1973; Ashmann and Bahre, 1977).

In California, a similar replacement of native plants by mediterranean invaders occurred 200 years later, induced by grazing, wood-clearing, and cultivation in the valleys and foothill slopes of the coastal ranges and the Sierra Nevada. Here, however, modern human impact never reached the same widespread catastrophic levels of denudation and sheet and gully erosion. In the extensive dense chaparral shrublands on steeper slopes and higher elevations in the coastal ranges and in Southern California, even after the displacement of the Indian populations, wildfires followed by deer grazing remained the main vegetation modifier. However, here also the dynamic equilibrium has been gravely distorted in recent times by destructive wildfires and uncontrolled urbanization.

In summary, because of its long and intensive human modifications, the SFZ in the Mediterranean has the greatest degree of structural and floristic diversity, the latter contributed chiefly by herbaceous plants, including many annuals. This has also induced a great faunistic diversity, especially of birds, reptiles, and insects. In Chile human impacts were longer and more intensive than in California, and therefore its SFZ is, in general, more open and more degraded. The California SFZ is the least diverse but the most affected at present by fires. Hence, a sequence in diversity values can be found in these three SFZ biomes according to the duration of the agropastoral biofunctions (Naveh and Whittake, 1979a).

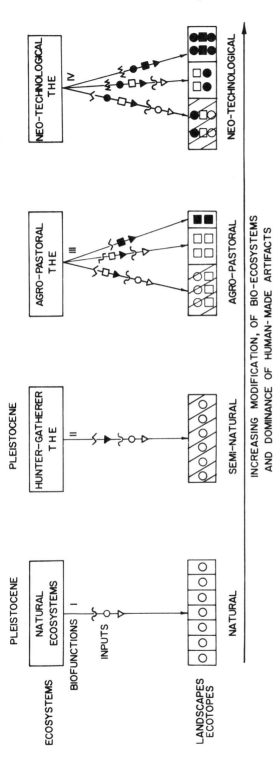

Figure 4-3. Evolution and degradation of mediterranean landscape ecotopes by energy, material, and information inputs from the geosphere, biosphere, and noosphere. (THE) = Total Human Ecosystem. Geosphere and biosphere inputs from natural and total human ecosystems: $(\sim\!\!\sim)$ = biological conversion of solar energy; $(\bigcirc)$ = natural material and organisms; $(\triangledown)$ = bio-physico-chemical information and control. Noosphere inputs from total human ecosystems: $(\sim)$ = technological conversion of fire energy; $(\urcorner\!\lrcorner)$ = technological conversion of muscle energy; $(\mathsf{/\!W\!\backslash})$ = technological conversion of fossil energy; $(\blacktriangledown)$ = cultural and technological information and control; $(\square)$ = agricultural material and organisms; $(\blacksquare)$ = rural artifacts (chiefly natural and synthesized). Landscape ecotopes: ($\boxed{\bigcirc\bigcirc}$) = closed forest, woodlands and shrublands of natural bio-ecosystems; ($\boxed{\bigcirc}$) = semi-open forests, woodlands and shrublands of semi-natural bio-ecosystems; ($\boxed{\bigtriangleup\bigcirc}$) = semi-open and open forests, woodland, shrublands and grasslands of semi-agriculture ecosystems; ($\boxed{\square\square}$) = terraced and cultivated fields and plantations of agric. bio-ecosystems; ($\boxed{\blacksquare}$) = farms, villages, roads, etc. of rural techno-ecosystems; ($\boxed{\bullet\square}$) = cities, factories, roads, etc. of urban-industrial techno-ecosystems. Multivariate state factor equations of biofunctions: (I) = $E_{s,v} = f(P,R,Cl_{dr,di}O_{gr} \dots T < 1,000,000)$; (II) = $E_{s,v} = f(H_{bu,hu,ga}P,R,Cl_{dr,fi}O_{gr} \dots T < 100,000)$; (III) = $E_{s,v,a} = fH_{ag\text{-}pa}(P,R,Cl \dots T < 100,000)$; (III) = $E_{s,v,a} = fH_{ag\text{-}pa}(P,R,Cl_{dr,di}O_{gr} \dots T < 100)$; (IV) = $E_{s,v} = E_{s,v,a} = fH_{ne,ag\text{-}pa}\,int.,ab,af,pr(P,R,Cl \dots T > 10)$. For an explanation of symbols of equation, see text, pp. 261–263.

Description of Major Vegetation Types

In contrast to California and Chile, where relatively little detailed phytosociological work has been carried out until recently, the Mediterranean has been studied extensively, first by Braun Blanquet and his geobotanical institute at Montpellier and later by many of his disciples all over the Mediterranean region. Much of this work has been summarized by Quezel (1977) and by Tomaselli (1977) in a special MAB report (UNESCO, 1977a).

In the Mediterranean, the zonal vegetation of the xerothermic subzone of the lower elevations in the coastal regions and foothills is called, in the Braun Blanquet terminology, *Oleo-Ceratonion* in the western Mediterranean and *Ceratonio-Pistacion* in the eastern Mediterranean. These are alliances composed of numerous plant associations forming parklike wood-, shrub-, and grasslands of scattered *Olea europea* (the wild olive) and *Ceratonia siliqua* (the very valuable fruitbearing carob tree) as the upper layer and *Pistacia lentiscus* (the mastic or lentisc tree) as the subdominant in the shrub layer. This vegetation type has no counterpart in California. But another important east Mediterranean type of this subzone, the deciduous woodlands of *Quercus ithaburensis*—the Tabor oak—in Palestine and *Q. macrolepis* in Turkey—the Vallonea oak—have a close overall resemblance to the deciduous *Q. douglasii*—the blue oak woodlands—in California. Both woodland types have a rich and highly diverse herbaceous understory and serve as most valuable grazing land for cattle and sheep.

The slightly wetter and cooler attenuated thermomediterranean subzone in the western Mediterranean is occupied chiefly by the *Quercion ilicis* alliance of forests and maquis shrubland associations, with *Q. ilex* as the original most important evergreen tree, now replaced chiefly by *Q. coccifera*. In the east Mediterranean, this subzone is occupied by the *Quercion calliprini* alliance with many different associations, all dominated by *Q. calliprinos,* sometimes with other evergreen sclerophyll codominants and many subordinates.

Quercus coccifera and the closely related east Mediterranean *Q. calliprinos* are the most widespread and typical components of these shrublands on all important soil types, covering an area of more than 2,000,000 ha, especially on terra rosa (derived from hard limestone and dolomite) and brown forest and rendzina soils (derived from soft limestone and mergel). *Quercus coccifera* rarely assumed a treelike habit, but *Q. calliprinos* is extremely polymorphic, and trees that have been protected for hundreds of years in cemeteries and sacred groves can reach huge dimensions—15-20 m in height, 2-3 m in trunk circumference, and 20 m in crown circumference. In general, however, both appear as tall shrubs, 2-5 m high, and when browsed heavily they form a lower, dense, intricate brush carpet, affording good soil protection. They have the typical dual root system of the "olive type" (Zohary, 1962), spreading horizontally in the shallow top soil layer close to the rock surface but also penetrating deep (sometimes 10 m and more) into rock fissures and crevices. In this way they can tap moisture surpluses stored in the rock layer after the rainy season for continued photosynthetic activity throughout the summer.

Like all other sclerophylls, they can regenerate vigorously after burning or coppicing from root suckers and shoots close to the ground, and these young lush sprouts and leaves are devoured by goats and cattle; but turning very soon into spiny hard leaves and rigid branches, they lose their palatability and are thereby protected from overbrowsing.

On lower elevations and in warmer winter temperatures *Pistacia lentiscus* is the most abundant evergreen sclerophyll shrub. Its prolific dense-leaved and highly polymorphic intricate shoots tend to form a compact cushionlike canopy, thereby shading and protecting the soil and creating favorable conditions for litter decomposition, humus formation, and biological soil enrichment. The survival potentials of this shrub in dry hot summers on rocky and shallow or sandy soils under constant defoliation pressures by fire, browsing, and cutting are even greater because of its lower palatability and high drought tolerance. Recent studies in Israel (Szwarcboim, 1978) revealed that a xeric ecotype from the drier borderlines of the SFZ, on Mount Gilboa, combines a remarkably active and passive resistance to drought and is thereby capable of photosynthesis and carbon fixation in temperatures up to 45°C with a very low transpiration rate and with low water potentials but with a high albedo (reflectibility) in the 500- to 700 nm region. In these aspects it very much resembles the most drought-resistant evergreen sclerophyll shrub in California, *Adenostoma fasciculatum*, or chamise (Rundel, 1977).

The xerothermomediterranean subzone contains some of the richest and most diverse floristic, faunistic, and scenic landscapes. In addition to above-mentioned vegetation types, it includes natural and planted pine forests, as well as wetland ecosystems. However, as the most densely populated and most intensively modified and transformed mediterranean region, it has suffered greatest destruction of its organic world and at present its ecosystems are more threatened than those of any other region in Europe and the Middle East.

The sclerophyll oak zone merges in Europe to the north and all around the Mediterranean on higher elevations with mixed deciduous and tall conifer forests in the accentuated thermomediterranean or submediterranean zone. But after the destruction of the original forest cover, sclerophyll shrubs have invaded this zone. This situation is very similar to that of higher elevations in northern California.

In general, the east Mediterranean SFZ has a more xerothermic nature and merges on its drier borders with semiarid Irano–Turanic steppe formations. In the ecotone borderline between both, in the 300- to 400-mm rainfall belt, dwarf shrubs and open woodlands replace sclerophylls. These communities are called batha in Israel, phrygana in Greece, and tomilarres in Spain. The most common dwarf shrub here, as well as in the lower degradation stages, is *Sarcopoterium spinosum* (thorny burnet), a very spiny, compact, and intertwined dwarf shrub about 0.5 m high. Like all others in this region, it is a drought-evading "heterophyllous evergreen," changing its large and lush winter leaves to small scalelike summer leaves, and thus becoming highly flammable. Its main tap root has many lateral branches adapting themselves to local soil and rock conditions and affording a very efficient soil protection, even on steep slopes. It is browsed by goats only during a short period in spring, and after being burned and cut it regenerates vigorously by resprouting from the rootstock and by fire-stimulated seed germination. The same is also true of all other Mediterranean dwarf shrubs, many of which are from the mint and from the rock rose (*Cistus*) families and have aromatic leathery heatherlike leaves that are detested by herbivores, including livestock. These plants are favored by both fire and grazing and they now dominate large areas in many Mediterranean countries. *Sarcopoterium* is one of the first woody pioneers on abandoned fields. In the past its advance was checked by its wide use as a favored fuel for backovens and lime kilns, but its dominance can now be observed in many sites as the first—and apparently also that last-stage of secondary succession. (On the problem of succession and climax in mediterranean plant communities, see pp. 302–305.)

The closest ecological counterpart of these dwarf shrub communities in the Mediterranean is the "coastal sage scrub" communities in the coastal region of central and southern California, as the most xeric mediterranean woody vegetation types, recently studied extensively by Westman, (1981). However, the most important formation of this xerothermomediterranean zone in California, as well as in Chile, is sclerophyll evergreen shrubland, or chaparral. It also extends into northern California in lower elevations, on drier sites and shallow soils, covering about 9% of the total area of California. In around 70% of these stands, *Adenostoma fasciculatum* (chamise), a highly flammable shrub with needlelike leaves, is dominant, sometimes forming dense pure stands on the drier sunny exposures. The north-facing and wetter sites are occupied by *Quercus dumosa*—scrub oak—and other co-dominants. The cooler and wetter accentuated thermomediterranean subzone, chiefly in the central coastal ranges and the lower western slopes of the Sierra Nevada, is occupied by broad-leaved sclerophyll forests and woodlands, mixed sometimes with conifers. These are almost completely missing in Chile, where they are dominated by evergreen oak species.

Because of the lack of quantitative information only general physiognomic and qualitative descriptions of these vegetation types are presented in a recent book on the terrestrial vegetation of California (Barbour and Major, 1977). Ecological comparisons of mediterranean ecosystems and landscapes in California and Israel can be found in Naveh (1967). Comprehensive comparisons between these vegetation types in California and Chile, from an evolutionary point of view of convergence, have been carried out in recent years as a United States–International Biological Program (USA/IBP). Its results have been summarized by Mooney (1977a), and further studies on convergence in mediterranean ecosystems have been carried out by Cody and Mooney (1978). Important information on mediterranean ecosystems, including comparisons of soil systems and animal biogeography, are contained in the above-mentioned volume on the origin and structure of mediterranean ecosystems (di Castri and Mooney, 1973). This book also contains the results of a comprehensive ecosystem study of the plant–litter–soil subsystem in a *Quercus ilex* forest in the south of France (Lossaint and Rapp, 1973). This has been followed up more recently by an even more comprehensive compilation of mediterranean-type shrublands (di Castri et al., 1981). Its contents are lucidly summarized by di Castri (1981), the chief editor of this volume. His conclusions on the coevolution of Mediterranean ecosystems and Mediterranean man, the atypical mediterranean nature of the South African–Australian complex, and the need for maintaining the great intrinsic ecological heterogeneity of Mediterranean landscapes through multipurpose management with new techniques are very similar to those expressed in this chapter.

Of special relevance for our discussion is the last part of the di Castri et al. volume, dealing with man and ecosystems and containing a very informative review of human and livestock impacts by Le Houerou (1981), one of the most knowledgeable Mediterranean ecologists. In the following chapter, Trabaud (1981), a leading Mediterranean fire ecologist from the important Mediterranean ecological research institute at Montpellier, summarizes the history of fire utilization in the Mediterranean and its ecological effects, as well as the results of his own revealing studies.

Hard and Soft Landscape Values

Direct Economic Benefits

Forestry Production. By far the greatest part of those ecosystems in which natural spontaneous floras and faunas have been retained (although, as shown above, under various kinds and intensities of modifications) are those unfertile sand dunes, undrained lowlands, and untillable uplands in which crop production is not feasible. These are mostly upland ecosystems, sometimes covering 50% and more of the total land area, but contributing very little to the national economy in monetary terms in comparison with the much more restricted cultivable land. Their extent in the Mediterranean Basin and major land uses are shown in Table 4-1.

The low productivity of these upland ecosystems is, above all, a result of severe edaphic and climatic limitation on phytomass production. This has been estimated recently by Lieth and Whittaker (1975) on a global basis, ranging for sclerophyll scrub from 250 to 1500 g/m^2 a year, as compared to 400 to 2500 for summergreen forest and 600 to 2500 for warm-temperature mixed forests. Mooney (1977b) has explained the inherent low carbon fixation capacity of the evergreen sclerophylls as an adaptation to the low-nutrient and moisture-limited habitats. Half, if not more, of this carbon is allocated to the extensive and deep root systems and much is "wasted" in high-energy terpene and other phenolin compounds in the leaves, which reduce their palatability for herbivores and make them highly combustible. Thus, although they stay green throughout the year, their productivity is much lower than that of the deciduous summergreen forests on their wetter margins. The average dry matter production for a *Q. ilex* forest in France has been estimated as 600 g/m^2 a year, although higher growth increments are achieved in the first years of regenerating after fire or cutting.

Table 4-1. Land Utilization in the Mediterranean Basin in 1976 (after Le Houerou, 1981)

Countries	Total area, Land area (10³ ha)	Area in Mediterranean								Non-Mediterranean land area			
		Total Mediterranean zone		Cultivated land area		Forests, maquis, and garrigues		Rangeland		Steppe and desert (too dry)		Non-Mediterranean zone (too cold)	
	10³ ha	10³ ha	% Total	10³ ha	% M.Z.	10³ ha	% M.Z.	10³ ha	% M.Z.	10³ ha	% Total	10³ ha	% Total
Algeria	238,714	9,287	3.9	6,500	70.0	2,424	26.1	363	3.9	229,427	96.1	—	—
Egypt	99,545	—	—	—	—	—	—	—	—	99,545	100.0	—	—
Libya	175,954	3,158	1.8	2,544	80.6	534	16.9	80	2.5	173,796	98.2	—	—
Morocco	44,630	13,798	31.0	7,830	56.7	5,190	37.6	778	5.6	30,832	69.0	—	—
Tunisia	15,536	8,190	52.7	4,410	53.8	530	6.5	3,250	39.7	7,345	47.3	—	—
Cyprus	924	855	100.0	432	50.3	330	38.5	93	10.9	—	—	—	—
Iran	163,600	20,000	12.5	4,000	20.0	14,000	70.0	2,000	10.0	143,600	87.5	—	—
Iraq	43,397	6,800	16.0	4,100	60.3	1,500	22.1	1,200	17.6	36,597	—	—	—
Israel	2,033	1,367	67.2	433	31.7	116	8.5	818	59.8	666	32.8	—	—
Jordan	9,178	1,400	14.4	1,175	84.0	125	8.9	100	7.1	8,318	85.6	—	—
Lebanon	1,023	1,023	100	348	34.0	570	55.7	105	10.3	—	—	—	—

Syria	14,418	6,567	35.6	5,260	80.1	457	7.0	850	12.9	11,851	64.4	—	—
Turkey	77,076	17,110	22.2	8,309	48.6	6,051	35.4	2,750	16.0	6,000	7.8	53,966	70.0
Albania	2,740	496	18.0	132	26.6	248	50.0	116	23.4	—	—	2,244	82.0
France	54,592	8,742	16.0	3,382	37.5	4,230	48.4	1,230	14.1	—	—	45,850	84.0
Greece	13,080	8,102	62.0	2,331	28.8	2,618	32.3	3,153	38.9	—	—	4,978	38.0
Italy	29,405	11,725	40.0	5,521	47.2	3,409	29.0	2,795	23.8	—	—	17,680	60.0
Portugal	9,164	5,656	61.5	1,800	31.8	3,156	55.8	700	12.4	—	—	3,528	38.5
Spain	49,957	31,700	63.5	16,000	50.5	9,200	29.0	6,500	20.5	—	—	18,257	36.5
Yugoslavia	25,500	2,380	9.1	800	34.3	900	38.6	630	27.1	—	—	23,210	91.0
Total	1,075,046	158,306	14.8	75,207	47.8	55,588	34.9	27,511	17.3	746,977	70.8	169,713	14.4

Source: FAO Production Year Book, 1977a.

Narrow-leaved evergreen pine trees play an important role in afforestation projects because of their hardiness and easy and rapid establishment and growth. In countries like Israel, Cyprus, Spain, and Greece, such upland pine afforestations have saved thousands of hectares from further depletion and degradation, provided labor for rural populations, and dramatically improved these denuded and barren mountain landscapes. In many cases they enabled the vegetative regeneration of the sclerophyll understory. This could be utilized, by proper forest management, for the creation of seminatural, mixed, and multilayered multiple-purpose forests (see below).

Morandini (1977), in discussing the economic role of forest production in the Mediterranean, rightly stressed the renewed economic importance of sclerophyll forests and shrublands for local firewood supply and, by mechanized wood-chopping, for commercial fiber production, which could pay for the thinning and opening of dense shrubland thickets into recreation forests and woodlands. He opposed the conservation policy of a total halt of intervention (out of entirely theoretical considerations of a hypothetical forest "climax"), and he stated that the natural place for intensive tree cultivation in this zone is *outside* the forest. Such plantations of high-yielding timber and fiber trees, like poplar and eucalyptus species, should be established in fertile and deeper soils in the lowlands or on gentle slopes of abandoned agricultural land, suitable for mechanized cultivation and for economic forest production. However, such production from monospecies plantations, even with potentially high-yielding conifers and eucalyptus trees on low-potential rocky and shallow or marginal upland soils in this zone, is highly doubtful. Unfortunately, this is not yet recognized by many foresters, who naturally would like to prove the important economic contribution of such forests to the national economy and/or to the departmental budget. This tendency is encouraged by the fact that forest departments, in general, are part of the crop-production-oriented ministries of agriculture and not of those oriented toward environmental protection and social services. The issue is further confused by the rather misleading definition of "exploitable forest," applied also in the SFZ to any tree-covered site in which "at least one industrial cutting during a rotation period can be made" (ECE/FAO, 1979) without discriminating between the more productive forests on deeper soils and in more mesic and cooler sites and those low-productive upland sites to which most forests in this zone are confined. For the same reason, data presented on forest production from Mediterranean countries can hardly be used for estimating their true economic value, as compared to alternative or complementary uses. Thus, for instance, in a recent review of forest and forest products in south Europe (ECE/FAO, 1979) by the Economic Commission of Europe, the net annual increment per hectare in cubic meters for all of France is 3.14, for Italy 2.12, and for Greece 1.73: but for all those countries in which most of the forest production is derived from such poor sites, like Israel or Turkey, no data on net production per hectare were presented. These countries were apparently excluded from the calculation of an average of 2.66 m^3 for all of south Europe, which is then used for prospects, policies, and decision making in future forest and timber production for all SFZ countries.

From the few available figures from these countries on actual wood production of such rocky uplands in the SFZ, including southern France and Corsica, it can be assumed that the annual wood increment does not exceed 1.0 m^3/ha and is often much lower. However, the current true deadlock in economic pine forest production stems from the fact that this low productivity is further reduced by steadily growing losses from wildfires, in spite of heavy expenditures for fire prevention and fighting, and from the new threats of air pollutants and pest damages, discussed later.

The same is also true for conifer forests in the drier parts of California, and it seems, therefore, that their importance lies much less in direct economic benefits and much more in their recreational, scenic, and other "soft" values.

Livestock Production. The most important direct benefit from these uplands is through conversion of their primary plant production into animal production by grazing and browsing. In the brush-dominated SFZ in the Mediterranean (as well as in Chile) the most important domestic grazers are goats. In 1976 these made up over 124,000 million head, as compared with approximately 1000 million sheep and 7000 million cattle. In California their role is still performed by wild ungulates, chiefly deer. According to French (1970), goats are the most efficient converters of fiber-rich and lignified woody plants, and they are able to consume daily up to 8% of their liveweight in dry matter, as compared with 3% or less by cattle and sheep. The Mediterranean black goats are especially well adapted to the meager and seasonable forage, scarce water supply, and rocky terrain, and can best take advantage of the prevalence of sclerophyll and thorny shrubs for milk, meat, and hair production. Thus, in *Q. calliprinos* shrubland in Israel, such a typical herd of Arab black Mamber goats, spending about two thirds of its grazing time on woody plants and regularly eating about 30 species (as compared to only 10 by cattle and even less by sheep), browsed chiefly after fire and only in the dry season, deriving 400 Scandinavian Feed Units (FU)/ha a year. (One FU is the caloric equivalent of 1 kg of barley, and this yield was about half the grain yield of an unfertilized, low-level dryland field, but only one fifth or less of a fertilized barley field.)

As discussed in a special symposium (FAO, 1965), in spite of low milk yields, these goats play an important role in the economy of these mountainous regions, and their density in relation to human populations is the highest in the world. Thus, for instance, in Greece, where sclerophyll shrub covers about 15% of the total land area, goat milk provides one sixth of the total animal husbandry and 65% of the family income in mountainous regions. In 1965 the net profit per goat was around $5.00. These range goats, being able to survive and produce in conditions no longer suitable for cattle and sheep, have been blamed as the main cause of Mediterranean land ruin, whereas in fact they happen to be only the last phase of the vicious cycle of land devastation brought on by indiscriminate burning, cutting, grazing, slope denudation, and cultivation. As stated by French (1970), this problem has been distorted by emotional reactions and misconceptions conducing to the conclusion that the only way open for rational land use is a complete ban on goats and the reclamation of Mediterranean uplands as closed forest vegetation—either by encouragement of natural revegetation or by afforestation, to which goats, without doubt, are an obstacle. On the other hand, systematic controlled goat grazing can be used to prevent the regeneration of shrubs from suckers and lower branches after cutting, thinning, and pruning of dense maquis shrub thickets, thereby converting them into attractive park-woodland recreational forests. This has been successfully done in Israel on a large scale by the Jewish National Fund's Forest Department in the admirable Goren Forest in western Galilee (E. Jospehi, personal communications, and our unpublished data, on file). The labor expenses for clearing were covered mostly by wood and charcoal production as well as by opening the dense woody canopy, dominated by tall *Q. calliprinos* shrubs. The carrying capacity for goats rose from less than one head per hectare per annum to more than three. At the same time, plant species richness of the formerly impenetrable, fire-prone maquis "climax" of less than 30 species (mostly woody) per 1000 m^2 has increased to more than 70, including many very valuable perennial pasture grasses and ornamental geophytes; in addition, high landscape and recreation amenities have been attained. (On the multiple-use benefits of such silvopastoral and recreational conversion, see also pp. 283–285).

One of the most promising methods of controlled, rational, and profitable utilization of these shrublands could be by improved Angora goat browsing in fenced subpastures. This has been tested successfully in Israel (Naveh, 1974a), where moderately grazed shrub pastures provided about 600 FU/ha a year, or the annual requirement of 1.8 Angora goats.

Ziani (1965) proposed the modernization of the traditional Dalmatian Zagor system of branch-lopping for stall feeding of cattle and goats by mechanized brush chopping and milling. This system should also be explored for developing new biochemical technologies for vitamin- and protein-rich human food and industrial products.

However, as discussed in detail by Le Houerou (1978), the greatest potentials for animal production are in the drier open woodlands and derived annual grasslands in this and the semiarid metiterranean zones. Therefore, the most attention for more rational management and improvement should be devoted to these grasslands. In such typical xerothermomediterranean open oak woodlands in Israel, with 550 mm average annual precipitation, the dry matter forage production ranged from 110 to 450 g/m^2 a year, with a six-year average of 240 g/m^2 a year and an average annual liveweight increase of close to 100 kg/ha, or 720 FU/ha. This output could be doubled in an economical feasible way by fertilizing and herbicidal weed control, but under the traditional system of heavy uncontrolled grazing these pastures produce only about 150 g/m^2 a year, or a caloric output similar to that of the goat-browsed shrublands, namely, around

400 FU/ha (Naveh, 1970). It is important to realize that similar and even higher pasture productivity can be expected in the semiarid mediterranean zone with only half the rainfall.

The productivity of the annual grasslands in California is very similar to that of their Mediterranean counterpart. In both cases, as we have seen, this varies very much from year to year, according to rainfall amounts and distribution, and their nutritional value in the dry season—after the drying out of the herbaceous vegetation—is very low. This great bottleneck in pasture production could be overcome by the enrichment of the herbaceous pastures with productive, protein-rich summergreen and drought-tolerant fodder shrubs. In this way the advantages of sclerophyll shrubs, like *P. lentiscus*, could be simulated, as regards their efficiency of energy, water, and nutrient tapping and their ameliorating effects on their habitat, but at a much higher level of efficiency by channeling solar energy into economically useful animal products. Work along these lines shows much promise in Israel (Naveh, 1975, 1979).

In California, the chief economic use of open woodlands and annual grasslands in the foothill zones is for beef cattle production in fenced, and in many cases very well-managed, ranches. On the other hand, in Chile, as in the Mediterranean, grazing of cattle, sheep, and goats is still on common unfenced grazing lands, with the exception of a few larger estates. Research efforts on the introduction of modern management and improvement methods have not yet changed the uncontrolled and mostly destructive land uses for animal production.

Miscellaneous Products. Another important means of economic utilization of these open areas, although they are decreasing steadily with the advance of industrialization and urbanization and the depopulation of mountainous regions in the western Mediterranean, is the production of cork from *Quercus suber* woodlands. This is mostly in combination with grazing and the use of coppice for fuel, charcoal production, tools, hedges, and the like. Many plants, among them the aromatic dwarf shrubs, are used as spices, balsam, healing teas, and so on. Also, many animals, especially birds, porcupines, wild boar, and deer, are a source of food and income as well as sport. The last is especially the case in California, where private ranches in the woodland and shrubland zones sometimes derive higher incomes from hunting fees than from cattle ranching. Some of the disastrous impacts of these uses will be discussed later.

Indirect Economic Benefits

Without doubt, the greatest indirect economic benefit derived from the SFZ stems from its recreational and scenic values, which are enjoyed each year by millions of tourists. This is true not only for the seashores but also for the uplands, with their extraordinary wealth of plant and animal life, especially in the spring, when hundreds of colorful flowering plants can be found (and unfortunately, are also picked, even sold, and their bulbs exported; Polunin and Huxley, 1966). These scenic and aesthetic values are very difficult to assess in monetary terms, although they do yield economic benefits.

This is even more the case with the great potentials of this zone in the procurement of genetic resources of valuable economic, medical, and ornamental species. A renewed process of domestication of these wild plants has been begun and will become increasingly important in the future. Thus, for instance, in the Weitzman Institute in Israel, drought- and disease-resistant and very protein-rich and wild biotypes of *Triticum dicoccoides*, the progenitor of wheat, are selected as genetic resources for cross-breeding and eventual domestication. There are also many highly valuable pharmaceutical and industrial compounds that are being used commercially. As Zohary (1973) has shown, the drier borders of the SFZ are specially rich in wild plants for human uses. He listed more than 100 species that have been used in folk medicine, some of which are still included in the pharmacopoeia. Another group of drought- and limestone-tolerant woody plants from all countries of this zone are presently being selected for their multiple-use benefit in soil protection and improvement, erosion and flood control, and landscape beautification, as well as for fodder and honey plants (Naveh, 1975, 1978).

Without doubt, future generations will be grateful for whatever we succeed in safeguarding from these untapped sources, not only in botanical gardens and collections but in their natural habitat, in which further evolution and speciation can proceed.

Free Ecological Services and Noneconomic Richness

In addition to these benefits, there are other ecological and socioeconomic "free services" that these natural ecosystems provide as long as their functional and structural integrity is assured. As Westman (1977) has shown, were it necessary to replace their functions or compensate for their destruction, doing so would cost millions of dollars.

These values are all connected with the capacity of these ecosystems to ensure environmental quality, stability, and health as part of their normal regulative and protective "life-supporting" functions. In the SFZ they are especially important because many of these ecosystems serve as watersheds to densely populated coastal regions and intensively cultivated and irrigated lowlands. We are not even aware of most of these functions, such as flood prevention, soil protection, filtering and breakdown of pollutants, and maintenance of natural balance in predator–prey (and pest) relations, until they are impaired, and only then can we estimate their "replacement values." They are also completely ignored, even in the most sophisticated cost–benefit analyses in engineering, drainage, and other development projects. Westman (1977) calculated the loss of natural pasturelands in the foothills of southern California, along the San Bernardino freeway, in terms of the resulting pollution damage or the cost of equipment to remove, each year, 440 kg/ha of carbon monoxide that had previously been absorbed by these grassland ecosystems, together with other pollutants, and in terms of lost soil-binding functions. In another example, he showed that the replacement value of ponderosa pine trees dying of severe damage from photochemical oxidants in the San Bernardino National Forest can be estimated by the amount of erosion and siltation they prevented. If this is similar to the erosion caused by the conversion of dense chaparral shrubland into open grass on steep slopes, amounting to 200 m³/ha a year, the cost of the removal of its sediments would be $122 million each year from an area of 4000 ha.

In a similar way, the highly beneficial "sanitary" functions of predators in the higher trophic levels of food chains in these mediterranean ecosystems that feed on agricultural pests can be appreciated only after these functions have been disrupted by shortsighted and wasteful nonselective pest control practices. Thus, Mendelsohn (1972) showed that in Israel the unwise use, against cyclic outbreaks of field mice, of wheat kernels coated with thalium sulfate in 1955–1956 exterminated most birds of prey and rodent predators by secondary poisoning. (Prior to this, the density per 10 ha of grain field amounted to 2 eagles, 10 buzzards, 4 kites, 4 harriers, 12 kestrels, and 79 voles per day!) This extermination boomeranged in that a massive increase of field mice and rodent pests occurred, causing much heavier damages than before. Nine years later, the wholesale extermination of jackals (suspected of transferring rabies), together with weasels, foxes, and wildcats, by baiting with chicken poisoned with fluoracetamide, also drastically reduced mongoose populations. Mongooses were the most effective killers of the Palestinian viper, and consequently, in the following years the number of people treated in hospitals for these deadly snakebites increased considerably. Furthermore, feral dogs (the chief transmitter of rabies), which do not depend on predation of wild animals, have now replaced these predators, and the unchecked populations of hares, field mice, and fruit-eating birds, after the final extermination of all 36 diurnal species of birds of prey by pesticides, are causing more damage than ever, in addition to heavy expenses for chemical control by airplanes and the like.

It is obvious that in spite of their low direct economic production values, these ecosystems have very high protection and regulation values, in addition to their great resilience. Thus, a free gift of nature, acquired during millions of years of evolution, can be rejected by a decision that takes a split second, or destroyed in a few minutes by the digging of a bulldozer or the spray from an airplane.

However, the extinction of wildlife, as well as the destruction of the last remnants of a *Quercus ilex* forest, a lowland meadow, or a swamp, all vital for the survival of rare and endangered plants and animals, or even the elimination of a unique geological formation or soil type, cannot and should not be evaluated in economic terms. As stated by Smith (1976): "The extinction of the unique, whether it be a treasured work of art or a species, is *irrevocable*." These noneconomic richnesses and their great aesthetic, spiritual, cultural, and scientific values are thus intrinsic qualities of these natural Gestalt systems (see Chapter 2) and their loss makes human society much poorer.

The SFZ in California, because of its great ecological diversity, has served as one of the most productive natural "field laboratories" for studies of the dynamics of speciation and evolution of wild plants, including plants with economic value (Raven and Axelrod, 1978). The same is also true for studies of wildlife and of manipulation of their natural habitat, including the national parks, which now serve as models for other countries with similar problems.

The educational benefits that can be derived from nature reserves in this zone have been demonstrated in a very convincing way in Israel: The establishment of field schools, staffed permanently by scientists, field guides, and instructors (suitable female high school graduates can serve in these field schools instead of in the army), in the vicinity of these nature reserves has turned them into main centers for field-oriented environmental education (Photo 1). Their various activities include short courses for organized groups, and summer recreation camps that combine natural history, archeology, and the like with hiking and recreation for adults, children, and families. All tenth graders in the country spend at least five days during their school year in such a field school, and for many of them this is their first opportunity to visit natural areas in a guided and inspiring way. In addition, many participate in natural surveys, especially birdwatching, carried out by the staff of these field schools. Thus, for instance, in a

Photo 1. Distribution of environmental education material and announcements on weekend activities of the Israel Society for Protection of Nature by the staff of the Tabor Oak Field School (written in Hebrew letters on the truck) along the highway.

recent special action, 200 such volunteers watched a pair of breeding rock eagles–the first of their kind ever to nest in the SFZ in Israel–for four months, from sunrise to sunset, on a pine tree above a frequently used trail leading to such a field school in the Judean hills. Five of these students carried out a special observation project for 1488 hours as part of their matriculation requirement in biology, and recorded some of the most exciting and scientifically important details on the feeding and nesting habits of these rare birds of prey.

Recently, in one of these field schools, a "Network for Observations and Information on Plants of Israel" has been established in which detailed phenological, botanical, and ecological information is collected by many volunteers all over the country and then computerized and stored at the Department of Botany of the Hebrew University, Jerusalem, for retrieval and further use by all botanists in Israel in their scientific work.

The difficult problem of how to include these "soft values" and their usefulness to human society in the chain reaction of decision making, from the public via the scientist and economist to the politician, has been treated in a very illuminating way by Ashby (1978). In an important lecture (mentioned in Chapter 2) he pointed out, rightly, that the moral choice between hard and soft values in environmental and nature protection, "between indulgence in the present and consideration of the future obliges people to strike a balance between counting what can be quantified and caring for what cannot be quantified." Therefore, instead of "stretching economic analyses to cover values that have to be stripped of some of their meaning, if they are quantified, these should be restricted to those values which are unquestionably quantifiable." These problems were discussed in Chapter 2 in connection with the self-transcendent openness of natural Gestalt systems. We cited also Egler's (1970) statement that "not only what can be counted, counts," especially in the computer age. But at the same time, all relevant information, even if not expressible in marketable commodity values or dollars in cost–benefit analyses, must be presented in a clear and illustrative way that can be comprehended by the public and the politician. For this purpose, nonmonetary and noneconomic richnesses should be expressed as functional parameters to which relative values can be attached in a matrix for decision making. How this can be applied to the conservation and management options of the SFZ will be discussed in the next sub-section.

The Need for Reconciling Clashing Demands and Attitudes

In most countries of this zone, the clashing demands of the traditional users of mediterranean uplands, the modern private and public "developers," and the classical conservation–protection fighters have led to a special version of the mediterranean "tragedy of the commons" (Hardin, 1968). In this, each side seeks to maximize its own profit or institutional interest, regardless of the fate of the "common" as a whole. Unfortunately, in many of these countries, lack of public awareness, understanding, and motivation is favoring those options which are short termed (from election to election) and money earning (for those with the most powerful pressure group) and therefore also the most harmful and irreversible.

For these reasons, it will not be enough to develop sounder conservation policies. These policies must be brought to the attention of the public, on both the local and national level, with the help of the mass media, schooling, adult education programs, volunteer groups, and the like as part of a major effort of environmental education. They must be supported by a powerful but trustworthy lobby that is respected not only in the professional but also in the public sense because of its lack of personal interests but highly motivated involvement.

In contrast to California, where great advances have been made in this respect, most Mediterranean zone countries are still lagging far behind. Thus, for instance, the great majority of those scientists who should know and care most, and therefore could also be the most influential, namely the biologists, are molecular biologists with reductionistic approaches to nature, who know (and care) more and more about the structure and function of DNA but less and less about the structure and function of nature in action in the field and its fate. The few conservation-minded terrestrial biologists are still mostly split into the old-fashioned, tragic organismic compartments of botanists, zoologists, and so on, and most of them, especially the phytosociologists of the Braun Blanquet school, are clinging to the classical one-sided protection–conservation philosophy, discussed below. In addition, the unfortunate language and communication barrier cuts off the English-speaking scientific community, preventing cross-fertilization and the adoption of new approaches, similar to those of their northwest European colleagues and well demonstrated in The Netherlands (Bakker, 1979), as mentioned in Chapter 1. There is, therefore, urgent need for the education of a new breed of field-oriented and conservation-minded *ecosystem biologists*. At the same time, however, it is hoped that the latter will not be misguided by those who regard computerized ecosystem models not as an important tool for integration and interpretation of field data but as a more prestigious and convenient substitute for the systematic and tedious collection of field data, on which these models must be based (see also Chapter 2).

Without doubt the most important professional group dealing with these uplands in their day-to-day work and in decision making for their future are the foresters. But unfortunately, as Nicholson (1974) has pointed out in an important article on forestry and conservation, until very recently (and in the Mediterranean even today) much of the training of professional foresters has been inadequate to equip them to appreciate the broader role of forestry and to communicate with others interested in land use. His statement that "the harmonization in forest exploitation with ecological principles has much to go" is especially true in the Mediterranean and, as regards fire and fuel management (see below), in California as well. Many of the most influential foresters in the Mediterranean still cling to an almost mystical belief that the only rational and economically sound alternative to the neglect and decline of these uplands and to "unproductive" evergreen woodlands is their conversion into dense monocultures of pines, along the patterns of the worst European forestry practices. A similar critical view has been expressed by Leopold (1978, p. 109) on American forestry. "It is troubling to note that many current silvicultural procedures being applied with much enthusiasm can, if applied too extensively, impoverish native faunas and for that matter, floras as well. In our zeal to make forests pay, we stand to lose wildlife diversity and other intangible values of the forest—and ultimately perhaps, nontangible ones too."

Unfortunately, such foresters also have their counterpart "polarizers" among the most outspoken conservationists. These "are carried away by their own emotions or dogmatic beliefs at the expense of environmental interests which they purport to champion." Nicholson pleads for the "*integrators,* who are strong pragmatists seeking to judge by the scale of the results obtainable, by substituting collaboration for embattled deadlock, and which course will most effectively secure the integration of conservation objectives with social and economical objectives." In the Mediterranean, more than elsewhere, there is an urgent need for a joint strategy between forestry and conservation for the wise and balanced long-term management of these resources. This can be achieved only if on both sides the more farsighted and broad-minded elements in professional leadership struggle courageously and tenaciously against narrow and divisive sectional views and aims.

In conclusion, there is urgent need for an integrative approach to Mediterranean land use, aiming at a reconcilation between the need for conserving the biological diversity and productivity of these upland ecosystems and the socioeconomic needs of its inhabitants and the national economy. This goal requires the cooperation of those who care for, those who deal with, and those who live from its resources. Its practical goals should be the preparation and implementation of closely interwoven networks of multiple land-use patterns. In their choice, most relevant variables from the following main domains should be considered.

1. The *bioecological* domain, related to those physical, chemical, and biological processes that ensure ecosystem function and structure and thereby the highest attainable productivity, diversity, and stability.
2. The *socioecological* domain, related to environmental quality and to sociohygienic, psychological, and cultural requirements that ensure the greatest possible overall benefit for the largest possible human population from a national and regional point of view.
3. The *socioeconomic* domain, related to the direct economic benefits to be derived from these uplands as pastures, forests, and recreation parks after their reconstitution and under rational management.

As a practical example of this integrative approach to multipurpose land use, some simple models are presented in Figures 4-4-4-6 for the evaluation of multiple benefits and their mutual influences on east Mediterranean semiagricultural upland ecotopes. In these, two major block variables have been used: (a) production functions, acting as positive and amplifying feedbacks; and (b) regulation and protection functions, acting as negative feedbacks and constraints.

Within the first group, two major types of variables can be distinguished:

1. Production functions based on the channeling of biological productivity and natural renewable resources into economic products that are transferable into market goods and therefore expressible directly in monetary terms as plant and livestock products and water yields. In the case of forest and other plant products, the primary production is utilized; livestock products of milk, meat, mohair, wool, and the like must be channeled efficiently into secondary production. Increased water yields can be derived from surplus soil and rock moisture reaching catchment areas if not utilized by deep-rooted summer-growing woody vegetation.

2. Production functions that can be achieved by "optimization of natural landscape values." Such values may be of a purely bioecological nature (e.g., biotic diversity or wildlife), or they may constitute a complex and closely interwoven mixture of bioecological, geomorphological, and human-perceptual processes of scenic diversity, attractiveness, and accessibility, which can be enhanced by ecological management and physical installations. Only a few of these variables can be channeled into marketable goods with direct monetary values, such as income from tourism, entrance fees for parks, or fees for camping grounds or hunting permits. One of the greatest threats facing the uplands and their natural and scenic landscape value is the tendency of over-enthusiastic outdoor recreation economists and planners to regard these monetary benefits as the main object and not as a desirable side product, and to therefore attempt to increase these benefits by overdevelopment and commercialization which destroys natural beauty and solitude and replaces them with crowded, noisy, and ugly mass outdoor recreation areas. The latter could be provided much more cheaply (from the overall social and ecological cost–benefit viewpoint) in specially designed recreation parks and planted forests close to the urban centers.

Within the second group, two major types of variables can be distinguished. The first comprises regulation and protection functions based on natural, inbuilt biophysical information of a negative feedback loop that ensures self-regulation and dynamic stabilization. These operate chiefly through the vegetation canopy and its "living sponge" in the direction of moderation of the physical environment and reduction of the kinetic energy of radiation and wind and water. This results in an improvement of the microclimate and environment and at the same time reduces the adverse effects of man-created disturbances of these landscapes. It is obvious that these functions cannot be fully expressed in monetary terms, although (as mentioned earlier) they can prevent severe longer-term financial losses and repercussions. They are an indispensable part of the noneconomic richnesses that are generally neglected in cost–benefit analyses; they are lumped together as *environmental and watershed protection.*

On the other hand, as mentioned in Chapter 2, the effectiveness of these information fluxes is much greater than the energy consumed thereby, and therefore attempts at expressing these functions in energetic terms may be misleading. These cybernetic regulation functions are the qualitative structural counterparts of the quantitative energetic aspects and should be treated as such.

As will be discussed below in more detail, another very important regulation function in mediterranean forest-, shrub- and woodlands is fulfilled by the fire cycles. These have been greatly distorted by pastoral land use through over-grazing and heavy defoliation pressure, which has reduced the herbaceous under-story canopy and fuel. Conversely, complete protection from fire has increased fuel and fire hazards, which have been accentuated by the planting of dense, highly flammable, and fire-vulnerable pine forests.

Fire risks will also be increased by recreational uses, but these can be reduced by management regulation functions (i.e., through controlled burning and fuel and vegetation management). It is therefore important to evaluate these variables as resistance to fire hazards in the different land-use and management options.

Having defined the main functional variables operating as either amplifiers or constraints, it is necessary to evaluate their intensity by weighing each in relation to the other variables involved in specific land-use goal(s) and their mutual in-fluences and compatability. As a first approximation, this can be done by a simple scaling system of the relative multiple-use benefits (Figure 4-4) in com-bination with a cybernetic model that distinguishes the high, moderate, and low positive and negative influence of each variable on all others (Figure 4-5). These scalings have been based on the following considerations.

For economic production functions, the actual expected economic benefits— as far as known—can be used as a quantitative guideline for relative scaling. This has been done in the case of livestock production from upland pastures, for which previous studies (Naveh, 1970) have shown that by intensive ecological management these outputs can be raised by 300% in comparison with degraded grasslands. It can be assumed that by replacement of low-value indigenous shrubs with more palatable and productive fodder shrubs and trees which have shown much promise in multilayered seminatural woodlands and shrublands, these outputs can be raised and additional 200% or more. This is shown in Figure 4-4 in the rise for livestock production from the A systems to the D_3 and E_3 pasture systems. On the other hand, in dense maquis and planted forests, no profitable livestock production can be expected. Such production will also be limited in D recreation and multiple-use systems, but it will be higher in the seminatural (E) systems because of the planted fodder shrubs and trees, which are also highly valuable for browsing wildlife.

Forest and plant production, as another example, will be naturally highest in the planted forest (C) systems. In the recreation and multiple-use forests and woodland systems also, some economic benefit can be derived from wood and charcoal by coppicing, and this could pay for much of the removal of the undesirable understory in E_1 systems and even more in E_2 systems, where industrial, pharmaceutical, balsam, honey, and spice plants can be planted in combination with ornamental and cover plants such as *Lavandula officinalis* and *Rosmarinus officinalis* (see Photos 2-6).

Recreation and landscape amenities are very low in depleted upland ecosys-

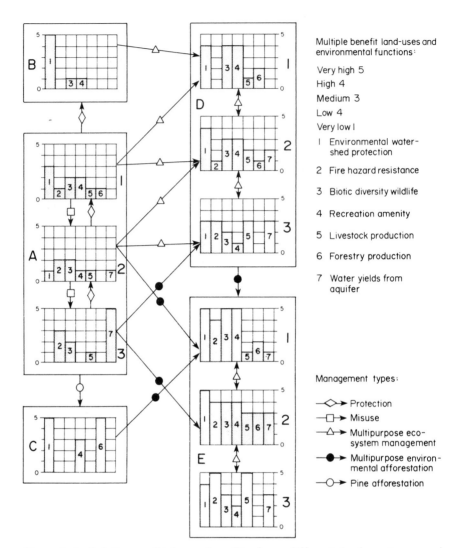

Figure 4-4. Relative multiple-use benefits from different land-use types and management systems (Naveh, 1979). (A) Degraded natural uplands: (1) = degraded maquis; (2) = degraded open shrublands; (3) = degraded annual grasslands. (B) Protected maqui reserves. (C) Planted pine forests. (D) Improved natural uplands: (1) = recreation forests, woodlands and parklands; (2) = multipurpose woodlands and parklands; (3) = pastures. (E) Converted semi-natural uplands: (1) = semi-natural recreation forests and parklands; (2) = semi-natural multipurpose forests and parklands; (3) = semi-natural multi-layered fodder-tree-shrub-grass pastures.

tems and in dense, impenetrable, and inaccessible maquis thickets. In dense pine afforestations, they can be increased considerably—as is done in Israel—by thinning, the development of roads and special facilities, and so on, but the encouragement of mass recreation can increase fire hazard, and this will act as a severe constraint on further development. As shown above, dense maquis can be converted into highly valuable natural recreation forests (D_1 and D_2 systems), but highest benefits will probably be achieved in E_1 and E_2 systems by planting ornamental and shade trees in seminatural parks, at the same time reducing fire hazard.

As mentioned earlier, water yields from aquifers, reaching catchment areas from a surplus of soil and rock moisture, can be increased considerably. These values are therefore highest in heavily grazed and shrub-depleted A systems, intermediate in open recreation and pasture systems, and lowest in denser forests and maquis (B and C). Biological production functions, especially biotic diversity and wildlife, are the main assets for outdoor recreation and landscape amenities. They are low both in depleted uplands and in dense and monotonous maquis thickets, and they are missing completely in densely planted pine forests. They can be raised considerably, however, by ecological management (e.g., burning, grazing, coppicing, and thinning of natural forests and woodlands) and will reach highest values by the introduction of additional diversity in multilayered and seminatural recreation systems (E_1 and E_2) and lower values in those systems devoted chiefly to livestock production (D_3 and E_3).

Photos 2-6. Protected, degraded, improved, and converted maquis shrublands in Israel. **Photo 2.** In front: Um Rechan Forest Reservation in Samarian mountains, after complete protection for 60 years. 3–5 meter high, dense and almost inpenetrable, highly inflammable and stagnating shrubland, dominated almost exclusively by *Quercus calliprinos* (40–50% cover) and *Phyllirea media* (20–30% cover) with 14 subordinate shrub and dwarfshrub species. Very few herbaceous plants have remained together with the lower shrubs and dwarfshrubs, near rock edges. In the rear: Much more open and lower shrubland outside the reservation with different intensities of grazing, coppicing and burning. The crest of the hill to the right was planted with dense pine forest.

Photo 3. Across the fence of the reservation, near the Arab village of Um Rechan. Only few, heavily browsed and cut shrubs are visible. (The tree with bare branches is an almond tree.) But many more, stunted shrub remnants can be spotted only by close inspection between rock outcrops. These would regenerate, if human pressures would be eased. Note, that in spite of heavy grazing, there is a dense, low grass cover with many flowering geophytes.

Photo 4. Multiple-use recreation forest in Upper Western Galilee, created from dense maquis thicket, like in Photo 2, by thinning, pruning and preventing regeneration from stumps and suckers by controlled goat grazing. (For further details, see also p. 274.)

Photo 5. Multiple-use recreation and fodder forest, created from degraded shrubland on shallow and rocky Pale Rendzina soil in Lower Galilee by planting of limestone-and drought-tolerant shrubs and trees with multiple-use benefits. In left front: remnants of original plant cover of thorny *Sarcopoterium spinosum* dwarfshrubs, replaced to the right by *Rosmarinus officinalis*, an ornamental, honey and medical cover plant from the Western Mediterranean. In the center: *Pistacia atlantica,* a local shade and fodder tree, planted only 5 years ago. In the background to the left: *Atriplex nummularia,* a very palatable, Australian fodder shrub; to the right: *Cotoneaster franchetti,* a very valuable ornamental and fodder shrub. The plot is fenced by *Cupressus horizontalis* trees.

Environmental and watershed protection is dependent on a dense and vital vegetation canopy which can prevent water, soil, and wind erosion, absorb dust, filter aerosols and gaseous air pollutants, and reduce noise and other detrimental environmental influences. This protection will be highest, therefore, in undisturbed natural and planted forests. It can reach similar values as a result of biological and managerial measures to prevent erosion, especially in vulnerable spots in recreation parks near campgrounds and along roadsides, in multilayered V_1 and E_2 systems in spite of intensive use for recreation. At the same time, however, these values will be lower in natural pastures, especially if heavily grazed and depleted, as in the A_2 and A_3 types.

Resistance to fire hazards is inversely related to the density and amount of vegetation cover and litter and to their fuel properties. It also is affected by man-induced fire risks, which will be highest in intensively used recreation areas and in populated regions. It is therefore lowest in highly flammable dense pine forests and maquis and highest in denuded I systems. It can be raised considerably also in E systems by replacement of xerophytic and highly flammable native woody plants by more mesic and lusher shrubs and trees and by creating special

Photo 6. Fodder shrub-parkland, created in Jezreel Valley, four years ago, from semi-arid grassland on poor basaltic soil by planting drought resistant and productive, leguminous *Cassia* and *Acacia* species from Australia. The prostrate plant, spreading on road bank in front is *Myoporum multiflorum*, a low-flammable, ornamental and honey plant, suitable as a fire break plant.

fire protection zones in which the vegetation is kept low through controlled grazing and herbicidal treatments. The mutual influences of these variables and their compatibility in multiple land use is shown in a cybernetic model in Figure 4-5; these are summarized as a sensitivity matrix in Figure 4-6. A sound mathematical treatment of such cybernetic feedback models has been proposed by Riggs (1977).

As proposed by Vester (1976), the summarizing of these values from left to right for the active sums (AS) and from top to bottom for the passive sums (PS) enables quantitative assessment of these interactions. Thus, the variables with the highest AS exert the strongest influence on all others, regardless of how they are influenced by the others. In our case B and E and A and C act most strongly in a positive way and F in a negative way. Those that are most influenced (i.e., with the highest PS) are D and again F in a negative way. Further, the quotient $Q = AS/PS$ can be used for defining the most active and relatively most successful variants (i.e., those exerting the greatest influence on all others, but being least affected by them), which in our case C are in a positive way and D in a negative way. Those with the lowest Q, which are the least compatible and influential, are F, which has no positive influence, and C, which has no negative influence. Those with the greatest P values, which are the critical variables, exerting the greatest influence and simultaneously being most affected, are B and E in a positive way and F in a negative way. Variables E and F have the highest overall Q and P values. Those with the lowest P values—called by Vester "buffering elements" or "latent elements," are again F and G.

It is obvious that those variables that act as management regulation functions have the most beneficial effects and are, therefore, the most positively active and competitive factors. This is true not only for B but also for E, under the condition that grazing be used as a regulative function for B and C. This is in contrast to D and F, which cannot be used in the same way, and which are therefore the most negatively active competitors.

The main conclusion from these hologram models for a multiple-use strategy

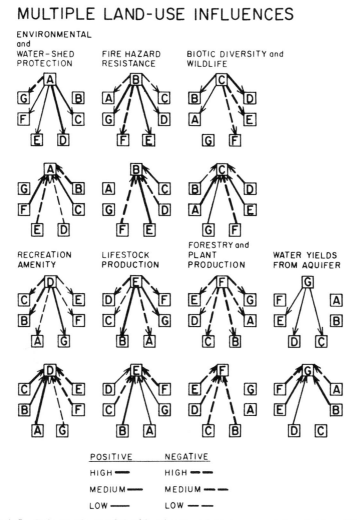

Figure 4-5. Cybernetic model of land-use and environmental variables. [Refer to legend designations for A, B, ... G in Figure 4-6 (Naveh, 1979).]

EFFECT FROM↓ON	A	B	C	D	E	F	G	AS +	AS -	Q +	Q -	TOTAL
A	•	0	2	3	1	0	-3	6	3	1.2	.6	1.8
B	-2	•	-1	2	2	-3	3	7	6	1.3	.1	2.4
C	1	2	•	3	-2	-2	0	6	4	1.5	.3	1.8
D	-1	-3	-2	•	1	-2	1	2	8	.2	1.6	1.8
E	-2	3	1	-3	•	-2	2	7	6	1.4	1.2	2.6
F	3	-3	-3	-2	-3	•	-3	3	14	0	1.5	1.5
G	0	0	0	1	1	0	•	2	0	0.3	0	0.3
PS +	5	5	4	9	5	0	6	-	-	-	-	-
PS -	5	6	6	5	5	9	6	-	-	-	-	-
P +	30	35	24	18	35	0	12	-	-	-	-	-
P -	15	35	24	40	30	126	0	-	-	-	-	-
TOTAL	45	71	48	58	75	126	12	-	-	-	-	-

```
A = Environmental Watershed Protection
B = Fire Hazard Resistance
C = Biotic Diversity Wildlife
D = Recreation Amenity
E = Livestock Production
F = Forestry Production
G = Water Yields from Aquifer

Influences:
3 high
2 medium
1 low
0 none

AS = active sum
PS = passive sum
Q = AS:PS - active and competitive variables
P = AS×PS - critical and buffering variables
```

Figure 4-6. Sensitivity matrix of land-use and environmental variables (Naveh, 1979).

that will ensure highest overall benefits and will be flexible enough to leave options for the future is that multiple-purpose recreation systems are best, especially those based on intensive improvement and conversion by environmental afforestation to create multilayered, rich, and diverse plant communities. On the other hand, a land policy directed exclusively to forestry will have the most pronounced inhibiting effects on all other options, with the possible exception of unilateral environmental protection, which can be attained in a different way by multipurpose afforestation, as described above.

Moreover, recreation, because of its potential great negative effect, should be handled with care. It will conflict not only with forestry and livestock production, but also with protection functions to which priority should be given in any overall landscape development considerations. Grazing, on the other hand, can be a very effective regulative tool in this respect, but if it is not controlled effectively, it will also have adverse effects on all protective functions.

It is also obvious that fire hazard resistance, because of its important active role, should be taken into account in any rational Mediterranean land-use policy. In contrast, water yields, as the most passive element, cannot be considered a major land-use goal, but only a desirable by-product in certain multipurpose land-use strategies.

Neotechnological Landscape Degradation and Its Prevention

General Trends in Degradation

The Mediterranean Basin is not only one of the most threatened of all Mediterranean-type climate regions, but probably, after the tropical forests, the most critical part of the world environment. The threats to the Mediterranean "microcosm" have been described vividly by Henry (1977) and by others in a special issue of *Ambio* (1977) on the Mediterranean.

As has been shown earlier, the cycles of the present neotechnological degradation functions are characterized by their accelerating speed, measured in a span of only years; and they are driven by deviation-amplifying mutual causal relations of spiraling positive feedbacks by exponentially growing populations and their consumption, expectations, and technological powers, coupled with land-use pressures of rapidly growing tourism. Thus, in the Mediterranean Basin, between 1950 and 1976 the total population increased from 94 to 220 million, and tourism is increasing at an annual rate of 7%, with more than 100 million visitors each year flocking to the most attractive, and at the same time scenically and biologically most sensitive, regions (see below). According to Tangi (1977), this tourism pressure on the Mediterranean now represents about one third of all international tourism pressure.

Although there are differences, and even conflicting trends, in different countries and regions, the general direction is toward the disruption of the formerly man-maintained dynamic agropastoral equilibrium that has contributed so much to the biological diversity, productivity, stability, and scenic attractiveness of these seminatural landscapes. This fact, the speed and extent of these modifications, and their mostly irreversible nature dictate the urgency of conservation measures in the SFZ.

These trends can be summarized under the following headings.

1. Intensification of the traditional and modern agricultural–pastoral upland uses by rootgrubbing of shrubs and trees and ploughing out of steep marginal land, combined with increased grazing pressures on diminishing and declining pastures by greater livestock herds, sustained with supplemental fodder; in many places, too, increased hunting, woodcutting, and the like.

In several (chiefly European Mediterranean) countries, however, there is an opposite trend:

2. Depopulation and abandonment and neglect of agricultural and pastoral uplands.
3. Intensification and expansion of agroindustrial land use by heavy inputs of fossil fuel, chemical fertilizers, pesticides and herbicides, large-scale land clearing, drainage, inundation, and irrigation schemes with far-reaching impact on natural lowland and upland ecosystems.
4. Expansion of urban–industrial sprawl, including mining and highway construction, into adjacent but also more remote uplands.
5. Explosion of mass recreation and tourism, leading to large-scale and mostly uncontrolled development of resort areas, camping sites, second houses, and tourist installations in scenically and ecologically most sensitive zones. The tourist pressure in the Mediterranean is the highest in the world. Thus, for instance, each summer in Majorca there are 15,000 visitors per square kilometer.

As a result, the following threatening alterations are occurring:

1. Rapid loss of open, unspoiled landscapes, with natural and seminatural ecosystems, their soil, plant, and animal life, and production and protection-regulation functions, and their replacement by built-up areas and urban-industrial wasteland, with far-reaching consequences for overall environmental quality, stability, and health.
2. Functional, structural, and visual degradation of the remaining open landscape, and especially its biological impoverishment and the ecological disruption of its natural ecosystems by accelerated erosion, soil, air, and water pollution, and neotechnological despoliation, combined with the creation of "monoculture steppes" and forests and monotonous and highly flammable shrub and weed thickets, leading to a steady increase in devastating wildfires and the costs of their extinction and damages.

The overall result is not only a steep decline in biological and economical production but also the loss of organic variety, unspoiled natural wealth, solitude, and beauty, and the noneconomic richnesses as the major assets of these Mediterranean landscapes. This is also indicated in Figure 4-3 by the increasing dominance of neotechnological urban–industrial artifacts of technoecosystems, replacing bioecosystems and leading to a rapid vanishing of the open Mediterranean landscapes.

Unfortunately, not enough figures are available for exact predictions and extrapolations, which would show in a much more dramatic way—in each country and in each location—what can be expected in the next 20–40 years if these trends continue unhampered without efficient countermeasures. This could be done, possibly, as a *Redbook of Threatened Mediterranean Landscapes*. Only a few examples of the pressing problems in different countries and regions, and some promising ways to solve them, can be shown here.

Effects of Increased Traditional
and Neotechnological Pressures

The impact of combined heavy agricultural and pastoral pressures on soil–vegetation systems in the xerothermomediterranean zone (the southern Judean hills) and the thermomediterranean zone (western Galilee) in Israel, which are typical of very extensive areas in the Middle East, North Africa, and Chile, have been described in detail by Naveh and Dan (1973). The cutting and uprooting of the most resilient sclerophylls, like *Quercus calliprinos* and *Pistacia lentiscus* (see above), and of hardy perennial grasses, like *Poa bulbosa* and *Hordeum bulbosum,* all of which afford most effective soil protection on steep and rocky slopes, even under heavy grazing, and the cultivation of shallow soil patches on nonterraced slopes have induced renewed and accelerating erosion–degradation cycles. These are leading to bare soil and rock deserts, with *Asphodelus microcarpus* (the hardiest nonpalatable and nonburnable geophyte remnant) as the last plant between rock outcrops. The harsher and more fragile the environment, the more rapid, far-reaching, and irreversible are these anthropogenic biofunctions. In general, most threatened are the open woodlands, shrublands, and grasslands on the drier ecotones of the SFZ. These contain probably the floristically and faunistically richest and most resilient mediterranean ecosystems and are serving as highly valuable drought-resistant genetic pools of immense value for future human uses (see p. 276).

The only way to interrupt this vicious cycle is by providing rational alternatives for higher production and incomes from much better-managed lowland fields (for crop production) and upland pastures (for animal production), along with forests and coppices for fodder and fuel. Only very dramatic results of manyfold increases in yields, together with suitable legislative and administrative measures and the constant assistance of devoted and motivated field research and extension workers, will be able to overcome the immense ecological, cultural, and social constraints. That such dramatic rises in production can be achieved has been shown in Arabic villages of the Samarian and Judean hills (Briegeeth, 1981) on the West Bank in Israel, in North Africa (Le Houerou, 1978), in Greece (Liacos, 1977), and in Corsica (Etienne, 1976).

The adverse impact of hunting in the Mediterranean in general, and especially in the western parts, is already causing serious reductions in many bird and mammal species. This process began in the years when hunting turned from an aristocratic privilege into a popular mass sport. In recent years, however, the shooting of game, netting of birds, and so on have become a major menace to wildlife, especially in Italy, where the density of eight hunters per square kilometer (with 2 million hunters) is the highest in the world (as compared with 3.65 in France and only 0.41 in Great Britain). This situation has been described in detail by Cassola (1979), Renzoni (1974), Cassola and Lovari (1976), Lovari (1975), and Lovari and Cassola (1975) from Siena, who are among the few biologists in the western Mediterranean devoting themselves to the cause of nature conservation and public education.

As in all other outdoor recreation activities, but with a more direct destructive impact on natural ecosystems, the deviation-amplifying positive feedback chain leads from more people with more income, more leisure time, more motor vehicles, more highways (hence, more mobility), and more powerful and precise firearms, to greater destruction potentials if these are not sufficiently counterbalanced by deviation-counteracting negative feedbacks of hunting laws and their enforcement and supervision and by public education and pressure. According to Cassola (1979), even the recent, much improved hunting laws in Italy still have very serious drawbacks. The survival of some threatened animals, like the Sardinian mouflon and the Abruzzo chamois, could be ensured within protected national parks, but for most others the future holds little cause for optimism. This is especially the case because shooting interacts synergistically with other (above-mentioned) Mediterranean neotechnological modifications. Thus, the reduction in the size of wilderness areas is reducing the survival chances of the greater carnivores, like wildcats and wolves. (Until recently, these, as well as weasels, otters, and owls, have been considered "pest animals" and could be hunted in Italy all year round!) The disruption of predator–prey relations and the massive secondary poisoning by pesticides described earlier prevail in all industrialized countries of the SFZ. An additional cause for the extinction of species of local origin, such as the grey partridge in Italy, is the introduction of more aggressive exotic game animals, such as pheasants and Middle European wild boar species. Similar problems have also arisen in California in the wake of the introduction of exotic species, such as feral hogs and wild boars (Barret, 1978).

It is unfortunate that the current Italian hunting laws provide huge sums, derived from hunting fees, for hunter associations (making them even stronger pressure groups) but nothing for the promotion of nature conservation nor for environmental education, especially for schools (Cassola, 1979).

Examples of Successes in Public Education
for Conservation

How effective children can be in nature conservation campaigns, if they are made a part of well-coordinated and well-promoted efforts, has been demonstrated by Israel's success in saving its most endangered wild flowers from extinction (Alon, 1973). This campaign was initiated by the Society for the Protection of Nature and was supported by the Ministry of Education (which also provides the salaries of the guides in the field schools). Special teaching activities and curricula were developed for the kindergarten to high school levels which involved the pupils in activities for saving the wild flowers. These campaigns are renewed every year with the beginning of the winter flowering season. Very attractive posters with all protected wild flowers (as well as animals) are distributed by the Society to schools, public offices, and the like, and can be bought at a nominal price. The Society for the Protection of Nature in Israel has thousands of enthusiastic members, all subscribing to the semiprofessional monthly journal *Nature and Land* or to its special edition for youngsters (an English edition is also available), in which the various weekly and monthly outdoor activities are announced. Hundreds of nature lovers of all ages take part in these activities, which range from hiking and excursions by car and bus to nature reserves and other scenic natural and/or historical and archeological sites, to bird-watching, to camping at combined recreation and instruction camps in the field schools. But all these successes in public conservation education would be in vain without the creation of a strong legal and administrative governmental infrastructure for nature protection and its support by an influential parliamentary lobby that cuts across all political parties. The exciting story of the untiring efforts of the well-organized and highly motivated Nature Conservation Authorities in Israel to enforce the laws protecting endangered flowers, animals, and landscape values, the declaration of over 100 nature reserves, and their management, development, and protection against the mounting pressures of a small, densely populated, and rapidly developing country like Israel, has been told recently in a well-illustrated book by one of the pioneers in this field (Paz, 1981).

Another example of the kind of public education so badly needed in the Mediterranean has been provided in West Germany by an outstanding and highly successful book, *Save the Birds—We Need Them!* (Stern et al., 1978) followed more recently by another similarily successful book on saving wildlife, in general (Stern et al., 1980). Showing the importance of birds and their habitats as indicators of the quality of life, this book effectively combines 450 artistic color photos and illustrations with an inspiring, instructive text. It is an example of how ecology should be presented to the public, and demonstrates that, because of the intricate interdependence of life—people, birds, open landscapes, and unspoiled ecosystems—nature conservation is the concern of everybody. It was sponsored by Lufthansa, which donated 285,000 DM and thereby made possible a low selling price. In addition, the authors donated .50 DM from every book sold, and a total of 500,000 DM were made available for the purchase of the Meggendorfer lakes and their conversion into a unique educational bird refuge. Since its appearance in August 1978, it has sold more than 300,000 copies and was on the best-seller list of *Der Spiegel* for a whole year.

The Threatened Mediterranean Wetlands

As has been shown so convincingly in the books by Stern et al. (1978, 1980), the major cause of the extinction of birds and wildlife in general is the loss and despoilment of their living space and natural habitats. Of greatest importance in this respect are the wetlands and their adjacent terrestrial systems. According to a MAB report (UNESCO, 1979), there are still about 1,000,000 ha of natural and seminatural wetlands in the SFZ of the Mediterranean, and 70% of the European Palaearctic waterfowl species spend their winters, together with many other bird species, in these wetlands. However, an International Union for the

Conservation of Nature and Natural Resources (IUCN) survey (Carp, 1977) has shown a very alarming picture. With the exception of the Hule Reserve in Israel and the outstanding Camargue Reserve in France, all others are severely threatened by agricultural "development," drainage, pollution, uncontrolled hunting, and the like. Thus, in one of the most important and largely intact wetland systems in southwest Europe, the Marismas del Quadalquivir in Spain, out of an area of 250,000 ha, only 3000 are protected within the Donanan National Park. This is a most important wintering place for several European duck species, geese, and other birds, including the rare Spanish imperial eagle, and is also the last refuge for the Spanish lynx. The park, established and studied by a team of devoted ecologists, is now gravely endangered by agricultural and tourist developments in bordering areas as well as by unwise road planning, pesticides, and the like. Here, as in the Mediterranean uplands, careful and well-balanced multiple-use management, with broad buffer zones keeping out undesirable external influences on one hand, and on the other retaining favorable internal ones [as proposed by Morzer Bruyns (1972)], is vitally important. Spain has the great advantage over many other Mediterranean countries (including California) of having a central governmental agency–the ICONA–for nature conservation. But this organization needs stronger political backing and public support, and greater funds and more professional staff and field supervisors, in order to fulfill its important task. In Spain, more than in most other Mediterranean countries, the need for landscape-ecological integrated approaches to conservation planning and management of natural resources is more and more recognized, in both professional and academic circles (Viedma et al., 1976). It is therefore hoped that this important park and its surroundings will be saved.

The Hule Reserve in Israel is another example of the difficult problems facing the conservation of wetlands in rapidly developing Mediterranean countries. This small remnant of 40 ha from an area of 6200 ha drained between 1951 and 1957 (the other swamps were drained even earlier) was saved only after great efforts by a small group of devoted biologists who at the time founded the Society for the Protection of Nature, described above.

The original swamp and lake were one of the biogeographically most unusual biotopes in the world, with extremely rich and diversified faunas and floras. Israel has already lost 20 plant and 20 animal species from these wet habitats, and 20 more fish, amphibia, reptiles, and waterfowls and 23 plant species growing in these wetlands and along the rapidly vanishing or polluted riverbeds and winter rain ponds are threatened by extinction. [Forty more animal species are endangered in Israel–chiefly by pesticides–and 50 more plants (Paz, 1981).] Because of its small size, the lack of a freshwater supply, and the inflow of polluted water from the artificial fishponds and intensively cultivated fields, the wildlife of this reserve has suffered badly. Therefore, the Nature Conservancy Authority decided to close it to the public and to carry out a very skillful biological and hydrological reclamation, which is now complete (Paz, 1975). In addition, in the northern coastal plains, the artificial fishponds of the collective settlement Maagan Michael were declared a bird refuge, together with the Taninim River Nature Reserve, which is at the mouth of Israel's last nonpolluted creek and which has unique riverbed fauna and flora. An adjacent field school serves as a major focus for field studies and environmental education in wetland conservation. In recent years, both numbers and species of wintering birds are increasing again, using these man-created and -protected ecosystems as breeding grounds.

This nature reserve was saved from the pollution and despoilment that would have been caused by the erection of a power station near the river mouth by the first decision in land use in Israel, determined by ecological considerations. The planning team, including a wildlife ecologist, convinced the authorities to prefer the already polluted outlet of the Hadera River as the final location, in spite of the opposition of the town of Hadera, by using a matrix of real relative values instead of the standard cost–benefit analyses in their argument (Hill, 1972) (see Photos 7 and 8).

Industrial Landscape Degradation in Coastal Tourist Centers

The undesirable side effects of uncontrolled and unwisely planned industrial development are already causing severe damage to aquatic and terrestrial systems in some of the biologically and scenically most sensitive coastal landscapes in the SFZ. Much of this could have been avoided by more careful planning, zoning, and landscape restoration and revegetation. Some of the worst examples can be found on the 200-km-long Adriatic coast of Yugoslavia, considered until recently one of the least despoiled and most attractive Mediterranean coasts. One would

Photo 7. In the Polag Nature Reserve, near Netanya, one of the last, unspoiled spots along the shores of the Mediterranean coast in the Sharon Plains, with flowering *Pancratium maritimum*.

Photo 8. Polluted shores and sea by oil spills and plastic waste. In the background, the newly erected electric power station, near Hadera.

have expected a country with strong central governmental powers to be more successful in reconciling the demands of industrial development with those of environmental quality and landscape values, especially since the latter serve there as the basis of a flourishing tourist economy and an important source of income and foreign currency.

Some huge industrial complexes are already established, and others, which will need increasingly larger amounts of fossil energy, are planned in these sensitive coastal stretches. For this purpose more oil drillings will be added to the already existing 26 coastal oil drillings near the tourist island Brac, in Sipan near Dubrovnic and Dugi Otok, and the pollution of the Mediterranean Sea near the coast by crude oil transport to North Adriatic ports will increase further. (In 1978 only about $30,000 was paid in penalties in 30 cases of oil spills.) The industrial sprawl from Rijeka has spread along the coast and the bay of Martiniscica, and has already despoiled some of its most attractive stretches. Even at the most beautiful bay of Yugoslavia, Bakar (in spite of the opposition of the population), 64 coke ovens and a chimney 252 m high belonging to the thermal power station are despoiling the landscape and causing heavy air pollution. At present, according to reports of visitors (Carp, 1979), the uncontrolled air pollution at Spit, Pula, Rijeka, Sibenik (here a pine forest is in progressive stages of air pollution destruction), and Togir is almost unbearable. Even the Party newspaper, *Vjesnik,* admitted: "A life in smoke, dust and noise."

In spite of recent investments of millions of dollars for resort hotels at the island of Kirk, the greatest petrochemical complex in Yugoslavia is planned in collaboration with the Dow Chemical Company, which can take advantage of the lack of pollution control regulations and/or their nonenforcement. A concrete bridge 1.3 km long is planned from Kirk to the mainland—not for tourists, but to connect a pipeline for the transport of crude oil. In addition, as in most other Mediterranean countries, there is a great amount of unplanned housing development in protected landscapes (about 30,000 illegal houses along the Adriatic coast).

The proposed *Redbook of Threatened Mediterranean Landscapes* should present and publicize all these examples, and many others, with illustrations and explanations. New methods developed for the assessment of scenic resources in California and presented at the conference on national landscapes at Lake Tahoe in 1979 (see Chapter 3) could be most helpful. Case studies should be included showing that industrial despoilment, pollution, and destruction of natural ecosystems cannot be accommodated with a flourishing tourist economy.

Recreation Pressures

Also greatly endangered by the combined impact of environmental pollution and commercialized mass recreation and tourism are the island ecosystems from Majorca all across the Mediterranean Sea to Cyprus. Vegetation on coastal cliffs and island rocks—as well as on those of the mainlands—is probably the least affected by man in the Mediterranean and therefore deserves special protection. Many rare animals, such as the monk seal and Eleonora's falcon, depend on these habitats. However, in the case of the rocky islet of Dragonera, near Majorca, "developers" established a gambling resort. Unfortunately, efforts to prevent this did not bear fruit.

However, the mere creation of a nature park by no means ensures the protection of the natural and scenic values for which the park was intended. Sometimes, the development of roads, and of parking, picnicking, camping, lodging, sport, and swimming facilities in parks or around their commercialized and uncontrolled fringes, can endanger these values even more. This is especially the case in many Mediterranean countries, where park authorities cannot control the onslaught of thousands of (mostly undisciplined) visitors, not seeking nature for its own value, but rather as a nice place for picnics and barbecues. Therefore, when nature reserves and parks are designated in order to protect outstanding natural values, the authorities should realize that nature conservation cannot be reconciled with motorized mass recreation, and that for this purpose other outlets must be found, such as planned recreation parks and forests or artificial lakes and dams. Yosemite National Park in California is an example of the problems encountered by mass recreation, and the decision there to close major parts to motorized traffic is a timely, but probably not sufficient, one. In certain cases, even stricter limitations on the daily numbers of visitors must be imposed, and such decisions need great political, public, and administrative backing, which is not easily found in the Mediterranean. The above-mentioned *Redbook of Threatened Mediterranean Landscapes* could also show how soon these parks will become overcrowded "outdoor recreation slums"—the fate that has already befallen most shores and beaches of the Mediterranean.

A satisfactory solution to these problems can probably be found only within comprehensive landscape master plans. Such a plan was prepared in 1964 for the island of Crete by a team of Cretan and Israeli planners and specialists, under the inspiring leadership of the great landscape ecologist and planner, the late A. Glickson (1970). This plan—although never fully realized—can serve as a model for integrated Mediterranean coastal and upland development and conservation. Among its basic targets was the reconstruction, by soil conservation measures and afforestation, of the uninhabited mountain landscape and the creation of a continuous landscape belt of varying width all along the ridges and watersheds of Crete. Along this Cretan Trail some camping grounds and accommodation for tourists in villages would be established, but no private development of villages would be permitted in the whole backbone of the island, which would thereby be turned into a kind of national park. Motor vehicles would be permitted only wherever north–south roads intersect the landscape system, but the rest would be accessible only to foot travelers, pack horses, and mules. Glickson's multiple-purpose approach, stressing the cultural importance of the re-creation of the indigenous mountain landscape, has been adopted in the model of upland-ecosystem planning and management, presented in this chapter.

Threats from Air Pollution

The potential damage to conifer forests by sulfur dioxide has already been recognized and is being studied in the Mediterranean SFZ (UNESCO, 1977b). This is not the case with photochemical oxidant pollutants, even though they are much more dangerous and constitute an alarming threat to mediterranean ecosystems. The special meteorological and topographical conditions favor the creation of photochemical smog and its chief phytotoxic component—ozone—in the densely populated, industrialized, and heavily motorized coastal mediterranean regions, and its transport with the sea breeze far inland, to mountainous and forested slopes. Here, these pollutants can cause heavy damage to sensitive trees, especially conifers, but may have much more far-reaching and cumulative effects on the ecosystem as a whole, especially if combined with other pollutants and with dust. Sulfur dioxide, as well as fluorides, on the other hand, are emitted from distinct industrial sources that can be traced, monitored, and controlled in a much more efficient way.

Thus, these photochemical oxidants, especially ozone, have already been recognized as a major threat to the future of the mixed conifer forests in Southern California. In the San Bernardino and Angeles forests, in the Los Angeles "smog basin," many thousands of *Pinus ponderosa* trees are dying or declining as a result of the combined stress of ozone and secondary pest attacks, chiefly pine bark beetles. The annual mortality exceeds 3% and the reduction in wood production is estimated at about 83%. Also, on the western slopes of the southern Sierra Nevada facing the San Joaquin Valley, up to 100 km from the nearest urban smog source, more and more severe damage to pine trees can be observed wherever the average daily atmospheric ozone concentration exceeds

0.05 ppm. Here, the typical injuries of chlorotic mottle on the needles, followed by general chlorotic decline of the whole tree from the lower branches upward, are generally combined with pest infestations by aphids and pine needle scales (Miller and McBride, 1975; Pronos et al., 1978).

In the metropolitan centers of the Mediterranean coast in Israel, photochemical smog often causes atmospheric ozone concentrations above 0.1 ppm and these are increasing from year to year. As in California, these air pollutants, generated under the influence of high solar radiation during most months of the year and especially in the summer, are transported downwind toward the Judean hills and reach Jerusalem in the early afternoon. These polluted air masses apparently are trapped in the Shaar Hagai canyon, and near the road leading to Jerusalem even higher concentrations of ozone have been measured. These have resulted in widespread chlorotic decline of the planted *Pinus halepensis* trees and—in combination with heavy infestations by *Matsucoccus josephi* scales—in almost total mortality of injured trees. In Israel, about 60% of all planted forests are composed of this, the only indigenous pine species, and progressive decline and mortality can now be observed in many forests. On these trees, as well as on *P. pinea, P. brutia,* and *P. canariensis* trees, which are more resistant to *Matsucoccus j.* scales, the phenomenological and histological symptoms of chlorotic mottle and decline observed by our study team show striking similarity to those observed in the above-mentioned forests in California. Controlled fumigation tests have induced these symptoms with ozone concentrations of 0.05 ppm and 0.1 ppm (Naveh et al., 1980). Because of its heavy resin exudations, *P. halepensis* is resistant to bark beetle attacks. This is not the case with *P. canariensis,* and indeed, in the coastal region of Israel, where atmospheric ozone levels are highest, groves of these ozone-stressed pines are now heavily damaged by *Orthomucus erosus* bark beetles. It would therefore be wrong to assume that other coniferous species in the Mediterranean will not succumb, sooner or later, to similar secondary pest attacks by scales, mites, and/or beetles, which are favored when chemical sprays, dusts, and the like weaken their predators. Ozone might also predispose these trees to additional stresses, such as drought, disease, salt spray, and other pollutants, and increase their damages manyfold. As reported by Tomaselli (1977), there is already widespread damage to pine trees on the Italian coast and in France; *P. pinea* is succumbing to *Matsucoccus* infestations. Along the major roads in central Italy and Toscana, cypress trees are dying.

Westman (1978), in the first phytosociological study of coastal sage dwarf shrub communities in California, found a significant reduction of broad-leaved herbs in communities exposed to smog in the Los Angeles mountains. There is therefore an urgent need for early recognition and monitoring of present and potential impacts of air pollutants as part of integrated, long-term ecosystem studies in the SFZ. Such studies are already being carried out in Southern California, but unfortunately they could not include control plots from nonpolluted sites because none remain in Southern California.

The great vulnerability of dense conifer forests to air pollutants and pest outbreaks may not only severely affect their future productivity and reduce their commodity and recreation values, but may also further increase their fire hazards (see below). This makes the need for a reorientation toward more stable multilayered and multipurpose Mediterranean forests even more important.

Problems of Dynamic Conservation Management

The Effects of Noninterference, Abandonment, and Complete Protection

The complete abandonment and cessation of cultivation, woodcutting, and grazing, resulting from the depopulation of rural uplands in many European countries in the SFZ, especially in France and the Mediterranean islands, is leading, in its early stages, to a recovery in height and density of the herbaceous and woody plant cover. Very soon, however, aggressive tall grasses and perennial thistles crowd out the smaller herbs, and a dense, species-poor, and highly combustible weed thicket establishes itself in open woodlands and grasslands.

In shrublands, the vigorous vegetative regeneration of sclerophylls from stunted shoots and almost imperceptible rootstocks is followed, in general, by the gradual enclosure of the shrub canopy and the almost total suppression of the herbaceous understory. After the initial high photosynthetic rates of the regenerating shrubs has slowed down, the dense 3- to 6-m-high shrub thicket gradually becomes stagnant and even senescent and very fire prone, because of the accumulation of dry and dead branches and undecomposed litter. On Mount Carmel in Israel, for instance, such undisturbed one-layered tall shrub communities are dominated almost exclusively by *Quercus calliprinos,* with a few codominant shrubs and subordinate dwarf shrubs and very few shade-tolerant perennial grasses and geophytes, chiefly near rock outcrops and shrub edges. In comparison with disturbed, semiopen, and multilayered shrub communities, their floral diversity is much lower (21 plant species per 1000 m as compared with 120 species) and their species richness and abundance of birds, rodents, reptiles, and insects is also much lower (Naveh and Whittaker, 1979a, b). Zohary (1973), after extensive studies in the SFZ of the Near East concluded that such undisturbed *Q. calliprinos* "climax" communities are extremely monotonous and species poor. From the more humid Adriatic *Q. ilex* forests, Horvat et al. (1974) reported similar effects: After the cessation of intensive use by rural populations 40 years ago, they have lost their open appearance and their floral and faunal richness. Thus, not only by excessive defoliation pressures, but also by their total suppression, the man-maintained equilibrium between the woody and herbaceous strata is lost, together with the biological diversity of these seminatural ecosystems.

This re-dominance of the woody components after abandonment or protection should not be confused with the deterministic, directional process of "progressive succession towards a self-perpetuating quasi-original maquis or forest climax" by what Egler (1954) called the classical "relay floristic succession to climax" model, and what Connell and Slatyer (1977) called the "facilitation succession model." This re-dominance is the result of highly stochastic multivariate biofunctions, in which, in addition to initial site conditions and fluxes of climate, the "initial floristic composition" (Egler, 1954), especially available rootstock and seeds, as well as active involvement of birds, rodents, ants, and other insects and human modifications by grazing, fire, cultivation, planting, or abandonment, are most important. We described such ecotope state factor equations in Chapter 2, and in this chapter we have described the process of evolution of cultural Mediterranean landscapes. In these specific cases, however, several processes may occur side by side:

1. *Autosuccession* [as first defined by Hanes (1971) for postfire chaparral regeneration], namely, the above-mentioned vegetative regeneration and—in favorable moisture and shade conditions—the germination of existing sclerophylls and of fire-stimulated dwarf shrubs.
2. *Arrested succession,* described by Connell and Slatyer (1977) as the "inhibition model."
3. *Direct ecesis* of sclerophylls in abandoned fields, in the shade of neglected orchard trees, grapevines, or planted pine trees, and the like, without any hypothetical "intermediate" successional stages of batha or garigue.

Similar autosuccession and arrested successions are also typical of the chaparral and coastal sagebrush communities in California. In the latter, fire, nonstable germination, and growth-inhibiting allelopathic agents have been shown to be involved in certain cases (Naveh, 1960, 1967; Muller et al., 1968). The same might also be true in the Mediterranean, but this possibility requires careful study.

Tomaselli was one of the most important and active Mediterranean plant ecologists and died recently in a tragic automobile accident, together with his co-workers from the University of Pavia. In the above-mentioned important MAB–UNESCO report on the conservation of Mediterranean forests and maquis, Tomaselli (1977) defined the ultimate goal of nature conservation in the SFZ as the reattainment of such a maquis climax. He therefore based his recommendations for a conservation policy of noninterference and complete protection on a hypothetical model of "possible progressive evolution from bare soil to this climax." Unfortunately, he did not provide any substantial documented evidence or proof from systematic studies of such a process in the SFZ. The same is also true for his claim that thinning and opening dense maquis shrub by cutting, grazing, and burning has many detrimental effects on the soil, the microclimate, and microorganisms. Not adhering to any such preconceived succession-to-climax dogmas and taking into consideration our present state of knowledge, we can conclude that, contrary to climax theories, closed and undisturbed sclerophyll shrub ecosystems in the Mediterranean (and this is also true for California) are neither diverse nor productive nor stable, and they can be perpetuated and rejuvenated only by fire or cutting and thinning. As dense, impenetrable, and highly flammable thickets, their economic, biological, scenic, and recreational value is very low. They also have little historical–ecological value as "natural, quasi-original" climax communities. The existence of such natural (i.e., primeval) maquis climax is highly doubtful, since people had already modified these ecosystems in the middle Pleistocene, even before the present Mediterranean climatic patterns were established. At most, they could be regarded as the "potential natural vegetation" sensu (Tüxen, 1956) that would establish itself when man suddenly stops all interference. As such they are of scientific and practical value as indicators of ecological site potentials, but they have no intrinsic value that would justify a general noninterference conservation policy, as suggested by Tomaselli and many phytosociologists from the Braun-Blanquet school in the Mediterranean. The same is true, of course, for the lower undisturbed and arrested successional stages as well.

In general, the decision on the most suitable methods for conservation management of these seminatural upland ecosystems should be made on the basis of broader bioecological, socioecological, and economic considerations, and as part of general landscape master plans for each site and location. Thus, there is no reason at all to oppose, for the sake of "nature conservation," the conversion of suitable sites of nonproductive and species-poor *Cistus* dwarf-shrub lands, covering thousands of hectares in Sardinia and elsewhere, into productive *Pinus radiata* forests if this is done in an ecologically sound way and has socioeconomic justification. In fact, in Sardinia this planting was carried out in combination with reseeding of annual legumes for soil protection and improvement. In suitable sites their productivity is much higher than that of Mediterranean conifers, but inflammability is considerably lower. There is also no reason to oppose the use of other "exotic" species for commodity or other purposes, as long as very strict precautions are taken to prevent their spread as undesirable weeds (the same is also true for Mediterranean plants), and if this is not done within nature reserves and other protected areas. In these, *any plant*—including Mediterranean conifers or even sclerophyll trees—that is not autochthonous and spontaneously growing should be considered exotic and therefore undesirable.

As mentioned already, the drier ecotones of the SFZ deserve high priorities for conservation of their great genetic potentials and biological diversity. These can be ensured only by continuation or simulation of the same defoliation pressures of grazing and fire under which they evolved. Thus, it would be a great mistake to attempt to restore the "original Ceratonio–Pistacion climax" of Mount Gilboa in Israel, or similar areas elsewhere, by noninterference conservation of the present "degraded" open shrublands. These are mosaics of a drought-resistant *Pistacia lentiscus* ecotype and other shrubs and dwarf shrubs with an extremely rich herbaceous canopy, with a very high floristic diversity of over 170 species per 1000 m^2 and a very high faunistic diversity (Naveh and Whittaker, 1979a). In this case it was equally wrong to "rehabilitate" great parts of this mountain by afforestation with *Pinus halepensis,* not only for ecological reasons of losing these unique ecosystems for conservation, but also for economic reasons. Under the prevailing semiarid xerothermomediterranean conditions of low rainfall, high summer temperatures, and frequent dry and hot sharav winds, the prospects for economic wood production from these trees in the near future and before the next raging wildfire—in spite of expensive fire prevention measures—are very slim. Fortunately, the declaration of a nature reserve to protect the attractive, endemic "iris of the Gilboa"—*Iris haynia*—saved at least part of this mountain from "rehabilitation." Here, the Nature Reserve Authorities rightly prevented the smothering of the scattered iris plants by taller aggressive grasses and herbs, and ensured their reproduction, together with the reduction of fire hazards, by allowing strictly controlled cattle grazing.

In the wetter and higher-elevation ecotones of the SFZ, the conservation of the last remaining *Quercus ilex* forests is of no less importance. But here too, a strict noninterference-conservation policy can sometimes be more harmful than useful. Thus, at Supremonte, near Orgosolo, Sardinia, Susmel et al. (1976), in one of the few integrated Mediterranean ecosystem studies, showed that the unique *Q. ilex* forest, with trees up to 1000 years old, has maintained its energetic and metabolic balance and is reproducing itself in spite of swine grazing by traditional pastoralists. These consume about 50% of all acorns and 7% of primary production; cattle and sheep, grazing chiefly in adjacent degraded maquis shrubland and subalpine meadows, consume only 1%. The chief detrimental factors endangering the forest are fire and woodcutting. The removal of the shepherds and of their livestock would not only mean an irreversible loss of cultural and biological diversity, but would create a hostile population and thereby increase manyfold the fire hazards and uncontrolled woodcutting. Susmel also emphasized the importance of the shrublands and grasslands for the conservation of the forest and suggested improving the pasture production of the meadows and restricting the afforestation of suitable (and not only Mediterranean) forest trees outside the forest to small, carefully selected sites with highest forest potentials.

The Role of Fire in Conservation

In a summary of the effects of fire on Mediterranean ecosystems, Naveh (1974b, p. 431) stated: "Fire has apparently served as a most beneficial tool in the skilled hands of our Mediterranean ancestors, but it has been abused greatly by later generations, and it is being now neglected and rejected. It is up to the ecologist and the enlightened land user to show how fire can be turned from a curse in Mediterranean lands to a blessing."

Because of the destructive combination of fire and uncontrolled grazing in recent times, the important role of fire as a selective force in the evolution of Mediterranean ecosystems and in the maintenance of their productivity and diversity and its potentials for Mediterranean wild-land management has been overlooked completely. When fire, like goat grazing, has been judged only from its ill effects when abused, it has been regarded mostly as an entirely negative degradation factor, leading away from the "sclerophyll forest and maquis climax" and therefore being detrimental to nature conservation as well (Tomaselli, 1977).

Robertson (1977) has vividly described the changes in upland uses and the resulting increase in fire intensities in southern France, which is typical of large regions in the SFZ. Formerly, little fuel for hot and frequent fires was left from the intensive agricultural upland use for vineyards, vegetables, orchards, olive and sweet chestnut groves, and the like, in combination with the coppicing and barkstripping of oaks, cutting of pines, grazing and browsing by sheep, goats, cattle, and pigs, and the use of "small fires" to burn weeds and stubble fields. This is still the case in densely populated Middle East and North African uplands, where these and other heavy defoliation pressures (described above) almost completely prevent fires.

In the last 30 years, however, with the depopulation of rural areas and the abandonment of fields and pastures, followed by the above-described reencroachment and piling up of dry and dead wood from senescent trees, great amounts of fuel are accumulating without barriers to the spread of fires. At the same time, the spectacular increase in recreation and tourism, and the resulting pressures for wild-land development, have increased manyfold the number of accidently, and also intentionally, ignited fires. In 1973, Le Houerou (1973) estimated that each year close to 200,000 ha of Mediterranean forests and shrublands are burned, causing direct damages of at least $50 million, but much more in ecological and social costs. However, under present policies of total fire suppression, and notwithstanding the great efforts and expenses in fire prevention and fighting, the number and extent of, and damage caused by, fires increase from year to year. Thus, in Italy between 1965 and 1973, the number of fires increased from 2360 to 6092 and the burned forest and shrubland areas from 23,500 to 77,400 ha (Susmel, 1973). In Israel in recent years, the number of pine trees burned has almost exceeded those newly planted, and in many places in the SFZ, each pine forest is apt to burn every 10 years. In the "red belt" of the pine forests of the mountain slopes along the Riviera in southern France, an average of 26,000 ha of forest have been destroyed in recent years, losing thereby about 4% each year. Any tree's prospects of reaching the age of 20 before being hit by fire are only 60%, in spite of one of the most advanced and best-organized fire prevention and firefighting systems, including 12 huge Canadair airplanes, each carrying 5 tons of water from the sea to the fire site, and special well-trained local firefighters. Thus, in August 1979, the newspaper *Le Monde de Paris* reported that "a red sea" had flooded the French Mediterranean coast during the mistral, burning down 40,000 ha of pine forests. The local authorities, to whom thousands of hectares of forest belong, can no longer afford to use expensive labor for clearing the dry understory, and prescribed burning for these purposes is not yet being considered. At the same time, on the Costa Brava in Spain, near the tourist villages of Llorat de Mar, the pine forests of the nature park were burned down, apparently in order to obtain this land cheaply after the destruction of the nature park for construction of hotels and bungalows. In France, too, forest fires clear the way to open the protected "green belt" to construction and development.) This fire killed 22 Spanish summer vacationers, and in the following week, eight other fires were set in the same region. In 1978, there were 8331 wildfires—45% set intentionally by arson-ists—in Spain on an area of 430,000 ha. According to the Minister of Agriculture, at this rate no forests will be left in Spain in less than 20 years. In addition to the traditional reasons for such fires, wildfires have turned into good business for land developers and for paper manufacturers (especially near Valencia and in Majorca), who can acquire the charred stems cheaply and replace them with fast-growing eucalyptus trees.

In California most of the wild land, covered with highly combustible chamise shrubland intermixed with conifer forests and grassy woodlands, covering exten-sive areas on steep and inaccessible slopes, are owned or managed by state or federal governmental agencies. However, instead of learning from the Indians how to use low-intensity fires judiciously and in harmony with nature, the European fire suppression philosophy was adopted very early and has not been changed—in contrast to the Australian Forestry Departments, which in recent years have developed advanced methods for controlled burning that are applied widely and with great success. That the fire suppression policy is failing com-pletely, in spite of more sophisticated tools, was demonstrated in a dramatic way during the first Mediterranean symposium on fire and fuel management, in August 1977, in Palo Alto (Mooney and Conrad, 1977), when some of the worst wildfires California ever experienced raged not far away, in the central Coast Range. They proved the warnings of Biswell (1977), the inspiring pioneer and leader in fire ecology and its application in mediterranean land management, that as long as forest departments direct all their thoughts and efforts only towards fire extinction, and not toward fuel management and reduction by pre-scribed fires as well, and as long as not even a half dozen people qualified for useful prescribed burning can be found in California, the severity of these wild-fires and the costs of preventing them will continue to rise. His work in Califor-nia and that of his students in the Mediterranean have already proved that prescribed burning is an important tool not only for reducing fire hazards and the costs of fire control, but also for increasing timber production in forests and for improving livestock grazing, forage production, and water flow for springs in shrubland and woodlands. It can also be a most beneficial tool for conservation management by manipulation of wildlife habitats and increased biotic diversity and scenic values. Thus, in the California chaparral shrublands, the summer density of deer rose from 30 per square mile in untreated closed shrubland to 98 in the first year after burning and to 131 in the second; with gradual reencroach-ment of the brush canopy, it dropped again to 84 in the fifth year. The ovula-tion rates of adult does were twice as high in the burnt and opened chaparral. The importance of the change from a suppression to an ecological fire phi-losophy in wildlife conservation was demonstrated in this conference by Walter (1977) for birds and by Lillywhite (1977) for small vertebrates. Both presented results on the indirect benefical effects of habitat improvement through the rejuvenation of the Mediterranean sclerophyll ecosystems and their rise in pro-ductivity and structural and floristic diversity. They stressed the potentials of prescribed burning rotations in time and space to encourage a mosaic of different stages of postfire regeneration, stimulating the natural ones in "fire-generated" (pyrogeneous) habitats in California and the Mediterranean. As in the case of grazing, highest diversity and productivity seem to be ensured only by a certain optimal frequency and intensity of burning, and not by extremes of non-use or overuse. These are important lessons, to be translated into actual conservation–fire management. For the assessment of the effects of such practices on different species, Walter (1977) proposed useful tolerance models. For the description and prediction of postfire successions, a new classification scheme was developed in Australia (Noble and Slatyer, 1977) that could replace the deterministic, classical succession concepts (see above) by more dynamic cybernetic approaches.

Whether fire can be replaced by other mechanical or chemical means for opening the brush canopy and reduction of fuel and the ecological effects of these measures needs careful study in different habitats and conditions. In the case of the striking increase of flowering geophytes after burning Mediterranean shrublands, it seems that in addition to the increase in light, fire also directly stimulates the germination and regeneration of these plants and removes heat-nonstable chemical inhibitors that accumulate in the unburned litter and duff (Naveh, 1973).

In certain conditions, where such means are necessary because of great fire hazards, the use of heavy machinery on steep slopes for total brush conversion—as done in California on a large scale—can be very detrimental to soil stability and to wildlife. It is regrettable that public agencies are still reluctant to accept ecologically sounder practices in land and fire management. Thus, the California State Forest Department continues to spend large sums for sowing the European aggressive *Lolium multiflorum* pasture grass as an erosion control measure after fire, in spite of its adverse effects on local spontaneously germinating plants—including shrubs and conifers—and its inefficiency in preventing postfire runoff and erosion.

Thus, even in California there still seems to be a great communications gap between academic research workers and teachers (some of whom have world capacities in their field) and the practical public and private land managers. On one hand, the latter do not know, or are not willing to make use of, relevant new research information, and on the other hand, researchers do not know about the pressing problems of these land managers and do not contribute to their solution or do not present their results in an interpretable way, applicable to practical landscape management. To overcome this gap, new approaches and tools for public and professional conservation education and decision making are needed here, as well as in other countries of the SFZ.

However, impressive progress in the field of fire management for conservation can be noted already in national parks. Thus in Yosemite National Park, Sequoia National Park, and Kings Canyon National Park, integrated fire management is being implemented (Parsons, 1977). It is aimed at minimizing the social losses but maximizing ecological gains by employing various combinations of fire suppression and prevention, fuel management, controlled burning, and land-use planning. In the highest subalpine elevations, all fires are left to burn, and in the middle elevations of the mixed *Sequoiodentron giganteum* conifer forests, prescribed fires are used to reduce fuel to a point where natural fires can be allowed to burn, in order to restore the natural pre-European wilderness state, when frequent and widespread surface fires kept these forests open and parklike.

In Southern California, an integrated system for fire behavior information in shrubland management (Kessel and Cattelino, 1978) is applied. It is based on computerized methods of gradient modeling, linking site inventory of vegetation, fuel and the like with cultural and hydrological information in order to simulate fire behavior and its consequences, using Landsat imagery for site potential inventories. As reported in detail in a very readable book by Kessell (1979), these integrated methods of gradient modeling for fire and resource management have been developed and refined chiefly in Glacier National Park in west Montana as an ecologically sound and administratively feasible plan for restoring fire to its natural, important role in this park, after its complete suppression since 1930. Gradient modeling itself is an outstanding example of "technological transfer" by employment of an ecological method and its conceptual philosophy, as developed in basic research, into an important holistic tool in landscape management. It is presently applied by Kessell for fire management in Australia as well, and it is hoped that it will soon reach the Mediterranean, where the need for more enlightened fire management policies, as part of integrated multiple upland-use patterns, is increasingly being acknowledged.

An important change in the attitudes of foresters has been initiated by a recent FAO–UNESCO technical consultation on forest fires in the Mediterranean (FAO, 1977a). The presentation of models of fire analyses in land-use planning matrices (Derman and Naveh, 1978) and the inclusion of fire hazard resistance in multiple-use benefit matrices (Naveh, 1978) stimulated much interest among the forester participants in this conference, especially among those attempting to solve the severe fire problems as part of integrated programs for the renovation of upland agriculture and restoration of the social and ecological role of Mediterranean natural areas. This is also the case in France, where a special interministerial mission is charged with this task (Robertson, 1977).

As a practical example of such fire analysis models in the regional planning process of east Mediterranean uplands, Figures 4-7 and 4-8 are presented. The major planning tool in this respect is a fire analysis model (FAM) to assess the probabilities of causing and spreading wildfire and its implications on the landscape (Figure 4-7).

The basis of the model is the description of the planned land uses and the inventory of natural and man-made components (Stage I). This should make possible a factual prediction of what will happen to the land use as a result of a fire event or its prevention (Stage II). A further stage is the prediction of probabilities of fire ignition and spreading. This stage is also based on the components of the land-use inventory, as was shown in Stage I, and on the recognition of the features of the planned land-use activities. The probabilities of fire ignition and spreading serve as a basis for the evaluation of the role of fire in

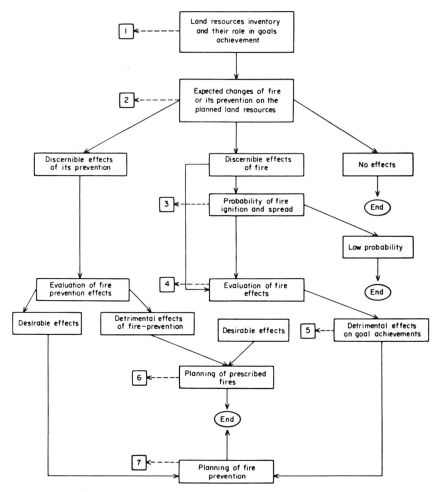

Figure 4-7. Fire analysis model in regional land-use planning (Derman and Naveh, 1978).

regional planning (Stage IV). The results of this evaluation may range from planning some installations for fire prevention or control to the change of the land use.

At first sight, Stage III-the prediction of fire ignition and spreading probability—should precede Stage II—the factual prediction, because obviously if there is no chance of fire, there is also no reason for its discussion. However, in our procedure, Stage III followed Stage II because of the above-mentioned beneficial implications of fire or its prevention. Therefore the effect of fire or of its prevention has to be assessed in the regional planning in each case, and not only as a result of a high probability of its ignition and spreading.

At the same time, as shown in the model, the absence of fire effects this procedure making Stage III superfluous, which would not have been the case in the apparently obvious order.

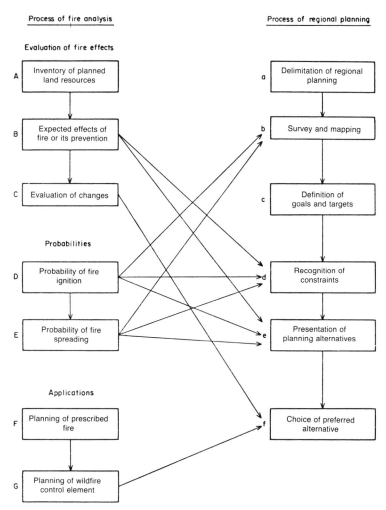

Figure 4-8. Fire analysis in the stages of regional planning (Derman and Naveh, 1978).

However, if fire has important effects, the different causes leading to fire ignition, as distinguished from its spreading, should be considered thoroughly. If the prevention of fire has significant effects on land use, such as increasing the probabilities of fire ignition and spreading by fuel accumulation in maquis and chaparral, the planner has to express these changes in planning parameters.

In general, the desirable results of fire and detrimental results of its prevention will lead to a similar conclusion, namely, the necessity of prescribed fires and therefore the need for physical planning for them. The option of detrimental results of fire or desirable results of its prevention will also lead to similar conclusions: the need for education of the public, for supervision, legislation, the planning of specific land-use allocations, and assisting in fire prevention, including planting of fire-resistant trees and shrubs.

In stage b of the R.P.–survey and mapping, stages D and E of the FAM must be carried out. In stage d (recognition of constraints), both the expected effects of fire or its prevention (B) and the probability of its ignition (D) and spreading (E) should be considered. In stage e (presentation of planning alternatives) the same factors should be considered, and in stage f (the choice of preferred alternative), evaluation of changes caused by fire (C) and planning of wildfire control elements (G) should be considered. Thus, if our FAM shows a very high probability of fire ignition and spreading and highly detrimental effects on pine forests, a different less vulnerable alternative should be chosen for afforestation, or an entirely different land use should be planned. This would be the case, for instance, on the eastern slopes of the hills of Galilee in Israel.

In Figures 4-9 and 4-10 are presented examples of matrices for the evaluation of the fire results on these uplands, including both direct and indirect immediate and long-range effects on natural and man-made landscape variables and their effects on land use. Since each element of the land-use inventory differs in its contribution to the realization of the planning goals and targets, the data collected with the help of these matrices should then be scaled and ordinated according to their positive or negative value in achieving these goals and targets.

It is obvious that throughout the whole procedure the dual value composites of positive and negative effects resulting from both the fire and its prevention have to be taken into consideration, and landscape planning and management have to be designed accordingly.

The burned changed components / components	Hydrology	Lithology	Soil	Vegetation	Wild life	Air quality	Noise	Micro-climatology	Engineering constructions	Adjacent land-use
Hydrology	—	O	□	□	△	△	O	O	□	□
Lithology	O	—	□	□	△	△	O	O	△	△
Soil	O	O	+	□	□	□	O	O	□	□
Vegetation	O	O	□	+	□	□	O	O	△	□
Wildlife	O	O	□	□	+	□	O	O	△	□
Air quality	O	O	△	□	△	+	O	O	□	□
Noise	O	O	△	□	△	△	—	O	□	□
Micro-climatology	O	O	□	□	△	□	O	—	△	□
Engineering constructions	O	O	□	□	△	△	O	O	+	△
Adjacent land use	O	O	□	□	□	□	O	O	□	+

△, No effect.
—, Not burned directly.
O, Not affected directly, unburnable.
+, Direct effect.
□, Indirect effect.

Figure 4-9. Matrix of interrelationships between fire and land-use components (Derman and Naveh, 1978).

The changed component \ The burned components	The changed element	Vegetation	Soil	Wildfire	Hydrology
Vegetation	Plant destruction	□	□	□	O
	Plant regeneration	□	□	□	O
	Germination of dormant seeds	+	□	△	O
	Flowering	□	△	□	O
	Resprouting	□	△	□	O
Soil	Acceleration of soil erosion	□	+	△	O
	Destruction of soil's profile A.	□	+	△	O
	Changes in chemical composition	□	+	△	O

Figure 4-10. Example of possible ecosystem changes as a result of fire (Derman and Naveh, 1978).

The resolutions of this FAO meeting, recommending for the first time the use of controlled burning and grazing for fuel management in the Mediterranean Basin, as well as of the above-mentioned symposium on fuel and fire management of Mediterranean ecosystems (Mooney and Conrad, 1977), could serve as important landmarks for the much-needed changes in attitudes toward fire. Specially relevant in this respect for the Mediterranean are the promising results of the pioneering studies by Liacos (1973, 1977) in Greece on prescribed burning to reduce fuel and fire hazards in pine forests, and the already mentioned important fire ecology studies in southern France by Trabaud (1981). Stock should also be taken of the outstanding work carried out elsewhere in this field, especially in other Mediterranean-type climate regions in South Africa and Australia. Much of this work has been reported in the proceedings of the annual meetings of the Tall Timber Fire Ecology Conferences in Tallahassee, Florida. Two very comprehensive books have been published on the state of research and practice in fire ecology and management on the fire-swept continent of Australia (Gill et al., 1981; Luke and McArthur, 1977). The review of these books is outside the scope of this chapter, but two main conclusions can be drawn from them:

1. The Australian achievements in fire ecology research and management bear witness that fire can be turned, with the help of thorough research and judicious and skilled management, from a terrible master into a good servant.
2. In designing large-scale and sophisticated technologies of prescribed burning rotations, great care should be taken not to completely wipe out the fire-induced biotic diversity and the natural feedback information maintaining it.

Conservation Management of Oak Woodland and Chaparral in California

In providing these notes of concern about what might be done in the Mediterranean-type vegetation of California, as contrasted with current practice, one should be aware of an increasing number of individuals and their writings who represent the leading edge of potential change in vegetation management practices in that state. In the proceedings of an important conference on the ecology, management and utilization of California oaks (Plumb, 1980), and in the proceedings of a symposium on living with the chaparral (Rosenthal, 1974), it is not difficult to detect the emerging thinking and forces that represent advanced attitudes on these matters.

In the first of these two conferences, the California oaks are considered from the vantage points of the fire resistance of oak seedlings, prescribed burning in California oak management, oaks and their place and role in the California landscape, measures to offset detrimental effects of urban development on oak trees, and management implications for birds and mammals of California oak habitats. Also included are ecological relationships between southern mule deer and California black oak, livestock utilization of the oak woodlands, and effects of the blue oak on rangeland forage production. Oak woodland–grasslands occupy about 10 million acres out of a total of 68 million acres of wildlands in California, and of these blue oak (*Q. douglasii*), a deciduous white oak, is the most important oak in the drier foothill woodland regions. At this oak timbers conference, Holland (1980) and Holland and Morton (1980) reported on the value of these oaks for livestock, in addition to their importance in ensuring soil fertility, stability, a favorable microclimate, and vital habitats for wildlife. They pointed to the detrimental effects on their integrity of large-scale and indiscriminate removal from foothill rangelands. Clear-cuttings are carried out widely by farmers, following shortsighted recommendations of farm advisors based on other studies claiming that pasture output, spring water, and stream flow can thereby be increased. However, Murphy (1980), superintendent of the Hopland Field Station of the University of California in the Coast Range foothills, who was himself involved in these studies, came to the conclusion that range operators should carefully assess the changes and land-use benefits that will result from tree removal before deciding on such ventures.

With woodlands of Engelmann and coast live oaks, an analysis of 15 vegetation map survey profiles in California of stands containing one or both of these species would tend to indicate that Engelmann oak grows more frequently in or next to more fireproof habitats. The woodlands with Engelmann oak are quite rare. Many have already been, and continue to be, lost to both agricultural and urban development. Protection of the few remaining Engelmann oak woodlands is identified as a high-priority item for oak management in Southern California.

In the Sierra Nevada, California black oak is an important species in mixed-conifer and ponderosa pine forests. Tappeiner and McDonald (1980) reviewed the occurrence of this oak relative to site considerations, species composition, and stand structure, and to its response to fire and its role as a nurse crop for conifers. They provided recommendations for managing this oak for wildlife, wood, and aesthetics in the Sierra Nevada, and envision growing it in groups that range from 0.5 to 10 acres. On poorer sites the California black oak is able to maintain itself, while on better sites treatment is seen as necessary to counter-act the trend toward conifers.

Litton (1980), writing from the perspective of a well-known landscape architect, long active in considering scenic landscape planning methodologies, and who is affiliated with the University of California at Berkeley and with the Pacific Southwest Forest and Range Experiment Station of the United States Forest Service, stresses the oaks as being important visual elements in the California landscape. Statewide distribution, historic association, and variety of visual roles played are cited to suggest their significance. Further, much of the character of the local and regional landscape is identified with different species of oak and their visual arrangements. It is strongly contended that conservation of that character helps to maintain scenic amenity. Litton (1980, p. 162) notes that

> the maintenance of local or regional landscape character built upon native oaks represents an extremely important and more effective means of conserving scenic amenity than any attempt to provide a man-made substitute. The original landscape of an area is normally an extensive combination of regionally typical landforms, vegetation patterns, and water forms. To this will be added new land use patterns, either urban or rural; but in usual situations no overall man-made substitute takes the place of the earlier landscape. Physical signs of human activities, including clearing, grading, construction and new planting are more apt to be seen as fragmenting rather than integrating. The use of introduced trees is not automatically condemned, because it can be both efficient and attractive; but the sensitive use of indigenous oaks, along with their perpetuation in place is an important means of integrating new development with the broader surrounding landscape.

Oaks are discussed by Litton in regard to highways, urban forests, and recreation needs; and a plea is registered for landscape research on native oaks. Such research would address how to maintain oaks in urban places, whether forms of recreational silviculture can be developed for perpetuation of oak woods, how oaks are commonly appreciated in certain California landscapes, and whether assembly and dissemination of current information about oaks would be useful to citizens' groups, urban planning departments, environmental planners, and landscape architects. In the case of recreational silviculture—that silviculture suited to landscape resources—experimentation is viewed as best starting in rural and wildland areas. Bringing together the four professional disciplines of the plant ecologist, forester, park administrator, and landscape architect is seen as necessary (Litton, 1980, p. 166).

> The social scientist is called upon to determine public values and also to communicate with a public which can be hostile to misunderstood manipulations in park landscapes. It takes time for an interdisciplinary research and planning team such as this to integrate their varied capabilities and to identify goals. Silvicultural manipulations take time to show effects. A long-term policy is needed to assure that the undertaking is not scuttled by neglect or adoption of new, short-view policies.

With land development seriously encroaching on oak woodlands in Southern California, Rogers (1980) explored measures to offset the negative effects of urban development on oak trees. Particularly identified are effects of construction on oak trees with regard to loss of air, water, nutrients, roots, and/or top foliage. Rogers sees solutions centering on drip-line preservation during construction, special procedures for grading cuts and fills around oak trees, the use of specialized soil/air drainage systems, and compensatory pruning and fertilization of trees that are impacted. He further regards implementation of a complete program as needed to protect the oaks, and this includes presite planning, horticultural evaluation, and data collection design, as well as preparation of trees prior to, during, and after construction around them.

A symposium on living with the chaparral sponsored by the Sierra Club, the California Division for Forestry, and the United States Forest Service was held at the University of California at Riverside in 1973. Topics discussed included California wildlife and fire ecology, hydrology of chaparral soils, fuelbreaks and type conversions, the impact of fire suppression on Southern California conifer forests, and the effects of wastewater on chaparral. Also covered were the role of herbicides in chaparral management, fire and chaparral prior to European settlement, a history of forest fire control in Southern California, the consequences for human life of managing or not managing chaparral, and public concern over the consequences of type conversion. In a paper dealing with a decision-analysis approach, D. Warner North, a member of the Decision Analysis Group at the Stanford Research Institute, reported on the application of decision analysis to the problem of fire protection for a wildland area (North, 1974). A contract for such investigation was made by the Forest Service, with initial effort turned toward a representative area in Southern California, a 65,000-acre area of watershed covered with old, heavy chaparral vegetation. Most of the area had not had a major fire since 1932; the Forest Service officials regarded it as one of their most serious fire protection problems.

Analysis focused on fuel management aspects of fire policy, with an objective being to demonstrate the use of decision-analysis methods in developing a comprehensive economic assessment of fire policy alternatives.

The decision-analysis model had two parts. A conflagration model provided the extent of large fires occurring over time for a given fire protection policy, "conflagration" being used to mean fires covering 1000 acres or more. A damage-assessment model determined the losses resulting from these fires in monetary terms. This included watershed damage and direct damage to structures and other improvements, and effects of wildlife, aesthetics, recreation, and so on.

Three fuel management alternatives were looked at in the representative area: no fuel modification; the Forest Service plan for a system of fuelbreaks; and a third possibility, in which fuelbreaks would be expanded from several hundred feet to a width of approximately 0.5 mile. (The rationale for such expanded fuelbreaks was that they would increase the chance of stopping a fire under severe wind conditions; the feasibility of expanded fuelbreaks was not investigated in detail, the alternative being included to determine whether fuel management on this scale could be "economically justified.") A conclusion reached, while many aspects required more refinement, was that extensive fuel modification appears to be economically justified in the area concerned. Indications were that conventional fuelbreaks return benefits well in excess of their costs.

Following this study in the initial representative area of the Matilija and San Antonio Creek drainages in the Los Padres National Forest, near Ojai, decision analysis was to be applied to a new representative area in the Santa Monica Mountains. The investigators' hope was to "contribute to rational planning in protecting Southern California from wildland fires and to establish on an even broader basis the advantages of using decision analysis in forest management" (North, 1974).

It would appear that this report and the method used dent some popularly held perceptions about the present value of fuel modification costs and the expected present value of fire damage and of suppression costs under the three regimes of no fuel modification, conventional fuelbreak system, and expanded fuelbreak system. It should be expanded to other alternatives, including multilevel, multipurpose vegetation patterns investigated in other Mediterranean-type climates which are described below.

In concluding, we cite from the opening notes by T. L. Hanes, one of the most actively engaged scientists in ecological research in the chaparral (Hanes, 1974, p. 5). "Chaparral needs periodic fire to ensure its vitality. Also, fire in the chaparral is inevitable. Fire exclusion from the chaparral does not prevent fire, it only forestalls it. A better understanding of the role of fire in the ecology of the chaparral is imperative if man is to live with the chaparral."

Nature Parks and Reserves: Requirements and Opportunities

From the list of recognized protected areas of IUCN, it is evident that in some countries in the SFZ, especially in the Levant, there are no protected areas at all and there is apparently little interest in establishing them. In others there are only a few and none in the SFZ proper. Thus in Chile, for instance, Rundel and

Weisser (1975) stressed the importance of the new Campana Park in central Chile for the protection of mediterranean sclerophyll vegetation types.

Also, several important mediterranean ecosystems in California are not sufficiently protected; for example, coastal sage dwarf-shrub land, which has been reduced to 10-15% of its original size and is suffering from photochemical oxidant air pollution in the Los Angeles smog basin. Of great historical and biogeographical importance as the center of evolution of southeastern vertebrata are the *Quercus douglasii* and other oak woodlands in the coastal foothills (J. Davis, personal communication). These, as already mentioned, are under the greatest pressures from urbanization and are being rapidly lost. At the same time, the remaining ones are undergoing undesirable changes due to clear-cutting. The preservation of these diverse, attractive, accessible mediterranean landscapes in California is further endangered by the failure of these oaks to reproduce; and there is not one sufficiently large nature reserve in which conservation management is carried out to solve these problems. The Hastings Natural History Reserve of the University of California Natural Land and Water Reserve System (see below) is the only protected area in the central and southern foothills, including the major evergreen sclerophyll vegetation types and oak woodlands, in which these problems could be studied more intensively if sufficient funds and staff were available.

In the Mediterranean, Italy is of greatest biogeographical importance as a bridge between Europe and Africa and as an important south–north and east–west route of bird immigration. Lovari and Cassola (1975) and Cassola and Lovari (1976) pleaded for the protection of the few wetlands remaining and of the unique relic populations of Mediterrean plants and animals from the Quaternary Ice Age, localized in isolated mountain and island habitats. Also of great importance in this respect is Sardinia, for which Cassola and Tassi (1973) prepared a detailed proposal of natural areas requiring protection. The final establishment of the Gennargentu National Park as the largest wilderness area where the last Griffon vultures, mouflons, and monk seals still live, and which also includes the above-mentioned oldest Supremonte *Q. ilex* forest, is of greatest priority in Italy (F. Cassola, personal communication). Unfortunately, competence disputes between the State and the regional governments do not favor a rapid and satisfactory solution to these problems, and these natural ecosystems and their plants and animals are not represented either in the parliaments or in the cabinets of the State or the region.

For scientific purposes, and especially for a better understanding of the structure and function of these ecosystems plus an assessment of their "life-supporting" free services, not only unique and rare but *representative* sites of sufficient size are most important. In the Mediterranean SFZ these can be ensured only within protected areas and nature reserves (and even here one often encounters unpleasant suprises). Also, in California, such greatly needed study areas (see below) are rapidly vanishing even in the extensive and more remote wildlands. A detailed description of research needs on natural areas in California has been provided by Cheatham et al. (1977). For this purpose the University of California has initiated a special program of Natural Land and Water Reserves Systems (NLWS), acquiring reserves with nonstate funds, private gifts, and land-use agreements to provide protected aquatic and terrestrial ecosystems for research and study. Under the dynamic leadership of its field representative N.U. Cheatham, these reserves have become models of well-functioning ecosystems for conservation and research and have been specially cited by the Soil Conservation Society of America as "each being a natural plant community management in a wise and judicious manner." Some of these 18 study areas—near campuses—presently serve as the most important outdoor classrooms and living laboratories for ecological courses. By far the greatest (22,680 ha) is the Santa Cruz Island Reserve at the Channel Islands, covered with woodlands, grasslands, and chaparral and containing important endemic and relic plant and animal species. A major menace to conservation efforts on these islands are feral goats. This problem has been only partly coped with by fencing, but recent attempts to find a more efficient eradication program have been stopped because of the opposition of "nature lovers" (who love the feral goats more than the nature that is being destroyed by them) and the public outcry they have raised. Presently, however, none of these or any other reserves are being used for a comprehensive interdisciplinary and integrated ecosystem study, which could provide still lacking information on the structure, function, and dynamics of these sclerophyll ecosystems, as affected by protection, grazing by livestock and/or wildlife, fire, cutting, and so on. This information is urgently needed for the conservation management of these ecosystems and for the breaking down of the ivory towers of discipline-oriented botanical, zoological, pedological and similar research.

Another very significant nongovernmental initiative in California and in the SFZ at large is carried out by an impressive group of 300 professional scientists and dedicated amateurs, sharing common concern for the rapidly vanishing and declining natural and seminatural landscapes of California. This is the *Inventory of California Natural Areas* (Hood, 1977) by the California Natural Areas Coordinating Council. Its purpose is to serve as a guide for a comprehensive and coordinated effort to protect a full sampling of these landscapes "for the use and enjoyment of future generations." It is a first statewide catalog, all inclusive and down to earth, with a sound ecosystem approach and not confused by preconceived succession-to-climax theories. It would therefore serve as a model also for other countries in the zone. It aims to include all natural and man-modified areas from two major categories: Those that are *unique* or of particular scientific and educational interest with rare or endangered species, populations, and geological, pedological, and historical features; and those that are *representative* of the various biotic communities found in California (Table 4-2).

In addition, The Nature Conservancy of the United States prepared, in 1975, an important document for the Department of the Interior on the preservation of natural diversity and recommended a coordinated national program to maintain diversity. It also mediated a national ecological diversity inventory in California. It is hoped that all these efforts will finally be pooled in order to have the greatest possible impact on decision making. Not less important is their coordi-

Table 4-2. Example of Inventory Sheet from Inventory of California National Areas

---

BIG TUJUNGA WASH   190265 BE

Chaparrel, coastal sage scrub, riparian

Los Angeles   34°17' N   118°20' W

Sunland $7\frac{1}{2}'$ T2N R14W

400 ha. (1,000 a.)   340–583 m (1,120–1,912 ft.)

United States Forest Service; Los Angeles County Flood Control District

Big Tujunga Wash is the flood plain, approximately 3 kilometers (2 miles) long and up to 800 meters (2,600 feet) wide, of the intermittent Tujunga Creek. The convergence of desert, chaparral, streamside woodland and freshwater communities provides a unique variety of plant and animal life found nowhere else in Los Angeles County. As occcasional flooding maintains the major part of the Wash as an open community, it is a habitat where natural hybrids and their offspring can compete with their parent species and become established.

The Wash is rich in plant life, including coastal sage scrub, chaparral and riparian, the latter, in fact, a desert-wash community. Among the dozen-plus species of trees found here are the black and Fremont cottonwood, *Populus trichocarpa* and *P. fremontii,* several willows, *Salix hindsiana, S. exigua* and *S. lasiolepis,* Coulter pine, *Pinus coulteri,* and California juniper, *Juniperus californica.* In the chaparral, chamise, *Adenostoma fasciculatum,* scrub oak, *Quercus dumosa, Ceanothus spinosus,* and mountain mahogany, *Cercocarpus betuloides,* are conspicuous. Prominent in the coastal sage are white and black sage, *Salvia apiana* and *S. mellifera,* California sagebrush, *Artemisia californica,* and *Erigonum fasciculatum.* Several cactus, *Opuntia occidentalis, O. parryi* and *O. ficus-indica,* are found here.

Three rare plants, *Chorizanthe leptoceras, Malacothamnus davidsonii,* and *Berberis nevinii,* have been reported from the area; the latter, which is common under cultivation, is known in the wild from only this locality. Some of the tallest known yuccas, *Yucca whipplei,* reaching a height of more than 6 meters (20 feet), are found in the rocky basins of the Wash.

Nearly 200 species of birds have been observed in the area, ranging from such typical water and shore birds as lesser scaup, *Aythya affinis,* redhead, *A. americana,* and dowitcher, *Limnodromus scolapaceus,* to such typical desert inhabitants as the cactus wren, *Campylorhynchus brunneicapillum,* and the roadrunner, *Geococcyx californianus.*

Among the numerous mammals that inhabit the area are the coyote, *Canis latrans,* gray fox, *Urocyon cinereoargenteus,* as well as a variety of rodents. Additionally, there are several desert insects, with disjunct populations, in the area.

Table 4-2. (continued)

Geologically, one of the more interesting features is the 1-meter (3-foot) high earthquake scarp which crosses the area. It was formed by the 1971 San Fernando earthquake. In the foothills to the north are outcrops of the Modelo and Fernando formations, the former an upper Miocene marine sedimentary formation with siliceous and diatomaceous shale, sandstone and siltstones; the latter dates to the Pliocene and is formed of marine sedimentaries, siltstones, conglomerates, and fine sandstones. Both formations are fossiliferous, with marine vertebrates and invertebrates.

Integrity: A portion of the area is diked; upstream are various camps and structures, including a road. The Wash is bounded on the south by a residential community, and a portion of Interstate 210 is slated to cross the area. Additionally, there are quarries on the perimeter and a few off-road-vehicle tracks in the Wash. A portion of the area was burned in 1975.

Use: Research, educational, observational.

---

nation with the activities of the different federal and state agencies that administer wildlands and natural ecosystems, especially the United States Forest Service. This, more than any other single agency, is responsible for California uplands and the remaining wilderness areas, and is presently preparing new directives for the protection, management, and extent of wilderness areas.

The most promising recent development on an international level is the adoption of the Man and Biosphere Project 8 on the establishment of a network of biosphere reserves in the Mediterranean (UNESCO, 1977b, 1979). The aim of these reserves is to conserve the diversity and integrity of biotic communities within natural ecosystems and to safeguard the genetic diversity of species, as well as to provide areas for ecological research and facilities for education and training. In the Mediterranean, these protected areas of land and coastal environments will include representative examples of natural biomes, unique communities and unusual features, examples of harmonious landscapes resulting from traditional land-use patterns, and examples of degraded ecosystems for their restoration. They should be large enough as a conservation unit and have adequate long-term legal protection. The emphasis on conservation of ecological diversity and cultural heritage, together with socioeconomic development, the inclusion of seminatural and degraded ecosystems and their restoration, multiple land use and its combination with research and training, is very much in accord with the approach presented here. It is hoped that these biosphere reserves will be realized as planned, that they will be fully coordinated with other international and local efforts, and that all other countries in the SFZ will also join this project.

The International Unit for Conservation of Nature and Natural Resources (IUCN), a network of governmental and nongovernmental organizations, scientists, and other conservation experts who joined together to promote the protection and sustainable development of living resources, has recently prepared a

"World Conservation Strategy" (IUCN, 1980). This important global document is now complemented by a series of more detailed source books for conservation strategies in the major ecosystems, including the SFZ. Much of the material presented in this chapter will be incorporated in this series and will (it is hoped) also reach those dealing with the practical implementation of some of the suggestions on conservation actions to which the concluding part of this chapter and of this book as a whole will be devoted.

Conclusions: Action Required

The Need for a Holistic Mediterranean Conservation Policy

The primary need in the SFZ is for a new holistic policy for the conservation and restoration of Mediterranean open landscapes as a whole. Its major goal should be the prevention of the accelerated and uncontrolled replacement and degradation of the biosphere by the technosphere along with its natural and seminatural ecotopes, thereby endangering the ecosphere as a whole and mankind as the critical link between bioecosystems and technoecosystems. Policy should be based on well-coordinated simultaneous efforts for better public and professional conservation education, more comprehensive landscape conservation planning and management, and more interdisciplinary, integrated ecosystem-conservation research.

This should be guided by system-oriented landscape ecological determinism aimed at relieving the present heavy and uncontrolled traditional and neotechnological pressures and reconciling the conflicting demands in these landscapes by optimization of their value for long-term and overall benefit by ensuring the following functions:

1. Bioecological functions of natural and seminatural wetlands and upland ecosystems as last refuges for spontaneous organic evolution and as protective and regulative life-supporting and buffering systems for environmental and watershed protection.
2. Socioecological and cultural functions of scenic beauty, solitude, and wilderness and of preservation of unique natural and historical features for future generations, which are endangered now in the SFZ even in protected areas because of mass recreation and tourism pressures.
3. Socioeconomic functions of semiagricultural and agricultural upland ecosystems as a source of food and revenues from plant and animal products and tourism which can be ensured only by rational management on a long-term sustained basis.

A Redbook of Threatened Mediterranean Landscapes

In order to identify the major issues of landscape and ecosystem conservation and their cause and implications in different countries in the SFZ as reflected by the extent and speed of landscape degradation, the issuing of a special *Redbook*

is suggested. It should provide factual updated information on the present conservation status (as demonstrated by case studies on successes and failures) and
should serve as a monitoring and prediction tool regarding the fate of open
mediterranean landscapes and as a basis for conservation measures and decision
making in each country. This book should be not only a scientific document but
also an important tool for public conservation education. It should therefore be
prepared in an appealing and convincing way by documenting and illustrating
the interdependence between the welfare of the population and that of its open
landscapes, in a way similar to that used in the successful *Save the Birds—We
Need Them!* (Stern et al., 1978). It should also be published in the language of
each country concerned and made available to the public, schools, universities,
industries, businesses, politicians, and officials, as well as to foreign visitors and
tourists.

A special Conservation Fund for Mediterranean Landscapes could help in
financing this project and ensure its continuity. If only 10 cents were collected
from 90% of the 100 million visitors to the Mediterranean, directly and indirectly through travel agencies, airlines, shipping companies, local authorities,
hotels, and the like, the annual sum would amount to $9 million.

Interdisciplinary Ecosystem Education for Professionals
and Decision Makers

The success of this holistic conservation policy will depend, to a great extent, on
creating awareness, understanding, and motivation for the broader conceptions
of Mediterranean landscape conservation among all those who deal with these
landscapes at all levels of decision making. Above all, it is vital to bridge the
communication gap and educate a new breed of interdisciplinary landscape managers, replacing the present one-track-minded agronomists, foresters, noninterference–climax protagonists, graziers, recreationalists, economists, and the like.
This can be achieved only with the help of more efficient educational tools, such
as audiovisual aids, mass media, and high-level ecological, educational, and communication expertise, which should reach directly the main body of academic,
professional, practical, and political land-use planners, managers, and decision
makers, including junior staff and technicians. Of great importance in this respect are interdisciplinary workshops and intensive training programs using
environmental simulation games (Vester, 1976), which could help to create a
common pool of concepts and goals based on the principles outlined in
Chapter 2.

A General Conservation Strategy for Protection
and Dynamic Conservation

In order to apply this policy in actual conservation planning and management
strategy, landscapes and their ecosystems should be classified and assessed according to their inherent ecological and economic potential, the degree of
human modification and pressure, and their resulting specific conservation man-

agement requirements for protection, development, and rational utilization. For this purpose, the functional ecosystem classification described in Chapter 2 and the ordination model of landscapes (Figure 2-14) can be used as starting points, to be followed by more detailed classifications, in which special attributes of nature conservation potentials, such as diversity, rarity, endemism, and accessibility, should also be taken into account, as suggested by Goldsmith (1975) and Wright (1977).

Basing Mediterranean conservation strategy on this functional classification, we should distinguish between agricultural ecotopes (derived from agroecosystems), which are suitable for profitable, mechanized rational crop production by cultivation, irrigation, and so on; and all other nontillable and marginal ecotopes, to which our main attention is devoted.

As a baseline for landscape master plans, detailed conservation and protection strategies should be defined for the two major groups of natural and close-to-natural ("subnatural") and seminatural ecotopes:

1. *Natural and close-to-natural ecotopes*, in which both flora and fauna are spontaneous, and ecosystem structure and function have undergone very little human modification. These are very rare in this zone and have remained only on inaccessible sites, such as rock cliffs and steep mountain slopes. Their major conservation object is complete protection from any interference and disturbance of all their natural features within strict nature reserves. These few sites deserve highest priority in conservation measures, on both the international and national levels. Within the proposed *Redbook*, a special report should be given on their present status and on actions required to ensure their full protection in each country.

2. In *seminatural ecotopes* flora and fauna are largely spontaneous, but here too domesticated animals may utilize a small part of primary production. As explained above, their structure and function have undergone far-reaching modifications through intensive and lasting human interferences, but in response to a certain optimal defoliation pressure they have maintained high biological diversity and a dynamic flow equilibrium. These are, however, lost, both by extreme noninterference and excessive overinterference. These ecotopes include the few remaining wetlands, riverside ecotones, and other coupling ecotones, along with aquatic ecosystems and broad-leaved sclerophyll forests and the much more extensive woodlands, shrublands, dwarf-shrub lands, and derived grasslands, which contain most of the natural Mediterranean plant and animal communities in different stages of degradation and regeneration.

The primary conservation object in all remaining wetlands, their ecotones and sclerophyll forests, and sufficiently large representative or unique ecotopes of all other soil–vegetation systems should be the perpetuation of all those natural features, patterns, and processes (including human interference, fire, and grazing) that ensure their structural and functional integrity and dynamic flow equilibrium, and thereby their ecological diversity and genetic resources, as well as their protection and regulation functions. The secondary object should be to fulfill the above-mentioned, mostly intangible, socioecological, aesthetic, and cultural functions, as long as these are not detrimental to the primary object. In contrast to the third category of semiagricultural ecotopes, their object is neither the production of harvestable resources nor the provision of recreational space— although these may be by-products of their conservation management.

These objects can be achieved only within strictly protected and controlled managed natural areas, as defined in the recent IUCN (1978) report. These can be managed nature reserves, nature conservation or wildlife sanctuaries, specially designated and protected parts of national parks, wilderness areas, or protected landscapes. Highest conservation priorities should be given to wetlands, coastal ecosystems and sclerophyll forests, the wetter and higher-elevation ecotones, the drier ecotones of the SFZ, and to sites with unique natural and historical features, including rock and soil types. In view of the almost total lack of scientific conservation–ecosystem research in this zone, specially designated scientific ecosystem research areas should be ensured in the vicinity of these nature reserves or within the biosphere reserves to provide the necessary information for the management strategies of conservation of biological diversity and ecosystem integrity. In addition, until better knowledge is available on the rational utilization of these seminatural ecosystems, larger areas should be protected as resource reserves with minimal interference, especially in regions in which agriculture activities have been abandoned. Within the *Redbook* of threatened Mediterranean landscapes, in addition to inventories on such threatened ecotopes, a special category should be devoted to those landscapes in which traditional Mediterranean upland uses are still intact and in which these seminatural ecotopes deserve special protection as part of natural biotic or anthropological reserves, with controlled minimal interference and more intensively managed agroecosystems (see below).

Multipurpose Management Strategies for Semiagricultural Ecotopes

Semiagricultural ecotopes include all seminatural, nontillable, and marginal ecosystems in the coastal lowlands and islands, especially in the mountainous uplands, which are not specifically mentioned in one of the other protected nature conservation categories and which can be utilized for agriculture and/or recreational purposes. Their major conservation object should be the achievement of maximal long-term combined ecological and economic benefits through controlled and directed manipulation of the soil–plant–animal complex, according to local site potentials and socioeconomic requirements. This can be achieved best within closely interwoven single-purpose and multipurpose land-use patterns with the help of two main ecotechniques:

1. *Ecosystem management* by manipulation and controlled utilization of the existing plant cover, such as cutting, thinning, pruning, chopping, coppicing, burning, grazing, selective chemical control, and, in pastures, fertilizing and reseeding. For conservation management of protected seminatural ecotopes, these techniques can also be partly applied, but only for ecological purposes, and not for agricultural, recreational, or other commodity purposes.

2. *Environmental afforestation* with drought-tolerant (and in many Mediterranean soils, also limestone-tolerant) trees and shrubs, which can fulfill multiple ecological, ornamental, and economic functions. This technique is essential wherever land denudation has reached an advanced stage, as well as in man-created ecotopes, such as roadside slopes, camping grounds, and recreation sites.

By these ecotechniques all these ecotopes can be converted into multilayered recreation forests, woodlands, and parklands. The more open sites can be converted into improved pastures or multiple-use forests, woodlands, and shrublands for maximum utilization compatible with recreation and production of wood and fodder for wildlife and restricted livestock grazing, with greater flexibility in utilization and management.

For decision making on the choice of these options and management strategies, multidimensional, hologram, and cybernetic models can be used in which the mutual influences of all relevant production, protection, and regulation function variables and their compatibility can be assessed and quantified. The simple qualitative models shown above can be transferred into more complex sensitivity matrices and used in computerized, dynamic, rational landscape planning projects described in the previous chapters. With the help of these models, the impacts of forestry, livestock, and recreational uses on protection and regulation functions and on biotic diversity and fire hazard resistance should be carefully weighed. Since the increase of water yields is in general only a by-product of certain (mostly nonconservative) land uses (denuding the slopes of woody and deep-rooted plants), it should not serve as a major land-use goal of these ecotopes in the SFZ.

Multiple-Use Management Areas and Priorities in Research

For multiple-use semiagricultural ecotopes, the newly proposed Multiple-Use Management Areas or Managed Resource Areas (IUCN, 1978) are especially suitable, and this conservation category should be promoted in the SFZ—on both the international and the national level—as a major vehicle for inducing

---

Photos 9-11. Ecology in Action-Integrated Approaches to Land Use and Conservation of Nature. Posters from an exhibit of the MAN and BIOSPHERE-UNESCO Program, inaugurated at the 10th anniversary conference of this program (in September 1981) in Paris. The aims of this international program of research and training are to provide the scientific knowledge and trained personnel needed to manage natural resources in a rational and sustained manner. This exhibit is a contribution to the effort to present scientific and technical information in a form which can be understood and used by planners, decision makers, educators and others. Photos by courtesy of MAB/UNESCO Secretariat, UNESCO, Paris. (The exhibit is obtainable from the MAB Secretariate, UNESCO 7, Place de Fontenoy 75700 Paris, France.)

Opening conservation to man

Is the best way to protect a natural area to seal it off in a "closed jar" from the outside human world? Sooner or later such a policy can destroy the area it was intended to protect. Ecological and sociological pressures - both inside and outside - eventually may shatter the reserve.

MAB emphasizes man's partnership with nature. A reserve is open and interacts with its region. The local people can be its guardians.

Synchronizing scientists and planners

...A TIME TO STUDY, A TIME TO ACT, A DATE TO BE KEPT...

The different time scales of scientists of different disciplines, and of scientists and planners should be harmonized. Remaining flexible and able to react to changing conditions is also necessary to deal with the moving targets of rapidly evolving land use problems.

Timing scientists and planners

Gearing scientists with each other

moving towards moving targets

Photo 10.

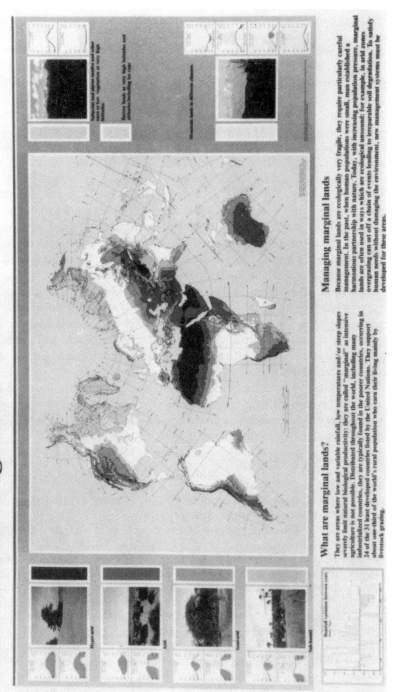

Photo 11.

lasting and far-reaching desirable changes in Mediterranean upland use, especially if they are combined with demonstration, research, and training. They could also become the intensively managed parts of natural forests, parks, and protected landscapes and biosphere reserves.

However, here also there is urgent need for interdisciplinary ecosystem research on the ecological and economic implications of different kinds and intensities of these land uses, and especially of recreational use, and of their combinations, to enable the quantification of above-mentioned models. Foresters should play an important role in applying their rich silvicultural experience to these new approaches, especially in the introduction and establishment of suitable low-flammable fodder and forest trees and shrubs and in the development of dynamic fuel management, including prescribed burning. [Additional subjects for research recommendations can be found in the recent MAB biosphere reserve report by UNESCO (1979).] The Scientific Committee on Problems of the Environment (SCOPE) projects on fire and on ecosystem modeling could play an important role in directing these research activities, but unfortunately only a few Mediterranean countries are members of SCOPE, and in others, like Israel, no funds are available for such studies.

Until funds and manpower for more comprehensive and sophisticated ecosystem studies become available, systematic coordinated field studies with uniform methods for the determination of plant and animal diversity along environmental, anthropogenic, and pyric gradients (Naveh and Whittaker, 1979b) can provide important and very much-needed information for conservation management practices.

In conclusion, the chief actions required in this area are in the realms of interdisciplinary system-oriented landscape-ecological education, planning, management, and research aimed at the conservation and restoration of the seminatural Mediterranean landscapes (see Photos 9-11).

References

Alon, A. 1973. Saving wild flowers in Israel. Biol. Conserv. 5:150-151.

Ambio. 1977. The Mediterranean (Special Issue) 6.

Ashby, E. 1978. Reconciling Man with the Environment. Stanford Univ. Press, Palo Alto, California.

Ashmann, H. A. 1973. Distribution and pecularity of Mediterranean ecosystems. In: F. di Castri and H. A. Mooney (Eds.), Ecological Studies, Analysis and Synthesis, Vol. 7. Springer-Verlag, Berlin, Heidelberg, New York, pp. 11-19.

Ashmann, H., and C. Bahre. 1977. Man's impact on the wild landscape. In: H. A. Mooney (Ed.), Convergent Evolution in Chile and California. Dowden, Hutchinson & Ross, Stroudsburg, Pennsylvania, pp. 73-84.

Bakker, P. A. 1979. Vegetation science and nature conservation. In: M. J. A. Werger (Ed.), The Study of Vegetation. Dr. W. Junk, The Hague, pp. 247-288.

Barbour, M. J., and J. Major (Eds.). 1977. Terrestrial Vegetation of California. Wiley, New York.

Barret, R. H. 1978. The feral hog on the Greek ranch. Hilgardia 46(9):283–355.

Biswell, H. H. 1977. Prescribed fires as a management tool. *In:* H. A. Mooney and C. E. Conrad (Eds.), Symposium on Environmental Consequences of Fire and Fuel Management in Mediterranean Ecosystems, August 1977, Palo Alto, California. USDA FS Gen. Tech. Rep. WO-3 4981, Washington, D.C., pp. 151–162.

Briegeeth, A. M. 1981. Pasture and range areas in Judea and Samaria. *In:* Proceedings of International Symposium on Pastoral Sheep Farming Systems in Intensive Economic Environments. Ministry of Agriculture, Tel Aviv, Israel.

Carp, C. G. 1979. Eine Küste wird verbaut. Stern 33:114–118.

Carp, E. 1977. Preliminary review of the wetlands of international importance in the Mediterranean region. IUCN, Morges (mimeo).

Cassola, F. 1979. Shooting in Italy: The present situation and future perspectives. Biol. Conserv. 12:122–134.

Cassola, F., and E. Tassi. 1973. Proposa per un sistema di parchi e reserve naturali in Sardegna. Bollettino della Societa Sarda de Sinze Naturali 7(12): 1–83.

Cassola, F., and S. Lovari. 1976. Nature conservation in Italy: Proposed national and regional parks and other areas deserving protection. Biol. Conserv. 9: 243–257.

Cheatham, N. H., W. J. Barry, and L. Hood. 1977. Research, natural areas and related areas, programs in California. *In:* M. G. Barbour and J. Major (Eds.), Terrestrial Vegetation of California. Wiley, New York, pp. 75–108.

Cody, M., and H. A. Mooney. 1978. Convergence versus nonconvergence in mediterranean climate ecosystems. Ann. Rev. Ecol. and Systematics 9:265–321.

Connell, J. A., and R. O. Slayter. 1977. Mechanisms of succession in natural communities and their role in community stability and organization. Am. Natur. 111:1119–1144.

Di Castri, F. 1973. Climatographical comparisons between Chile and the western coast of North America. *In:* F. di Castri and H. A. Mooney (Eds.), Ecological Studies, Analysis and Synthesis, Vol. 7. Springer-Verlag, Berlin, Heildelberg, New York, pp. 21–36.

Di Castri, F. 1981. Mediterranean-type shrublands of the world. *In:* F. di Castri, D. W. Goodall, and R. C. Specht (Eds.), Mediterranean-Type Shrublands; Ecosystems of the World, Vol. 11. Elsevier, Amsterdam, pp. 1–52.

Di Castri, F., D. W. Goodall, and R. C. Specht (Eds.). 1981. Mediterranean-Type Shrublands; Ecosystems of the World, Vol. 11. Elsevier, Amsterdam.

Di Castri, F., and H. A. Mooney (Eds.). 1973. Mediterranean Type Ecosystems, Origin and Structure; Ecological Studies, Analysis and Synthesis, Vol. 7. Springer-Verlag, Berlin, Heidelberg, New York.

Derman, A. D., and Z. Naveh. 1978. Fire analysis in land use planning in the Mediterranean region. J. Environ. Mgt. 7:253–260.

ECE/FAO. 1979. General review of the forest and forest products sector in south Europe, including trends and prospects for consumption and supply of wood and other forest products. Economic Commission for Europe, Timber Committee, ad hoc Meeting, Lisbon, September 1979 (mimeo).

Egler, F. E. 1954. Vegetation science concepts, I: Initial floristic composition— a factor in old field development. Vegetatio 4:412–417.

Egler, F. E. 1970. The Way of Science; A Philosophy of Ecology for the Layman. Hafner, New York.

Etienne, M. 1976. Etude sur les conditions d'amélioration des parcours dans les vallées du Golo (Corse). Group d'Etude de Herbages méditerranéens, Avril 1976, Paris.

FAO. 1965. Report on Goat Grazing Policies in the Mediterranean and Near East Regions. Expanded Program of Technical Assistance. FAO Rep. No. 1929, Rome.

FAO. 1977a. Production Year Book 1977, Vol. 31. FAO, Rome.

FAO. 1977b. Technical Consultation on Forest Fires in the Mediterranean Region. May 1977, St. Maximiene, France. FAO, Rome (mimeo).

French, M. H. 1970. Observations on the goat. FAO Agr. Studies No. 80 (Rome).

Gill, A. M., R. H. Groves, and I. R. Noble. 1981. Fire and the Australian Biota. Australian Acad. Sci., Canberra.

Glickson, A. 1970. Planned Regional Settlement Projects. Keter, Jerusalem.

Goldsmith, F. B. 1975. The evaluation of ecological resources in the countryside for conservation purposes. Biol. Conserv. 8:89–96.

Haas, G. 1966. On the vertebrate fauna of the Lower Pleistocene site Ubeidiya. Israel Academy of Science and Humanities, Jerusalem, pp. 33–68.

Hanes, T. L. 1971. Succession after fire in the chaparral of Southern California. Ecol. Monogr. 41:27–52.

Hanes, T. L. 1976. Vegetation called chaparral. Opening notes, Symposium on Living with the Chaparral, pp. 1–5.

Hardin, G. 1968. The tragedy of the commons. Science 162:1243–1248.

Henry, P. M. 1977. The Mediterranean: A threatened micro-cosmos. Ambio 6: 300–307.

Hill, M. 1972. Evaluation of Sites for Power Location. Center for Urban and Regional Studies, Israel Institute of Technology (Technion), Haifa, Israel (in Hebrew).

Holland, V. L. 1980. Effects of blue oak on rangeland forage production in central California. In: Proceedings of Symposium on the Ecology, Management and Utilization of California Oaks, Claremont, California. June 1979, USDA FS Gen. Tech. Rep. PSW-44, pp. 314–318.

Holland, V. L., and J. Morton. 1980. Effect of blue oak on nutritional quality of rangeland forage in central California. In: Proceedings of the Symposium on the Ecology, Management, and Utilization of California Oaks, Claremont, California, June 1979, USDA FS Gen. Tech. Rep. PSW-44, pp. 319–323.

Horvat, L., V. Glava, and H. Ellenberg. 1974. Vegetation Südosteuropas. G. Fischer Verlag, Stuttgart.

Hood, L. (Ed.). 1977. Inventory of California Natural Areas. California Natural Areas Coordinating Council, Sonoma, California (mimeo).

IUCN. 1978. Categories, objectives and criteria for protected areas. A final report prepared by the Committee on Criteria and Nomenclature, Commission on National Parks and Protected Areas. IUCN, Gland, Switzerland.

IUCN. 1980. World Conservation Strategy. Living Resource Conservation for Sustainable Development. IUCN, Gland, Switzerland.

Jenny, H. 1961. Derivation of state factor equation of soils and ecosystems. Soil Sci. Soc. Proc. 1961: 385–388.

Kessel, S. R. 1979. Gradient Modeling–Resource and Fire Management. Springer-Verlag, New York, Heidelberg, Berlin.

Kessel, S. R., and P. J. Cattelino. 1978. Evaluation of a fire behavior information integration system for Southern California chaparral wildlands. Env. Mgt. 2: 125–159.

Köppen, W. 1923. Die Klimate der Erde. De Gruyter, Berlin, Leipzig.

Le Houerou, H. N. 1973. Fire and vegetation in the Mediterranean Basin. *In:* Proceedings of the 13th Annual Tall Timber Fire Ecology Conference, Tallahassee, Florida, pp. 237–277.

Le Houerou, H. N. 1978. Plant and ecology applied to grazing land research, survey and management in the Mediterranean Basin. *In:* W. Krause (Ed.), Application of Vegetation Science to Grassland Husbandry. Dr. Junk, The Hague, pp. 217–274.

Le Houerou, H. N. 1981. Impact of man and his animals on the mediterranean vegetation. *In:* F. di Castri, D. W. Goodall, and R. C. Specht (Eds.), Mediterranean-Type Shrublands; Ecosystems of the World, Vol. 11. Elsevier, Amsterdam.

Leopold, A. S. 1978. Wildlife and forest practice. *In:* H. P. Browkow (Ed.), Wildlife in America. Council of Environmental Quality, U.S. Govt. Printing Office, Washington, D.C., pp. 108–120.

Liacos, L. A. 1973. Present studies and history of burning in Greece. *In:* Proceedings of the 13th Annual Tall Timber Fire Ecology Conference, Tallahassee, Florida, pp. 65–96.

Liacos, L. A. 1977. Fire management in pine forest and evergreen brushland ecosystems in Greece. *In:* Symposium on the Environmental Consequences of Fire and Fuel Management in Mediterranean Ecosystems, August 1977, Palo Alto, California. USDA FS Gen. Tech. Rep. WO-3, Washington, D.C., pp. 289–298.

Lieth, H., and R. H. Whittaker (Eds.). 1975. Primary Productivity of the Biosphere. Springer-Verlag, New York, Heidelberg, Berlin.

Lillywhite, H. B. 1977. Effect of chaparral conversion on small vertebrata in South California. Biol. Conserv. 11:171–184.

Litton, R. B. 1980. Oaks and the California Landscape. *In:* Proceedings of the Symposium on the Ecology, Management and Utilization of California Oaks, Claremont, California, June 1979. USDA FS Gen. Tech. Rep. PSW-44, pp. 161–166.

Lossaint, P., and M. Rapp. 1973. Soil vegetation relationships in Mediterranean ecosystems of southern France. *In:* F. di Castri and H. A. Mooney (Eds.), Mediterranean Type Ecosystems: Origin and Structure. Ecological Studies: Analysis and Synthesis, Vol. 7. Springer-Verlag, Berlin, Heidelberg, New York, pp. 199–210.

Lovari, S. 1975. A partridge in danger. Oryx 13:203–204.

Lovari, S., and F. Cassola. 1975. Nature conservation in Italy: The existing natural parks and other protected areas. Biol. Conserv. 8:127–141.

Luke, R. H., and A. G. McArthur. 1977. Bush Fires in Australia. Australian Govt. Publ. Service, Canberra.

Meigs, P. 1964. Classification and occurrence of mediterranean-type dry climates. *In:* Land Use in Semi-Arid Mediterranean Climates. UNESCO, Paris, pp. 17–21.

Mendelsohn, H. 1972. Ecological effects of chemical control of rodents and jackals in Israel. *In:* T. Farvar and J. P. Milton (Eds.). The Careless Technology. Natural History Press, New York, pp. 527–544.

Miller, P. R., and J. R. McBride. 1975. Effects of air pollutants on forests. *In:* J. B. Mudd and T. T. Kozlowski (Eds.), Responses of Plants to Air Pollution. Academic Press, New York, pp. 195–275.

Mooney, H. A. (Ed.). 1977a. Convergent Evolution in Chile and California Mediterranean Climate Ecosystems. Dowden, Hutchinson & Ross, Stroudsburg, Pennsylvania.

Mooney, H. A. 1977b. The carbon cycle in Mediterranean climate evergreen communities. *In:* H. A. Mooney and C. E. Conrad (Eds.), Proceedings of the Symposium on the Environmental Consequences of Fire and Fuel Management in Mediterranean Ecosystems, August 1977, Palo Alto, California. USDA FS Gen. Tech. Rep. WO-3, Washington, D.C., pp. 107–115.

Mooney, H. A., and C. E. Conrad (Eds.). 1977. Proceedings of the Symposium on the Environmental Consequences of Fire and Fuel Management in Mediterranean Ecosystems, August 1977, Palo Alto, California. USDA FS Gen. Tech. Rep. WO-3, Washington, D.C.

Morandini, R. 1977. Problems of conservation, management and regeneration of Mediterranean forests: Research priorities. *In:* Mediterranean Forest and Maquis: Ecology, Conservation and Management, MAB Tech. Notes 2, UNESCO, Paris, pp. 73–39.

Morzer Bruyns, M. F. 1972. Management of wetlands. *In:* IUCN 12th Technical Meeting, Banff, Alberta, Canada, September 1972, IUCN, Morges, pp. 277–293.

Muller, C. H., R. B. Hanawalt, and J. K. McPherson. 1968. Allelopathic control of herb growth in the fire cycle of California chaparral. Bull. Tory Bot. Club 95:225–231.

Murphy, A. H. 1980. Oak trees and livestock—management options. *In:* Proceedings of the Symposium on the Ecology, Management and Utilization of California Oaks, Claremont, California, June 1979. USDA FS Gen. Tech. Rep. PSW-44, pp. 329–332.

Naveh, Z. 1960. The ecology of chamise (*Adenostoma fasciculatum*) as affected by its toxic leachates. *In:* Proceedings of the A.A.A.S. Ecology Section, Eugene, Oregon, June 1960, pp. 56–57.

Naveh, Z. 1967. Mediterranean ecosystems and vegetation types in California and Israel. Ecology 48:445–459.

Naveh, Z. 1970. Effect of integrated ecosystem management on productivity of a degraded Mediterranean hill pasture in Israel. *In:* Proceedings of the 11th International Grassland Congress, Brisbane, Australia, 1970, pp. 59–63.

Naveh, Z. 1973. The ecology of fire. *In:* Proceedings of the Annual Tall Timbers Fire Ecology Conference, Tallahassee, Florida, March 1973, pp. 131–170.

Naveh, Z. 1974a. The ecological management of non-arable Mediterranean upland. J. Environ. Mgt. 2:351–371.

Naveh, Z. 1974b. Effects of fire in the Mediterranean region. *In:* T. T. Kozlowski and C. E. Ahlgren (Eds.), Fire and Ecosystems. Academic Press, New York, pp. 401–434.

Naveh, Z. 1975. Degradation and rehabilitation of Mediterranean landscapes. Landscape Planning 2:133–146.

Naveh, Z. 1977. The role of fire in the Mediterranean landscape of Israel. *In:* H. A. Mooney and C. E. Conrad (Eds.), Proceedings of the Symposium on the Environmental Consequences of Fire and Fuel Management in Mediterranean Ecosystems, August 1977, Palo Alto, California. USDA FS Gen. Tech. Rep. WO-3, Washington, D.C., pp. 299–306.

Naveh, Z. 1978. A model of multi-purpose ecosystem management for degraded Mediterranean uplands. Environ. Mgt. 2:31–37.

Naveh, Z. 1979. A model of multiple-use management strategies of marginal and untillable Mediterranean upland ecosystems. *In:* J. Cairns, G. P. Patil, and W. E. Waters (Eds.), Environmental Biomonitoring, Assessment, Prediction and Management—Certain Case Studies and Related Quantitative Issues. International Co-operative Publ. House, Fairland, Maryland, pp. 269–286.

Naveh, Z. 1982. Mediterranean landscape evolution and degradation as multivariate bio-functions—theoretical and practical implications. Landscape Planning 9:125–146.

Naveh, Z., and J. Dan. 1973. The human degradation of Mediterranean landscapes in Israel. *In:* F. di Castri and H. A. Mooney (Eds.), Mediterranean-Type Ecosystems: Origin and Structure. Ecological Studies: Analysis and Synthesis, Vol. 7, Springer-Verlag, Berlin, Heidelberg, New York, pp. 373–390.

Naveh, Z., and R. H. Whittaker. 1979a. Structural and floristic diversity of shrublands and woodlands in northern Israel and other Mediterranean areas. Vegetatio 41:171–190.

Naveh, Z., and R. H. Whittaker. 1979b. Measurement and relationships of plant species diversity in Mediterranean shrublands and woodlands. *In:* J. F. Grassle, G. P. Patil, and C. Taillie (Eds.), Ecological Diversity in Theory and Practice. International Co-operative Publ. House, Fairland, Maryland, pp. 219–239.

Naveh, Z., E. Steinberger, and S. Chaim. 1981. Photochemical air pollutants—a new threat to Mediterranean conifer forests and upland ecosystems. Environ. Conserv. 7:301–309.

Nicholson, M. 1974. Forestry and conservation. Environ. Conserv. 1:83–86.

Noble, I. R., and R. O. Slatyer. 1977. Post fire succession of plants in mediterranean ecosystems. *In:* H. A. Mooney and C. E. Conrad (Eds.), Proceedings of the Symposium on the Environmental Consequences of Fire and Fuel Management in Mediterranean Ecosystems, August 1977, Palo Alto, California, USDA FS Gen. Tech. Rep. WO-3, Washington, D.C., pp. 27–38.

North, D. W. 1974. A decision-analysis approach. *In:* Proceedings of Symposium on Living with the Chaparral. Sierra Club, San Francisco, California.

Parsons, D. J. 1977. Preservation of fire-type ecosystems. *In:* H. A. Mooney and C. E. Conrad (Eds.), Proceedings of the Symposium on the Environmental Consequences of Fire and Fuel Management in Mediterranean Ecosystems, August 1977, Palo Alto, California, USDA FS Gen. Tech. Rep. WO-3, Washington, D.C., pp. 172–182.

Paz, U. 1975. Nature protection in Israel—research and survey, Rep. 1. Israel Nature Reserves Authority, Tel Aviv, pp. 116–206.

Paz, U. 1981. Nature Reserves in Israel. Massada, Israel, 2 volumes (in Hebrew).

Polunin, N., and A. Huxley. 1966. Flowers of the Mediterranean. Houghton Mifflin, Boston.

Plumb, R. T. 1980. (Tech. Co-ord.) Proceedings of the Symposium on the Ecology, Management and Utilization of California Oaks, June 1979, Claremont, California, USDA FS Gen. Tech. Rep. PSW-44, Pacific Southwest Forest and Range Exp. Station, Berkeley, California.

Pronos, J., D. R. Vogler, and R. S. Smith, Jr. 1978. An evaluation of ozone injury to pines in the southern Sierra Nevada. USDA FS, Region 5, Rep. No. 78-1, San Francisco, California.

Quezel, P. 1977. Forests of the Mediterranean basin. *In:* MAB Tech. Notes 2, UNESCO, Paris, pp. 9-32.

Raven, P. H., and D. I. Axelrod. 1978. Origin and Relationships of the California Flora. University of California Publications in Botany, Vol. 72. Univ. of California Press, Berkeley, California.

Renzoni, A. 1974. The wild cat (*Felix sylvestris*) in Italy. Is the end in sight? Biol. Conserv. 6:11.

Riggs, D. S. 1977. How models of feedback systems can help the practical biologist. *In:* S. Levin (Ed.), Lecture Notes in Biomathematics, Vol. 13. Springer-Verlag, Berlin, Heidelberg, New York, pp. 175-205.

Robertson, J. M. S. 1977. Land use planning of the French mediterranean region. *In:* F. di Castri and H. A. Mooney (Eds.), Symposium on the Environmental Consequences of Fire and Fuel Management in Mediterranean Ecosystems, August 1977, Palo Alto, California, USDA FS Gen. Tech. Rep. WO-3, Washington, D.C., pp. 293-295.

Rogers, P. A. 1980. Measures that can help offset the detrimental effects that urban development has on oak trees. *In:* Proceedings of the Symposium on the Ecology, Management and Utilization of California Oaks, June 1979, Claremont, California, USDA FS Gen. Tech. Rep. PSW-44, pp. 167-170.

Rundel, P. W. 1977. Water balance in Mediterranean sclerophyll ecosystems. *In:* H. A. Mooney and C. E. Conrad (Eds.), Symposium on the Environmental Consequences of Fire and Fuel Management in Mediterranean Ecosystems, August 1977, Palo Alto, California, USDA FS Gen. Tech. Rep. WO-3, Washington, D.C., pp. 95-106.

Rundel, P. W., and P. J. Weisser. 1975. La Campana, a new national park in central Chile. Biol. Conserv. 8:35-46.

Rosenthal, M. (Ed.). 1974. Symposium on Living with the Chaparral. University of California, Riverside, March 30-31, 1973. Sierra Club, San Francisco, California.

Smith, R. L. 1976. Ecological genesis of endangered species: The philosophy of preservation. Ann. Rev. Ecol. and Systematics 7:33-55.

Stekelis, M. 1966. Archeological Excavations at Ubeidiya. Israel Academy of Science and Humanities, Jerusalem, pp. 1-32.

Stern, H., G. Thielke, F. Vester, and R. Schreiber. 1978. Rettet die Vögel-wir brauchen sie. Herbig, Munchen, Berlin.

Stern, H., W. Schroder, F. Vester, and W. Dietzen. 1980. Rettet die Wildtiere Pro Natur Verlag, Stuttgart.

Susmel, L. 1973. Development of present problems of forest fire control in the Mediterranean region. FAO, Rome.

Susmel, L., F.Viola, and G. Bassato. 1976. Ecologie della leccata del supramonte del supramonte di Orgosolo. Annali del Centro di Economia Montana delle Venezie, Vol. X, 1969-1970, Padova.

Szwarcboim (Shavit), I. 1978. Autecology of *Pistacia lentiscus*. Ph.D. Thesis, Israel Institute of Technology (Technion), Haifa, Israel (in Hebrew with English summary).

Tangi, M. 1977. Tourism and the environment. Ambio 6:336–341.

Tappeiner, J., and P. M. McDonald. 1980. Preliminary recommendations for managing California black oak in the Sierra Nevada. *In:* Proceedings of the Symposium on Ecology, Management and Utilization of California Oaks, June 1979, Claremont, California, USDA FS Gen. Tech. Rep. PSW-44.

Thrower, J. W., and D. E. Bradbury. 1973. The physiography of the Mediterranean landscape with special emphasis on California and Chile. *In:* F. di Castri and H. A. Mooney (Eds.), Mediterranean Type Ecosystems: Origin and Structure. Ecological Studies: Analysis and Synthesis, Vol. 7. Springer-Verlag, Berlin, Heidelberg, New York, pp. 37–52.

Tomaselli, R. 1977. Degradation of the Mediterranean maquis. *In:* Mediterranean Forests and Maquis: Ecology, Conservation and Management, MAB Tech. Notes 2, UNESCO, Paris.

Trabaud, L. 1981. Man and fire impacts on Mediterranean vegetation. *In:* F. di Castri, D. W. Goodall, and R. C. Specht (Eds.), Mediterranean-Type Shrublands; Ecosystems of the World, Vol. 11. Elsevier, Amsterdam.

Tüxen, R. 1956. Die heutige potentielle natürliche Vegetationskartierung. Angewandte Pflanzensoziologie 13:5–42.

UNESCO. 1963. Bioclimatological Map of the Mediterranean Zone. Arid Zone Research, UNESCO, Paris.

UNESCO. 1973. Programme on man and the biosphere (MAB). Expert panel on project 8: Conservation of the natural areas and of the genetic material they contain. MAB Rep. Ser. No. 12, UNESCO, Paris.

UNESCO. 1977a. Mediterranean Forests and Maquis: Ecology, Conservation and Management. MAB Tech. Notes 2, UNESCO, Paris.

UNESCO. 1977b. Programme on man and the biosphere (MAB). Regional meeting on integrated research and conservation activities in the northern Mediterranean countries. Final Rep., MAB Rep. Ser. No. 36, UNESCO, Paris.

UNESCO. 1979. Programme on man and the biosphere (MAB). Workshop on biosphere reserves in the Mediterranean region: Development of a conceptual basis and a plan for the establishment of a regional network. Final Rep., MAB Rep. Ser. No. 45, UNESCO, Paris.

Vester, F. 1976. Urban Systems in Crisis. Understanding and Planning of Human Living Space: The Biocybernetic Approach. Deutscher Verlag, Stuttgart.

Viedma, M. D., D. L. Leon, and P. R. Coronado. 1976. Nature conservation in Spain: A brief account. Biol. Conserv. 9:181–190.

Walter, H. 1977. Effects of fire on wildlife communities. *In:* H. A. Mooney and C. E. Conrad (Eds.), Symposium on the Environmental Consequences of Fire and Fuel Management in Mediterranean Ecosystems, August 1977, Palo Alto, California, USDA FS Gen. Tech. Rep. WO-3, Washington, D.C., pp. 183–192.

Walter, H. 1968. Die Vegetation der Erde, Bd. 2: Die gemässigten und arktischen Zonen. G. Fischer, Jena.

Walter, H. 1973. Vegetation of the Earth in Relation to Climate and the Ecophysiological Conditions. The English Universities Press Ltd., London.

Westman, W. E. 1977. How much are nature's services worth? Science 197: 960-964.

Westman, W. E. 1981. Diversity relations and succession in California Coastal Sage Scrub. Ecology 62:170-184.

Wright, D. F. 1977. A site evaluation scheme for use in the assessment of potential nature reserves. Biol. Conserv. 11:293-305.

Ziani, P. 1965. Principles and practices of goat raising techniques in Mediterranean countries. FAO Rep. No. 1929, Rome.

Zohary, D. 1960. Studies on the origin of cultivated barley. Bull. Res. Council Israel Sec. D, 9B:21-42.

Zohary, D. 1969. The progenitors of wheat and barley in relation to domestication and agricultural dispersal in the Old World. *In:* P. J. Uko and G. W. Dimpley (Eds.), The Domestication of Plants and Animals. Aldine, Chicago, Illinois, pp. 47-66.

Zohary, M. 1962. Plant life in Palestine. Roland Press, New York.

Zohary, M. 1973. Geobotanical Foundations of the Middle East. G. Fischer, Stuttgart.

Dynamic Conservation Management of Mediterranean Landscapes

Introduction

The update to this chapter could be summarized as the "bad and good news": the bad news is that the environment and landscapes of the Mediterranean sclerophyll forest zone (SFZ) have further deteriorated. The good news is that there is growing public awareness of these perils. There are more research activities and conservation management policies; and important international and national initiatives are instilling hope that at least in some of these countries this situation can be changed for the better before it is too late.

Continued Threats of Mediterranean Landscapes

The traditional and modern pressures on the open Mediterranean landscapes continue in SFZ countries. These pressures were described in Chapter 4 as neotechnological landscape degradation driven by exponential population growth, uncontrolled urbanization, and tourism.

This is especially true for the Mediterranean basin: according to the scenarios of the Blue Plan (1988), the total population around the Mediterranean basin will grow from 360 million to 520–570 million by 2025. Nearly two-thirds of this population will be in countries south and east of the basin, from Morocco to Turkey, where these populations will be almost five times more than in 1950. The main threats are in the coastal regions with the largest urban-industrial concentrations and the fastest-growing tourism. Here, population will increase from 140 million at present to 195–217 million by 2025, an increase of 45–62%. Its urban population alone will grow from 82 to 145–170 million. In addition to 170 existing energy and industrial installations of oil ports, refineries, and thermal power plants, 32 are in the planning stage and many more will be required by 2025, in addition to cement plants, steel works, fertilizer plants, etc. At the same time, tourist flow to the Mediterranean coast will increase (assuming the present annual growth rate of 2.2%) from 95 million to 220 million in 2025. According to other scenarios, the number of tourists will almost triple. About 40% of these holiday makers are merely seeking relaxation, sun, sand, and the sea, without caring much about natural and cultural values. Therefore, their accommodation close to the beaches is the main economic and cultural force driving further uncontrolled despoilation of the rapidly diminishing open shorelines and their intrinsic scenic and other "soft" landscape values.

Unfortunately, pressure on the Mediterranean coastline does not mean that pressure has been lifted from the inlands and uplands. The rapid loss of almost all natural sand dune habitats and wetlands and increased water pollution in lowland rivers is combined with a further decline of diversity and stability in the uplands. In California, drastic laws and efficient control measures have considerably curbed air pollution in the urbanized coastal region. However, the alarming rise of SO_2, NO_x, O_3 and acid deposition in the Mediterranean is playing a major role in the progressive decline of forests. This occurs not only in the coastal zone and close to the emission sources, but also farther inland in locations which are reached chiefly by phytotoxic photochemical oxidants, especially ozone (see Chapter 4, pp. 300–302). Thus, in 1987 82% of *Pinus pinea* trees were affected in Sardinia, and 78% of these trees in Tuscany. Other woody species, evergreen oaks, and other broadleaved sclerophyll maquis trees are deteriorating, so that more than 50% of all trees are affected by air pollution (Bussotti et al., 1991, 1992; Ferretti at al., 1992).

The rapid and careless shift from diversified and stable traditional agriculture to intensive large-scale, agro-industrial farming is combined with land abandonment and indiscriminate planting of pine and eucalyptus monocultures, creating large stretches of highly flammable vegetation. These developments have destabilized and despoiled rich and unique landscapes. Unfortunately, such practices are promoted by local governments and are heavily subsidized by the European Community (EC). As in California, these "high-tech" and high-input agricultural practices are increasing the rates of air, water, and soil pollution; soil erosion; salination; and siltation.

A typical example is Extremadura, one of the poorest regions in south Spain (CEPA, 1992). Here, the EC has allocated great sums from its European Structural Funds, in order "to develop and to bring this region into line with the rest of Europe." In reality, however, development has undermined the local economy, and threatens to destroy traditional agricultural practices and to degrade the natural resources and scenic and cultural assets of his region by uprooting many thousands of hectares of Mediterranean woodlands. In addition to the great investments in the lowlands, huge sums have been spent on large-scale afforestation projects, and on so-called "conservation and erosion control" projects. For these, heavy machinery and bulldozers are used to clear existing maquis vegetation and to build roads and terraces. Instead of controlling water run-off and soil erosion, they accelerate these by levelling out large areas and removing the protective natural vegetation cover, turning over the soil, and destroying the upper and richest profiles which had taken hundreds of years to build. As a result, more than 30 tons of soil per hectare are washed away every year, causing severe sedimentation.

We predicted 20 years ago (Naveh and Dan, 1973) that such neo-technological landscape degradation world result, in the more densely populated regions, in only a few islands of degraded open landscapes. Impoverished nature reserves and park islands would be turned—like most of the shores of the Mediterranean Sea and inland waters—into "over-crowded open-door recreational slums."

This "doomsday" scenario is indeed approaching year by year, and is affecting even the most remote mountainous uplands. Vos and Stortelder (1992) documented a rapid loss of organic, cultural, and scenic variety and a vanishing of the richest landscape units in the Solano basin, Tuscany. In their study, similar ecotopes were clustered spatially as recurring patterns of landscape units on maps of 1:50,000. These maps illustrate the changes which have occurred in the last 50 years and which are apt to continue in the next 50 years if no conservation measures are introduced.

As a result of large-scale emigration and agricultural intensification (encouraged by EC policies and subsidized wheat prices), the rich mosaics of coppiced and managed oak and chestnut forests, mountain grasslands, and the terraced mixed vine/olive and crop cultures have been replaced by intensive wheat and grape monocultures, pine plantations, and secondary scrub on abandoned pastures and fields. The fine-grained patterns have broken down into coarser patterns of fewer ecotopes and land units, with impoverished floras and faunas. In addition, the soils of arable slopes, cultivated now by heavy machinery, are suffering from erosion and are losing their fertility. Within 50 years, due to the spread of coniferous forests and secondary successions of scrub and montonous oak forests, five major landscape types will vanish altogether and the biological, cultural, and scenic richness of this region will be further impoverished.

Most severe is the situation in the fastest-growing east and south Mediterranean SFZ countries. Here, the major causes for land degradation and desertification are overgrazing and excessive removal of fuelwood. According to FAO studies, in the year 2000 about 140 million people in some of these countries will suffer a deficit of close to 30 million m^3 wood to supply their needs for cooking and heating. Taking into consideration that the wood production of undegraded forest ranges from 0.5 to 1 m^3 per hectare per annum (see Chapter 4, p. 273), this will affect very large areas and endanger almost all remaining forests and woodlands in these countries.

Threats of Global Climatic Changes

Threats to Mediterranean landscapes will be further aggravated by the predicted global heating and destabilization of rainfall and temperature regimes, resulting from increased levels of atmospheric CO_2 and other radiative important "greenhouse gases." In addition to the destructive impact on shorelines from rising sea levels, and the far-reaching effects on hydrological regimes and on agricultural crops, such climatic changes may have severe repercussions to the biological diversity and stability of inland and upland landscapes.

Research on the possible effects of climate change on SFZ ecosystems in California and their interaction with air pollution and fire frequency were conducted by Westman and Malanson (Westman and Malanson, 1991; Malanson and Westman, 1991).

These studies were based on the global climate models of the Geophysical Fluid Dynamics Laboratory and the Goddard Institute for Space Sciences' model, which predicted for California (assuming a doubling CO_2 concentration), a temperature rise of 4–6.5°C in January and 2–5°C in July, as well as an increase of 0.3–3.9 cm in precipitation in January, and of 0.06–0.6 cm in July, causing an increase in evaporation stress, particularly in the summer months. These predictions imply major shifts in ecotone boundaries and in distribution of all vegetation types, chiefly towards more xerothermic ones. It would also mean the displacement of subalpine forest/meadow by red fir (*Abies magnifica*) and mixed conifer forests of *Pinus ponderosa* and *P. jeffreyi* from lower elevations.

The gradient of elevated ozone levels with increasing temperatures—and of evapotranspiration stress—and their aggravated effects on coastal sagebrush vegetation in southern California could be used as an analog for the kinds of changes expected in this region and in other Mediterranean SFZ landscapes. On the basis of their combined field and growth chamber and simulation modeling, Westman and Malanson (1990) concluded that changes attributed to climatic variations alone were accentuated when the direct effects of climate change, increasing fire intensity, and ozone pollution were combined. These interactions—and the influence of rising CO_2 levels—together with species-specific responses to climate variables and individual species competitive capacities, may make the above-described predictions unrealistic.

Westman and Malanson (1990) made some important recommendations for practical conservation policies:

The opportunistic and resilient exotic annuals which invaded from the Mediterranean region (see Chapter 4, p. 264) could play an important stabilizing role during a period of rapid change by colonizing areas made bare by the death of shrubs less tolerant to new climatic conditions. Therefore they recommended a revision of current policies involving the active removal of these exotics in reserves. In order to ensure the survival of representative examples of major species assemblages through the current array of public parks, they recommended connecting corridors between existing protected areas which would aid in the migration of species to more suitable habitats under changing climate conditions. Expanding the area of parklands which are currently underrepresented will aid in ensuring the growth of new plant assemblages in regions that may currently be of less biotic interest. In view of the uncertain effects of climatic change, they pleaded for greater flexibility in the acquisition and management of parklands.

At the European conference on "Landscape-Ecological Impact of Climate Change" (Boer and de Groot, 1990), the major conclusion reached in a multidisciplinary workshop on the Mediterranean region was that temperature increases of 3–4°C would lead to increasing aridity with potentially far-reaching and wide-ranging environmental effects and mostly disastrous socioeconomic impacts. The aridity may increase even if the annual amount of precipitation remains unchanged or increases slightly, owing to (1) a higher evapotranspiration rate caused by higher mean temperatures, which could lead to an increase in potential evapotranspiration of 400 mm/year; (2) changing amounts and frequency of rainfall; and (3) vegetation and soil degradation processes which will lower the ability of the soil to retain available moisture.

These climatic stresses would even further aggravate the existing processes of landscape desiccation in this region. Considering that desert encroachment into the semi-arid Mediterranean region, induced by the heavy pressures of expontentially growing population, is presently estimated at more than 2% per annum, this would have catastrophic consequences (Le Houerou, 1988).

Assuming a general trend of landscape desiccation in the drier Mediterranean biomes, following climatic destabilization and a rise in summer temperatures and evapotranspiration, this could push the present subhumid biomes, like those of the Carmel seashore and mountain in Israel, as well as that of other Mediterranean mountain landscapes, closer to the much harsher, semi-arid biomes, and may push the latter into more arid semi-desert biomes. Such northward shifts of the desert have occurred apparently several times in the geological history of Israel and elsewhere. Previously, however, the transition periods lasted thousands of years, whereas presently this process could be condensed into one hundred years or even less. Most sensitive to these predicted climatic stresses could be the more demanding Eumediterranean herbaceous and woody plants in sheltered sites and more mesic slopes. These plants make up the evergreen and deciduous forests and maquis which are richest in arboreal species and distinguished by a wealth of rare and endemic herbaceous species.

On the other hand, in the drier xero-thermo-Mediterranean parts, especially in the semi-arid ecotones along the 400-mm isohyets, such as

Mt. Gilboa in Israel (see Chapter 4, p. 267), these harsher conditions and less predictable climatic regimes have favored the development of an inherently more drought-tolerant, xeric, and thermophyllous flora which will probably be better adapted to increasing climatic stresses. As shown recently by Merino and Villar et al. (1993) in the Donoma Park in southwest Spain, this flora could therefore be more resilient and elastic than that from mesic habitats. They also have many woody species in common with more mesic biomes. But, as described in Chapter 4, these had most probably undergone distinct intraspecific, ecotypic differentiation. It can therefore be envisaged that under increasing climatic stresses, these xeric vegetation types of the drier Mediterranean biomes may become the last refuges for Mediterranean plant and animal diversity.

Considerable research efforts have recently been made to study the ecophysiological effects of elevated CO_2 levels under changing moisture regimes, and more detailed regional climatic prediction models have been prepared in California and the other SFZ countries (Moreno and Oechel, (1993a). The outcome of these studies clearly show the chaotic and non linear behavior of atmosphere-biosphere interactions and their complex and mostly synergistic interactions with other environmental stresses, and that our present knowledge is not sufficient to attempt any robust prediction models. This led Naveh (1993) to state that the only certain prediction which can be made is that we will face a period of increasing uncertainty in spatial and temporal climatic trends and their ecological effects. We can also anticipate with great certainty that any increase in climatic stresses will further aggravate the process of overall landscape degradation to alarming degrees.

Naveh reached the following conclusions regarding conservation policies for Mediterranean landscapes:

1. These climatic uncertainties and their coupling with land degradation are an additional compelling reason to conserve and restore the health, integrity, and diversity of as many natural and semi-natural landscapes as possible, as outlined in Chapter 4. This would be the best insurance policy. It would ensure sufficient landscape connectivity and counter fragmenting, pollution, and scenic despoliation. It would provide options for sustainable life-support and for the redundancy of functional groups and vital keystone species, and afford the best chance for the survival of rich biotic communities.

2. Highest priority should be given to the protection of the last, richest plant refuges of the semi-arid ecotones in Israel and elsewhere, to preserve their genetic heritage as "drought reserve banks" for the restocking of more mesic sites undergoing a creeping process of desiccation. Sufficiently large areas with high natural and cultural values, serving as corridors and buffering zones with natural reserves and parks, should be declared as "protected, managed landscapes" or bio-

sphere reserves. For this purpose the recent IUCN publication on protected landscapes by P. H. C. Lucas (1992) is an important guide for policy-makers and planners.

3. There is an urgent need for the diversion of afforestation efforts on marginal uplands from dense and montonous coniferous and eucalyptus plantations to the establishment of parklike multiple layer forests, based on the selection and multiplication of beneficial keystone species from the drier Mediterranean biomes and ecotones. Of special importance are "drought-resistant forest reserves" in which the biological diversity and productivity of the Mediterranean arboreal vegetation and its rich herbaceous understory could be safeguarded.

4. To facilitate the creation of special climate stress refuges in which promising ecotopyes could be multiplied and stored in long-term seedbanks for future restoration projects, highest priority should be given to genetic and ecophysiological research on plant behavior under multiple stresses in combined field, greenhouse, and simulation modelling studies. These studies should be part of a long-term, well-coordinated and integrated Mediterranean research program, embracing all relevant biological and ecological complexity levels and scales from DNA and genotypes to ecosystems and landscapes. In these studies, much greater attention should be given to edaphic and biotic factors, to the root systems, and to their rhizophere and microflora. In landscape ecological studies attention should be given to the dynamics of patterns and processes, and to production, regulation, and carrier functions, as effected by different land-use regimes and under different scenarios of climate changes.

The Mediterranean Action Plan and its Blue Plan

As mentioned above, there are many signs suggesting a growing public awareness of pressing environmental problems and their alarming impact on human societies. These have already been many important initiatives from governmental and nongovernmental agencies to deal with these problems. As an example, we have chosen the Mediterranean Action PLan (MAP), approved in 1975 at a conference called by the United Nations Environmental Program and attended by sixteen Mediterranean countries. Despite geopolitical conflicts and abundant differences among many countries in the Mediterranean basin, the common threat of pollution of the sea brought them together in an effort to prevent its irrevocable degradation (Meith, 1985).

Although the Mediterranean Sea is the focus of MAP, the interconnected impacts of land-based activities, many of them far inland, and touristic and other pressures along the coastline clearly had to be addressed. The Mediterranean Action Plan calls for a series of legally

binding treaties to be drawn up and signed by the Mediterranean governments, the creation of a pollution monitoring and research network,
and a socioeconomic program to reconcile development priorities with a
healthy Mediterranean environment.

The Priority Actions Program consists of practical, specific activities,
including integrated planning and management of coastal zones; promotion of soil protection as the essential component of environmental protection in Mediterranean coastal zones; development of Mediterranean
tourism harmonized with the environment; environmental impact assessment in development of coastal zones; and balance between the hinterland and coastal zones. Case studies are prepared on rehabilitation and
construction of historic centers. The program is coordinated by a regional
centre in Split, Yugoslavia.

Research involving an assessment of the state of the Mediterranean and
identification of its major problems involved 83 laboratories from 16
countries and became known as the Mediterranean Pollution Monitoring
and Research Program, or MED POL. Included in MED POL's first
phase in 1975–1980 were baseline studies and monitoring. The second
phase of MED POL (1981–1990) dealt with comprehensive long-term
monitoring and research. These include monitoring sources of pollution,
nearshore and offshore areas, atmospheric transport of pollutants,
epidemiological studies related to environmental quality criteria, guidelines and criteria for land-based sources protocol, and pollution-induced
ecosystem modification.

The Blue Plan was initiated in 1979 as part of the socioeconomic component of MAP as the first prospective study in the relationship between
the environment and development in the Mediterranean, launched by all
countries within the region. Coordinated from a regional center in Sophia
Antipolis, France, it has reviewed present trends in socioeconomic development of the Mediterranean basin. The review analyzed freshwater
sources, industrial growth and industrialization strategies, energy, population movements, urbanization, rural development, tourism, and other
variables. It looked at scenarios for sustainable integrated social and economic development of the basin. Over the years, the Blue Plan has
studied a wide range of issues, including space use, urbanization, and
rural development; cultural heritage and cross-cultural relations; environmental awareness and value systems; tourism, space use, and the environment; systems analysis of development linked with environmental
protection; and preparation of alternative scenarios. A third phase, in
1987, involved synthesis of the entire exercise, and analysis of results by
the contracting parties (The Blue Plan, 1988).

A substantial volume of information was gathered about the Mediterranean region and a record of scientific and technical co-operation developed during the preparation of the Blue Plan. In addition to the main
report, specialized booklets deal with the evolution of hinterland and

mountain regions; the conservation of fragile areas, wildlife, and plant life; and the future of the islands and the forests.

If this important international and regional activity and its far-reaching conclusions are taken seriously by lawmakers, planners, politicians, and decision-makers, it may lead to practical action in all participating countries. The Blue Plan could thus serve as a valuable model of regional cooperation in issues vital to the ecology of landscapes and the inhabitants.

New Approaches to Education in Regional Development Planning

In most Mediterranean SFZ countries development planning has been concerned almost exclusively with efficient exploitation of natural resources for the country's economic needs, generally defined as quantitative growth of the gross national product. Finally, a broader and more qualitative perception of regional planning is emerging, in which the needs for socioeconomic utility—not only of the country as a whole, but also of the population directly affected—are balanced with concern about the environmental impacts of development. In this context, it is significant that a course on landscape ecology, design, and planning is part of the training program at the Mediterranean Agronomic Institute in Zaragosa, Spain. In addition, the Mediterranean Agronomic Institute of Chania in Crete is developing an interdisciplinary course in landscape ecology, in which the landscape ecologists involved in the Crete Red Book case study (see below) are taking an active part.

An innovative series of courses on techniques for ecologically based regional planning is being taught at the University of Haifa, Israel (Lieberman and Maos, 1992). These courses are offered in a cross-disciplinary Curriculum in International Development Planning for both local and overseas students with different backgrounds in natural sciences and humanities. Much consideration is given to sustainability and to conceptual and practical tools which landscape ecology provides for planning and management. Students are also familiarized with regional planning strategies and methods that have been employed in the Mediterranean basin and other parts of the world.

Specific examples of rural regional development planning, particularly in Latin America, look at agrarian reform and its consequences. Similarly, rural migration and accelerated urbanization are examined, and rural settlement and resettlement schemes are evaluated.

A major focus in this teaching effort is on the importance of realizing a new balance in ecological planning and management between the biosphere and the technosphere (see Figure 2-14, p. 84).

Advances in Dynamic Conservation Research and Management

The need for a shift in emphasis from passive protection of plant and animal species to dynamic conservation management of ecosystems and landscapes in order to ensure the continuation of all vital ecological processes has been endorsed as a major international conservation strategy (Ricklefs et al., 1984). This approach has now been adopted in the Mediterranean, as demonstrated in the first comprehensive publication on conservation of plants in this region (Gomez-Campo, 1985). Several recent ecological studies have emphasized the importance of fire, grazing, and human disturbances in the evolution of most Mediterranean landscapes, and in the maintenance of biological diversity and productivity. Some of these studies have been presented in the Fourth and Sixth International Conferences on Mediterranean Ecosystems (MEDECOS) (Dell et al., 1986; Arianoutsou, 1991), and in the Third European Fire Conference (Goldammer and Jenkins, 1990).

Further progress has also been achieved in actual conservation management. In Israel, for instance, controlled livestock grazing has been introduced as an important management tool in all nature reserves in the Mediterranean zone. Monitoring of these effects in 90 different sites in forests, maquis, open shrublands, grasslands, and wet habitats (Noy-Meir and Kaplan, 1991) confirmed our previous findings (Chapter 4, p. 302) on a much larger scale. It showed that light-to-moderate grazing in these reserves increases plant species richness and diversity. In wooded vegetation types the contribution of controlled goat grazing was most important. In protected and lightly grazed sites the frequency of fires was higher than in moderately and heavily grazed ones. Protected herbaceous plant communities—dominated chiefly by tall grasses—also suffered more from periodic outbursts of rodents.

Recent fire-ecology studies in Israel (Kutiel and Naveh, 1987, 1990), corroborated findings from California on the beneficial role that fire can have in Mediterranean shrublands and forests on nutrient cycling by mobilizing the nutrients tied up in the highly lignified wood and slowly decomposing litter and duff. The short-term post-fire nutrient flush is utilized by herbaceous fire followers for prolific forage and seed production, serving as an important link in the recycling of these nutrients.

In view of the increasing frequency, intensity, and destructivity of wildfires, the futility of attempts for passive fire extinction has been recognized in California, in Australia, and in South Africa. Active fire and fuel management have become the major challenge for wildland management. An important landmark was the publication of the book on prescribed burning in California wildlands by the late H. Biswell (1989). Blending knowledge of ecological principles with wisdom and expertise derived from long practical experience, this book is a vital source of information on fire management in California. It should also be used elsewhere by foresters, conservationists, and wildland managers dealing with Mediterranean landscapes who are ready to learn from the California experience. Fortunately, their growing numbers in the Mediterranean basin gives hope that controlled burning will soon be used widely as an efficient tool for dynamic conservation management. This is already happening in Israel, where it is used for the reduction of fuel and for the creation of broad grassy fuel breaks by the Jewish National Funds Forestry Division, and as an experimental tool in the Carmel National Park and Nature Reserve.

Advances in Holistic Landscape Conservation

As emphasized in Chapter 4, the ultimate success of holistic landscape planning and management policy will depend on the creation of awareness and understanding among those who care for, live from, and deal with these landscapes at all levels of decision-making. As a first step it is vital to bridge the communication gap between academicians and professionals, conservation-minded ecologists and production-minded foresters and agronomists, economists, and engineers. At the same time we need to educate a new breed of interdisciplinary landscape managers with broad ecological, economical, sociological, and technological backgrounds who can serve as "integrators."

In addition to the educational efforts, landscape ecologists in the Mediterranean have started to fulfill an important function in this process. In Italy, the Tuscan Regional Government has recognized the great value of the integrative landscape ecological study by Vos and Stortelder (1992) and has asked this team of landscape ecologists from the Dorschkamp Forestry and Nature Research Station in Wageningen, together with a team of landscape ecologists, planners, hydrologists, geographers, and biologists from the University of Amsterdam, to carry out a detailed environmental assessment of the impacts of the ambitious Farva River Barrage for water storage and irrigation of the Grosseto Plain, a project which had been fiercely opposed by naturalists and environmentalists. The comprehensive evaluation of all biological, ecological, hydrological, and scenic landscape attributes indicated that intervention on the Farma basin would cause serious damage to this unique ecosystem and would reduce landscape and natural resources which are considered important to the community as a whole (Pedroli et al., 1988). The Tuscan Regional Government accepted these conclusions and thereby prevented the implementation of a project which could have had severe environmental consequences.

The active involvement of landscape ecologists from the Mediterranean and other SFZ countries was very evident in three symposia on the future of Mediterranean landscapes at Montecatini in April 1992. The proceedings of this conference have been published in the July 1993 volume of *Urban and Landscape Planning* (Farina and Naveh, 1993). Most of these presentations reflected not only the great concern for the future of the endangered natural and cultural assets in the SFZ landscapes, but also offered innovative approaches and methods—mostly outlined in the supplement to Chapter 2—as well as proposals for integrated landscape planning, management, and restoration, which it is hoped, will lead to practical solutions.

The first Red Book case study, initiated by the IUCN-IALE Working Group, mentioned in the supplement to Chapter 1, has been carried out in Western Crete by a multinational team of landscape ecologists and planners, botanists, physical and cultural geographers, and range, forest, and remote sensing specialists. The study area covers 350 km² stretching from the touristically overdeveloped coast of the Aegean Sea, across the "White Mountains" and the world-renown Samaria Gorge National Park, to the relatively unspoiled coast of the Lybian Sea, and the small island of Gavdos.

The study included a detailed classification, mapping, and clustering of ecotopes and landscape units with the help of remote sensing and GIS, combined with intensive field work. The research team identified the major threats to the diminishing biological, historical, and scenic landscape values and recommended practical guidelines for nature conservation, forestry, livestock, agriculture, and touristic, as well integrated landscape planning and policy measures. These should ensure both effective preservation of overall landscape ecodviversity and socio-economic benefits from all vital "soft" and "hard" values of these highly scenic mountains.

In the introductory lecture to the Montecatini symposium, Z. Naveh (1993b), chairman of the Working Group and scientific coordinator of the Crete project, outlined the purpose and scope of these and other Red Books (which are now called Green Books) as innovative holistic landscape-conservation tools for which the Crete case study is serving as the first model:

1. In contrast to the IUCN Red Data Books for endangered plants and animals, dealing with taxonomic species level, these Green Books will deal with the concrete, physical landscape, at scales ranging from the smallest mappable landscape ecotope to regional landscape scales of 200–400 km². Most decisions on land uses and conservation measures are made on these terrain levels, which also determine the fate of plant and animal populations and their habitats. Therefore, the threats to such tangible and well-known landscape units have much more meaning and public appeal than threats to species or vaguely defined ecosystems. This will be even more evident if the Green Books

present relevant information not only on endangered natural assets but also on all other crucial issues and perils to cultural, historical, and scenic landscape assets which compose the total landscape eco-diversity.

2. In contrast to the IUCN Red Data Books and to the regular types of land studies and surveys, these Landscape Green Books should recommend alternative and more sustainable land-use practices and conservation strategies, including zoning and protecting, for such specific landscape unit within the regional landscape systems.

3. In order to avoid the "top-down" syndrome of conservation plans prepared by experts and imposed by administrators, efforts should be made for maximum involvement of the local populations from the early planning stages. With the help of open dialogue and by approaching conservation "from the roots," the fullest comprehension and cooperation should be ensured. Above all, these Green Books should demonstrate how demands to safeguard intrinsic biological and cultural "soft" landscape values can be reconciled with controlled utilization of "hard" values, which are vital for socioeconomic advancement.

4. To generate awareness of the dangers to these landscapes, concern about their future, understanding of the complex problems involved and their best solutions, and motivation for active involvement in their realization, these Green Books should not only be descriptive but anticipatory. Although based on scientific methods, this information should be presented to the planner, land manager and user, the political decision-maker, and the public at large in clear, nontechnical language. They should present different scenarios on the fate of these landscape units under alternative land-use options, with ample illustration by maps, diagrams, and photos.

5. In contrast to the fate of most scientific reports, which are at best filed away, Green Books should be used as practical background documents for planning, policy, and management guidelines by professionals, politicians, and the public. They should serve not only as a database and dynamic landscape model, but they should also lay out the foundations for long-term impact evaluation and forecasting.

6. Although containing information on present and anticipated environmental impacts, Green Books for landscape conservation differ from conventional "Environmental Impact Statements," which are merely defensive and passive reactions to planned development projects, limited to specific sites and their potential threats. Green Books should be active and positive initiatives to change existing undesirable trends and to prevent their continuation with the help of innovative, holistic, long-term, and multi-beneficial strategies on regional landscape scales.

7. Published under the auspices of IUCN in English, as well as by the local authorities in the language(s) of the country and/or the ethnic region covered, Red Books should be widely distributed to reach all

those concerned with these threatened landscapes. They have relevance for public and school education in conservation, as well as for
tourists.

8. According to the proposal of the Working Party (see supplement to
Chapter 1) these Green Books should be used as the major tool for
"rescuing" endangered landscapes from the Red Lists of Endangered
Landscapes that should be prepared according to the guidelines of
IUCN on a worldwide basis in each country.

The Tuscan Regional Government is supporting further landscape-
ecological studies, carried out by A. Farina (1993) and his co-workers at
the Aulla Natural Museum of History in the Lunigiana intermountain
riverbed, which will lead to further landscape protection at regional
scales. For this purpose the methodology for the Green Book concept is
further developed with the help of GIS and dynamic process-oriented
simulation models. Further case studies are planned in Portugal, Israel,
Slovenia, Ireland, Norway, and Costa Rica.

References

Arianoutsou, M. (Ed.). 1991. MEDECOS VI International Conference on
Mediterranean Climate Ecosystems, on Plant and Animal Interactions in
Mediterranean Type Ecosystems. Maleme (Crete) 23–27 September 1991.

The Blue Plan. 1988. Mediterranean Blue Plan Regional Activity Centre: Futures
of the Mediterranean Basin. Executive Summary and Suggestions for Action.
United Nations Environmental Programme, Mediterranean Action Plan,
Mediterranean Blue Plan Regional Activity Centre, Sophia Antipolis, France.

Biswell, H. 1989. Prescribed Burning for Wildland Management in California.
University of California Press, Berkeley.

Bussotti, F., M. Ferretti, E. Cenni, R. Gellinni, F. Clauser, P. Grossoni, and E.
Barbolani. 1991. New type forest damage to mediterranean vegetation in
Southern Sardinian forests (Italy). Eur. J. For. Pathol. 21:290–300.

Bussotti, F., R. Gellini, M. Ferretti, E. Cenni, R. Pietrini, and G. Sbrilli. 1992.
Monitoring in 1989 of Mediterranean tree condition and nutritional status in
southern Tuscany, Italy. Forest Ecology and Management 51:81–93.

Boer, M. M., and R. S. de Groot (Eds.). 1990, Landscape-Ecological Impact of
Climate Change. Proc. Eur. Conf. Lunteren, The Netherlands, 3–7 December
1989. IOS Press, Amsterdam.

CEPA—Coordinadora Extremena de Proteccion Ambiental. 1992. Dealing
with disparity. European Structural Funds in South West Spain. The Ecologist
22:91–96.

Dell, B., A. J. M. Hopkins, and B. B. Lamont (Ed.). 1986. Resilience on
Mediterranean-Type Ecosystems. Dr. W. Junk, The Hague.

Farina, A. 1993. A proposal for a simple, integrative Red Book procedure,
applied for threatened rural landscapes of the Northern Apennines (Italy).

Farina, A., and Z. Naveh (Eds.) 1993. Special Issue on "The Future of
Mediterranean Landscapes." Landscape and Urban Planning 24. July 1993.

Workshop of the IUCN-CESP Working Group on Threatened Landscapes, Montecatini, 1. May 1992. Special Working Paper, CESP-IUCN, Sacramento.

Ferretti, M., E. Cenni, B. Pisani, F. Righini, D. Gambicorti, P. De Santis, and F. Bussotti. 1992. Biomonitoraggio di inqinanti atmosferici: una esperienze integrate nella Toscana costieria. Acqua-Aria 8:747–758.

Goldammer, J. G., and M. J. Jenkins (Eds.). 1990. Fire in Ecosystem Dynamics. Mediterranean and Northern Perspectives. SPB Academic Publishing bv, The Hague.

Gomez-Campo, C. (Ed.). 1985. Conservation of Mediterranean Plants. Dr. W. Junk, The Hague.

Kutiel, P., and Z. Naveh. 1987. The effect of fire on nutrients in a pine forest soil. Plant and Soil, 104:269–274.

Kutiel, P., and Z. Naveh. 1990. The effect of wildfire on soil nutrients and vegetation in an Aleppo pine forest on Mt. Carmel, Israel. *In:* J. G. Goldammer (Ed.), Fire in Ecosystem Dynamics. Mediterranean and Northern Perspectives. SPB Academic Publishing bv, The Hague, pp. 85–94.

Le Houerou, H. N. 1988. Vegetation and land- use in the Mediterranean basin by the year 2050: A prospective study. UNEP (OCA)/WG.2/15.

Lieberman, A. S. and J. Maos. 1992. Regional Development Studies: Settling the Conflict between Development and Conservation. Proc. Israel Geographical Society, Beersheba, Israel, December 22, 1992 (Hebrew).

Lucas, P. H. C. 1992. Protected Landscapes. A Guide for Policy Makers and Planners. Chapman & Hall, London.

Malanson, G. P., and W. E. Westman. 1991. Modeling interactive effects of climate change, air pollution, and fire on a California shrubland. Climate Change 18:362– 376.

Meith, N. 1985. Mediterranean Action Plan. Mediterranean Co-ordinating Unit, Athens, Greece, and the Programme Activity Centre for Oceans and Coastal Areas of the United Nations Environment Programme.

Merino, O., and R. Villar. 1993. Vegetation response to climate change in a dune ecosystem in southern Spain. *In:* J. M. Moreno, and W. C. Oechel (Eds.), Anticipated Effects of a Changing Global Environment on Mediterranean-Type Ecosystems. In Press.

Moreno, J. M., and W. C. Oechel. (Eds.). 1993. Anticipated Effects of a Changing Global Environment on Mediterranean-Type Ecosystems. In Press.

Naveh, Z. 1993a. Conservation, restoration, and research priorities for Mediterranean uplands threatened by global change. *In:* J. M. Moreno, and W. C. Oechel (Eds.), Anticipated Effects of a Changing Global Environment on Mediterranean-Type Ecosystems. In Press.

Naveh, Z. 1993b. Red Books for threatened Mediterranean landscapes as in innovative tool for holistic landscape conservation. Introduction for the western Crete Red Book case study. Landscape and Urban Planning 24:241-247.

Naveh, Z., and J. Dan. 1973. The human degradation of Mediterranean landscapes in Israel. *In:* F. di Castri and H. H. Mooney (Eds.), Mediterranean-type Ecosystems. Origin and Structure. Springer-Verlag, Heidelberg, pp. 370–390.

Noy-Meir, E., and D. Kaplan. 1991. The Effects of Grazing on the Herbaceous Mediterranean Vegetation and its Implications on the Management of Nature Reserves. Interim Report to the Nature Conservation Authorities, Jerusalem (Hebrew).

Pedroli, G. M. B., W. Vos, H. Dijkstra, and R. Rossi. 1988. The Farma River Barrage Effect Study. Giunta Regionale Toscana. Marsilio Editori, Firenze, Italy.

Ricklefs, R. R., Z. Naveh, and R. E. Turner. 1984. Conservation of Ecological Processes. Commission on Ecology Paper No. 8 IUCN, Gland, Switzerland.

Vos, W., and A. H. F. Stortelder. 1992. Vanishing Tuscan Landscapes. Landscapes Ecology of a Submediterranean-Montane Area (Solano Basis), Tuscany, Italy. Pudoc Scientific Publishers.

Westman, W. E., and G. P. Malanson. 1991. Effects of climatic change on Mediterranean-type ecosystems in California and Baja California. *In:* R. L. Peters, and T. Lovejoy (Eds.), Consequences of the Greenhouse Effect for Biological Diversity. Yale University Press, New Haven, Connecticut.

Epilogue

Half a century ago, Tansley loosed the word *ecosystem* on an unsuspecting and unprepared world of biologists. Tho he defined it as an integrated unit composed not only of plants and animals, but also the climate, soil and other aspects of the environment (but not man), the idea was ahead of its time. A decade later, I applied the concept to Vegetation (the integrated mosaic of plant-communities in time and space). Since then, the subject has met with many vicissitudes. On the one hand, theoreticians have seen only the theory, yet fail to grasp its essential nature by identifying the larger social holons with the only wholes they know: their own individuality, and their own species Homo sapiens. On the other hand, the ecolometricians of the present have made ecosystem *analysis* fashionable by applying their highly developed numeracy (the quantitative analogue of literacy) to studying the parts, not the wholes. One character has compared this behavior to that of an intelligent young boy taking the family grandfather clock all apart, and diligently describing each weight and wheel, and even the interrelationships between them. But the whole clock remains to him a vitalistic, mystical, occult, spiritualistic, subjective, qualitative thing, unworthy of elegant scientific interest. Biologist William Morton Wheeler, General Ian Christiaan Smuts, and the Jesuit priest Pierre Teilhard de Chardin knew differently, even as philosopher Bertrand Russell, himself a mathematician, could frame the interests of scientists as philosophic "logicism".

And now, a half century later, Zev Naveh and Arthur Lieberman lift the curtain on a new drama that in my opinion will have far-reaching influence on the development of the academic ecologies: of the unintegrated "environmental science," of conservation, conservation ecology and land planning; of such special problems as human over-population, of natural resources, of environmental quality, pollution and degradation. Under the name of LANDSCAPE ECOLOGY (not to be confused with the esthetic fields of Landscape Architecture and Landscape Engineering of English-speaking countries), this volume presents a truly integrated concise view of that highest of all holistic phenomena, man-and-his-total-environment, our one Earth, the successful intelligent management of which is now at the crossroads.

It is an Epilogue in more senses than one, that I add a few comments about an Idea which has been understandably underplayed in Europe and Mediterranea, where man has man-handled the entire landscape for at least ten millenia; and underplayed in the Americas, where the pre-European Indian-influenced landscape has always been considered to be "pristine," "virgin," or "climax." It is also under-conceived by those who consider the landscape of our hunter-gatherer ancestors of a million years ago to be "natural". (A hungry hunter-gatherer, territorially limited, will eat the last bulb or the last live meat, even as a hungry sailor ate the last dodo, all blissfully unaware of Rare and Endangered Species.) And the idea is underconceived by the present fashionableness of the preservationists, despite the extraordinary worthiness of all their accomplishments to date. They have preserved, too often with no interest in further knowledge of the changing nature of what they have preserved, at the highest level of holistic integration of that Nature. And thus I briefly mention a "tool," a "device," a concept, a methodology, that is very much a part of Landscape Ecology: *An area that is kept as free as possible from human influence* can be called a NATURAL AREA (altho that term, like "landscape" and "ecology" themselves, has many other uses).

A **Natural Area,** be it one hectare or a million hectares, is thus a control, a standard, a common denominator, essentially a non-human area, which by comparison with all m a n-a g e d areas, urban, farm, forest, and range, allows us to judge and separate the role of man himself in the man-and-his-total-environment ecosystem. It is not to be supposed that a Natural Area can be entirely free of the influence of man. Air pollution and winds, water pollution and aquifers and streams, and all the past influences of man leave their fingerprints and their footprints. But without Natural Areas, *used* by man, we have no logical grounds for evaluating the *influence* of man. Nor should we emotionally assume that Natural Areas are always bigger, better, finer, more Edenesque, and more ideal than what man has done. Man-less nature never did produce the ethics, the esthetics and the logic of a civilization. Natural Areas should be established permanently by whatever legal and political safeguards a government allows. Time, endless time, is necessary. One does not "preserve" a live kitten or a pup, just as one cannot "preserve" an adolescent habitat that is an immature landscape (even tho it harbors a Rare and Endangered species). A Vegetation type merely 30 years old is a babe at breast. Think of a redwood forest over 2000 years old, the individuals themselves being mere vegetative stump sprouts from an older individual.

There are three major uses of Natural Areas. The first is for its **cultural** values. In this crowded world, wilderness even a mini-wilderness is no longer a dangerous and feared environment. It is a place of retreat, away from the crowds, to the home of our ancestors. Space, silence and privacy are rare and cherished commodities, where the contemplative in spirit can gain renewed strength in their intellectual and esthetic awareness of the natural—the better to continue their lives in the artificial haunts of man. The study of Natural History is one of the humanities, one of the graces of our age, a tenth muse of our civilization.

The second use of Natural Areas is for their **educational** values. Natural Areas are living outdoor museums, preeminent for the exhibition of the plant animal and mineral world about us. They should be treated with the same care and consideration as our historical monuments or exhibits in museum halls, many of which are under the guidance and care of trained leaders. Explanatory literature facilitates the communication of these educational values.

The third use of Natural Areas is for their **scientific** values. In laboratory research one manipulates alters and experiments with his materials under carefully controlled but highly simplified and unnatural conditions, under short-term grant-limited socio-economic restrictions. Currently one emphasizes elegant mathematical number-limited abstractions, arguments, interpretations and conclusions, which are logically very satisfying. But only in Natural Areas does nature have the freedom to develop, to change, in her own unpredictable aleatoric manner. Only in Natural Areas can man stochastically pursue the reality of Natural History and—understanding the better—gain greater wisdom from greater knowledge.

And thus we envision that Natural Areas (together with adjacent semi-Natural Areas, in which one factor or another is carefully and controllably modified) can play a most important role in a developing man-plus-environment Landscape Ecology. The more under-developed, the more over-populated a nation or area is, the more urgently it needs a series of Natural Areas, to see what Nature can teach us, in order to benefit our own life, and the unity of which we are a part.

Frank E. Egler
Aton Forest
Norfolk, Connecticut

Zev Naveh is Professor Emeritus of Landscape Ecology at the Technion – Israel Institute of Technology. Initially, he dealt with research in range ecology and management in Israel and also in East Africa. After 15 years at the Vulcani Agricultural Research Institute, he taught general ecology and landscape ecology at the Faculty of Agricultural Engineering and the Faculty of Architecture and Town Planning. His research activities covered fire ecology, restoration ecology, and landscape ecology. He has extensive international involvement in the practical applications of these fields and has received wide international recognition for his contributions. He is a distinguished member of the International Society of Mediterranean Ecologists (ISOMED). As a member of the Commission of Environmental Strategies and Planning of the International Union for the Conservation of Nature and Natural Resources, he has chaired a joint working group with the International Association of Landscape Ecology, on Landscape Ecology and Conservation. Prof. Naveh has published close to 200 scientific publications and is the author and coauthor of several books.

Arthur S. Lieberman is Professor Emeritus of Physical Environmental Quality at Cornell University. He has long experience in teaching, research, and extension work in regional and community resource development and has had close involvement with users of remote sensing techniques, land-use planners, and other practitioners and educators in these fields. His work in regional landscape planning emphasizes biophysical aspects of land-use determinations and information needs of physical planners. Currently, he serves as Resident Director of the Cornell Abroad Program in Israel, where he has coordinated a joint Cornell University-Haifa University teaching program in International Development Studies.

Index

Supplemental Subject Index